Food Industry and the Environment in the European Union

Practical Issues and Cost Implications

Second Edition

Edited by

Janet M. Dalzell, MIFST
Food Science Writer and Editor
Kent, United Kingdom
Former Senior Information Scientist
Leatherhead Food Research Association
Leatherhead, United Kingdom

AN ASPEN PUBLICATION®
Aspen Publishers, Inc.
Gaithersburg, Maryland
2000

Aspen Publishers, Inc., is not affiliated with
the American Society of Parenteral and Enteral Nutrition.

Library of Congress Cataloging-in-Publication Data

Food industry and the environment in the European Union: practical issues and cost implications / edited by Janet M. Dalzell.— 2nd ed.
p. cm.
Includes bibliographical references and index.
ISBN 0-8342-1719-8
1. Food industry and trade—Environmental aspects—European Union countries.
2. Energy conservation—European Union countries. 3. Packaging waste—Recycling—European Union countries. I. Dalzell, J. M. (Janet M.)
TD195.F57 F57 1999
338.1'94—dc21
99-048013

Orders: (800) 638-8437
Customer Service: (800) 234-1660

Editorial Services: Jane Colilla
Library of Congress Catalog Card Number: 99-048013
ISBN: 0-8342-1719-8

Printed in the United States of America

1 2 3 4 5

Table of Contents

Contributors

Clive V. J. Dellino, FInstR
Refrigeration and Cold Store Consultant
Norfolk, United Kingdom

Alan Gear, MSc
Chief Executive
Henry Doubleday Research Association
Ryton Organic Gardens
Coventry, United Kingdom

George Hazle
Exel Logistics
Bedford, United Kingdom

Martin Key
Key Environmental Management
Worcestershire, United Kingdom

Peter Page, MIMgt, MIInfSc, MIEx
Principal Consultant
The Packaging Technology Group
Pira International
Leatherhead, United Kingdom

Tony Palmer
Director and Editor
Cold Chain News
Norfolk, United Kingdom

Sibel Roller, MSc, PhD
Professor of Food Biotechnology
School of Applied Science
South Bank University
London, United Kingdom

Susan E. Selke, MS, PhD
Professor
School of Packaging
Michigan State University
East Lansing, Michigan

Jeremy Tait
Project Manager
Energy Efficiency Best Practice Programme
Energy Technology Support Unit
AEA Technology Environment
Oxfordshire, United Kingdom

Stephen Toogood
Effluvium, Ltd.
Cambridgeshire, United Kingdom

Philip Walsh
P. Walsh & Associates
Lincolnshire, United Kingdom

Andrew D. Wheatley, PhD
Professor of Water Engineering
Civil and Building Engineering Department
Loughborough University
Leicestershire, United Kingdom

R. White
Hampshire, United Kingdom

Preface

Since the first edition of this book was published in 1994, much has happened in the sphere of environmental issues as they affect the food industry. Whilst commercial genetically modified crops have been accepted by consumers in North America, consumers in Europe have been more hostile towards them. Many are unsure if scientists know enough about genetically modified foods, and this uncertainty, together with a dislike of the use of chemical pesticides and animal-growth-promoting drugs, has resulted in an increase in sales of organic products, a trend that looks set to continue.

Energy conservation and pollution control have been found by many in industry to be an excellent way of both reducing the impact of industry on the environment and saving a company substantial sums of money. The re-use or recycling of packaging waste has become mandatory in many countries.

The second edition of this book has incorporated these developments, and the impact of them on the cost and practicalities of implementing an environmental policy. I hope that this book will be of interest to senior managers and all in the food industry who are concerned about their environment and the impact they have on it.

Introduction

Environmental issues are of increasing concern to the food industry, as consumers and legislators have forced manufacturers to become more aware of the environmental consequences of their operations. Many food manufacturers have embraced these changes and have taken the initiative in improving their practices. For many companies, taking the environmentally sound option has become standard practice, whilst others have become aware of these issues through the introduction of legislation. The impact of a food factory on the environment is not limited to the waste materials that the processing operation generates; the treatment that the raw materials and ingredients have been subject to is also of relevance. Furthermore, many consumers are increasingly concerned about what happens to packaging waste and whether it is really necessary to have so much packaging in the first place. Whilst some reduction in packaging has occurred over the past few years, much of the packaging used is necessary if the product is to be protected and the shelf-life of the food enhanced.

Just as some food companies have integrated environmental issues into their company policy, many consumers buy products they consider to be environmentally sound, as a matter of course. This is particularly true for the younger consumer. One issue that is currently high on the European consumer's agenda is foods containing genetically modified ingredients. Whilst foods containing such ingredients have largely been accepted in the United States, many European consumers fear that genetically engineered crops and animals may play havoc with the gene pool and change the world irrevocably.

Packaging is another issue that is important to the environmental consumer, because it is so visible that the consumer is very aware of it, whereas the waste produced by a food factory is only brought to the attention of the consumer if there is a large spill that impacts dramatically on the rivers and is reported in the media. Each sector of the food industry produces its own unique mixture of waste materials; the dairy sector, for example, is a major source of industrial effluent in Europe. Dairy waste may have a drastic effect on the chemical oxygen demand of a waterway, and the cleaning agents that are used to remove the fat and protein deposits from processing equipment are highly alkaline and may raise the pH to toxic levels. It is in the interest of the manufacturer to treat waste to such standards that it is clean enough to release into waterways or sewers, as this avoids paying large sums of money for the waste to be treated by the local water company.

Money can also be saved by conserving energy. Additionally, this preserves the earth's natural stock of fuels and cuts down on emissions of carbon dioxide, a gas that has been linked to global warming. An environmental audit of a food factory will almost certainly identify areas where energy use can be reduced. Energy conservation is often the first step a company takes on the road to becoming more environmentally aware, as small, inexpensive measures can result in vast savings in energy and fuel bills.

The number of companies that have an environmental policy and are committed to producing food in an environmentally friendly way has grown over the years. There are still some companies that have not come to terms with environmental issues; this book should provide a practical and comprehensive introduction to this topic.

—Janet M. Dalzell

CHAPTER 1

Organic and Non-Organic Agriculture

Alan Gear

1.1 BACKGROUND

During the first half of the 20th century—indeed, for most of the period when Britain was a major economic power—government policy toward food was simple: It should be available at the cheapest possible price. Farming suffered as food—much of it capable of being grown in the United Kingdom—was imported from the Empire, where it could be produced at a much lower cost. Consequently, Britain entered the Second World War with its domestic agriculture in a seriously weakened state, a point driven home by the calamitous U-boat attacks on the Atlantic convoys that were its lifeline. Thanks to a combination of military successes and a drastic plan for home production, the country was not starved into submission. Nevertheless, the experience proved cathartic for government planners. The maintenance of an adequate food supply, through agricultural self-sufficiency, became a primary objective during the ensuing years of the Cold War.

The main vehicle for this policy—co-ordinated in England by the Ministry of Agriculture, Fisheries and Food (MAFF), and in the rest of the United Kingdom by similar government bodies—was a complex series of financial incentives that encouraged output. These included support of the market price, subsidies towards the cost of fertilisers, and capital grants for farm improvement. With no fear of temporary gluts driving prices down, it is hardly surprising that farmers responded enthusiastically.

Farmers were urged to specialise, backed by an ever-increasing effort invested in research and development. Gradually, the mixed family farm of pre-war years was replaced by all-arable or all-livestock enterprises. Nitrogen fertiliser use was to expand sevenfold during the next 40 years. Completely new types of animal husbandry, such as large-scale indoor pig and poultry units, came into being, superseding their more extensive outdoor counterparts.

During hostilities, the Germans had been working on organophosphate-based chemicals for use in biological warfare. It was found that these nerve gases could, with little modification, be pressed into service as insecticides. Similarly, DDT, which had been used extensively by front-line troops to control the lice infesting their uniforms, found a ready and expanding market as a novel pesticide. Other insecticidal organochlorine compounds—Dieldrin, Aldrin, BHC quickly followed, whilst new chemical substances were also developed to kill diseases and weeds.

Given such financial and technological support, output rose accordingly. Yield increases for cereals, for example, more than doubled. Though fertilisers played a major part in this success story, the role of the plant breeder in developing a succession of hybrid wheat varieties that responded positively to such chemical inputs was no less significant.

When, in 1973, Britain joined the European Economic Community (EEC), domestic farming support was replaced by the Common Agricultural Policy (CAP). Little changed, however, since Brussels' policy was also to encourage output throughout the territory of its European members. This was achieved by price support of the market, albeit in a somewhat different form.

By the mid 1970s, the United States inspired model of agricultural production—involving intensive use of chemical inputs, factory livestock systems and crop specialisation—had become the dominant world model. Not only was it applied extensively in the developed northern economies, it had also taken root in many third world countries. In the case of the latter, the spur had been not so much fear of food security in time of war but the spectre of failing to feed rapidly rising populations. If food supply was to keep pace with the birth rate, it was argued, output must rise accordingly. The outcome was Norman Borlaug's "Green Revolution", involving strains of high-yielding grains that responded to a diet of chemical fertilisers with improved irrigation and intensive use of pesticides. It, too, achieved impressive results but, ironically, overall levels of hunger, if anything, got worse.

However, by the end of the 1970s, questions were being asked about the thrust of agricultural policy. Concern over possible food shortages in the northern hemisphere had been replaced by anxieties over what to do about burgeoning surpluses. In the United States, this resulted in productive farmland being taken out of cultivation, whilst in Europe, immense sums of money were being expended to pay for the production, storage, and disposal of cereals, milk, wine, and other commodities produced in excess.

More alarming to some was the environmental damage caused by industrialised farming. As early as 1962, the American biologist Rachel Carson had warned, in her book *Silent Spring*, against the indiscriminate use of pesticides. It was not long before such predictions became reality in the United Kingdom, as birds of prey went into serious decline because of the use of organochlorine insecticides.

Better environmental monitoring throughout the 1980s revealed the widespread pollution of inland watercourses and underground aquifers by fertiliser run-off and pesticides. Many people lamented the destruction of the traditional English landscape, particularly in the midlands and the eastern counties, with its patchwork of small fields and hedgerows. These no longer made economic sense in an agriculture dominated by intensive cereal monocultures. The new landscape, sporting vast open prairie fields, gave little delight to the eye. It simultaneously deprived much of Britain's wildlife of its habitat.

Consumers, too, began to be concerned about the potential harmful effects on human health of food contaminated by pesticide, hormone, or antibiotic residues. A rise in vegetarianism, especially amongst the young, warned of a sea-change in attitudes towards the exploitation of animals kept under inhumane conditions.

By the end of the 1980s, the pressure to reform the CAP was irresistible. Something had to be done to curb the crippling financial burden caused by supporting surpluses, which in 1990 amounted to over 28.5 billion ECUs (£20 billion)—95% of the total CAP budget. Of this, more than 13 billion ECUs—46% of the total—was spent on the destruction and disposal of unwanted produce (European Community [EC], 1991).

Dairy quotas, to limit milk production, had been introduced a few years earlier, but the problem of over-production of cereals still posed a major headache. Pressure, especially from the United States, to bring European food prices more into line with world prices meant attacking the price support underpinning of the CAP. The method chosen to achieve this by then-EC Farm Commissioner Ray MacSharry, during negotiations over the Uruguay General Agreement on Tariffs and Trade (GATT) round, was to simultaneously take land out of production and reduce the guaranteed price paid for commodities. Farmers received alternative compensation via "set-aside" and "acreage" payments, respectively.

At last the deleterious environmental consequences of the previous policy were officially acknowledged, and new measures were proposed. Some individual schemes, such as management agreements for Sites of Special Scientific Interest or Environmentally Sensitive Areas, had already been introduced by the British government. In early 1992, the EC formally endorsed such policies under the Agri-Environment Programme and, for the first time, financial support was made available for environmentally friendly organic farming. Some European countries, such as Denmark, responded enthusiastically to these incentives, and their domestic production of organic food rose dramatically. In the United Kingdom, growth in sales of organic food rocketed in the wake of the calamitous bovine spongiform encephalopathy (BSE) or Mad Cow Disease epidemic that decimated the beef industry. Demand, though, was mainly met by importing produce from as far away as South America.

As the 20th century draws to a close, the prospects for farming appear polarised between those advocating sustainable agriculture in harmony with nature and those who predict an increasingly high-tech future of intensive agriculture driven by genetic engineering.

1.2 WHAT IS ORGANIC FARMING?

Throughout the past 40 years, the agricultural paradigm just described was enthusiastically promoted by government advisors and agrochemical interests and was taken up by virtually all farmers, except for a handful who chose to do things differently.

Inspired by the writings of agriculturists such as Sir Albert Howard (1940) and Lady Eve Balfour (1975), they took issue with the neo-orthodoxy that reduced plant nutrition to a mere consideration of supplying soluble nutrients in precise amounts, at the correct time, or that saw pest and disease control as a continual battle against nature. For such organic farmers, as they were to be called by the pioneering American agriculturist Robert Rodale, attention to the well-being of the soil as a prerequisite to plant growth was paramount. Upon its vitality depended the health of plants, and of the animals and humans that depended on plants for food. Or, as Lady Eve Balfour put it more eloquently, the health of soil, plant, animal, and man is one and indivisible.

In 1946, Lady Eve founded the Soil Association, an organisation dedicated to promoting a holistic vision of agriculture that, eschewing synthetic fertilisers and pesticides, sought to work in harmony with natural processes. Membership remained small during the ensuing decades, with little organic produce finding its way into the market. Indeed, throughout the 1960s and 1970s, it was nigh impossible to buy organic food, apart from the organic flour and cereals that were sold by whole food shops, specialist outlets, or direct from the farm. If you wanted to eat organic fruit and vegetables, you had little choice but to grow them yourself.

That a significant number of people opted to do just this was evidenced by the largely gardening membership of the Henry Doubleday Research Association, which from its beginnings in 1954 grew steadily to number around 6,000 by 1980 and 25,000 by 1998, making it the largest organic organisation in Europe. The first producer group was the Organic Growers Association, founded in 1980 by a group of mainly young vegetable producers, many of whom had settled in West Wales. British Organic Farmers followed a few years later. The market for organic vegetables grew rapidly, particularly after its take-up by the supermarkets, Safeway being the first to sell an organic range. To this day, a large proportion of domestic vegetable production derives from West Wales (though regrettably the vast majority is now imported), and the country's largest wholesaler is based on an industrial estate in Lampeter, Dyfed.

Other multiples followed the lead set by Safeway, though the recession in the early 1990s produced something of a hiccup in the steady growth in sales, and some players dropped out. Trade picked up again from 1995 onwards, as consumers sought "safe" food following the BSE scare. Most multiples increased both sales and range—organic dairy products and, to a lesser extent, meat proving especially popular. By the end of 1997, retail sales in the United Kingdom were growing at more than 25% a year and topped £250 million. Market forecaster Mintel ("Organic and Ethical Foods", 1997) anticipates that the UK market will grow quickly over the next five years, reaching £359 million at 1997 prices by the millennium, though industry insiders consider this to be a conservative estimate.

Elsewhere in Europe, demand for organic food is even more buoyant—some Scandinavian countries are predicting that by the year 2000, 20% of all food consumption will be from organically grown produce. Growth in the United States has also been rapid, with the emergence of entirely new chains of supermarkets selling all-organic produce.

For all that organic produce has been available for more than a decade, through outlets accessible to most people, there is still considerable public misunderstanding about organic food. Survey after survey confirms a public preference for the products of organic agriculture, at the right price. Mintel ("Organic and Ethical Foods", 1997) found that three-quarters of the population are sympathetic to the idea of buying organic products, with young consumers being the most committed. However, the term "organic" is at best understood hazily. Most people assume that food from organic farms is entirely free from pesticide residues, but there can be no guarantee that it does not contain unwanted pesticide residues that may have come via the soil, water, or even the surrounding air. Others believe that organic food is tastier and more nutritious than conventionally grown food, and indeed a case for this, albeit inconclusive, can be made. Some assume that all organic food must be suitable for vegetarians, misunderstanding the crucial role that livestock play in organic farming. By banning the use of genetically modified organisms in organic agriculture, a further strong argument is presented to the public that organic food offers the only safe alternative.

Before proceeding further, it is worth turning to two definitions of what constitutes organic farming. The first is supplied by IFOAM, the International Federation of Organic Agriculture Movements. It should be noted that these principles encompass a far wider brief than is usual when food production is considered, involving equity, environmental sustainability, and social justice.

IFOAM states that organic farming seeks:

- to produce food of high nutritional quality in sufficient quantity
- to work with natural systems rather than seeking to dominate them
- to encourage and enhance biological cycles within the farming system, involving micro-organisms, soil flora and fauna, plants, and animals
- to maintain and increase the long-term fertility of soils
- to use as far as possible renewable resources in locally organised agricultural systems
- to work as much as possible within a closed system with regard to organic matter and nutrient elements
- to give all livestock conditions of life that allow them to perform all aspects of their innate behaviour
- to avoid all forms of pollution that may result from agricultural techniques
- to maintain the genetic diversity of the agricultural system and its surroundings, including the protection of plant and wildlife habitats
- to allow agricultural producers an adequate return and satisfaction from their work, including a safe working environment
- to consider the wider social and ecological impact of the farming system

A rather more concise definition has been framed by the United States Department of Agriculture, as follows:

> Organic farming is a production system which avoids or largely excludes the use of synthetically compounded fertilizers, pesticides, growth regulators and livestock feed additives. To the maximum extent feasible, organic farming systems rely on crop rotations, crop residues, animal manures, legumes, green manures, off-farm organic wastes, and aspects of biological pest control to maintain soil productivity and tilth, to supply plant nutrients and to control insects, weeds and other pests.

How these principles are applied in practice is, perhaps, most easily understood by comparing a hypothetical organic farm with its non-organic counterpart.

One obvious difference is the absence of a large stack of chemical fertiliser sacks—on an organic farm, fertility is derived not from a bag but by mobilising soil micro-organisms. These, of which there can be as many as 5 billion in a few grams of soil, break down animal and plant wastes and, in the process, make the resulting nutrients available to plants. Organic fertilisers such as hoof and horn, bonemeal, and seaweed meal are occasionally brought in, but these are supplementary to, rather than a substitute for, on-farm-derived fertility.

This leads naturally to another significant difference: the presence of animals as an integral part of the farm, both as a source of income from livestock products and as recyclers of nutrients. An organic enterprise is almost invariably a mixed farm, employing a rotation of crops and livestock. Whereas the manure derived from conventional, intensive livestock units is viewed as a polluting waste product that incurs disposal costs for the farmer, to the organic farmer it represents the foundation of the farm's fertility.

Because organic regulations prohibit the intensive rearing of livestock, most animals are kept under free-range conditions, mostly outdoors. This necessitates the upkeep of stock-proof hedges and other field boundaries. In addition to their usefulness in preventing livestock from straying, hedges and accompanying grass headlands play an invaluable role in providing a habitat for many beneficial insects and predators. Consequently, ponds, hedges, woodland, and other conservation features are much more likely to occur on an organic farm, because they are such an essential part of a natural pest-controlling strategy. Though pesticides may be used occasionally, most organic farmers tackle pests, diseases, and weeds using environmentally benign alternative methods.

A fundamental yet less obvious difference lies within the soil itself. It is only when you closely examine the soil of most conventional arable farms that you realise how "lifeless" it has become. Depleted of organic matter, the countless micro-organisms have all but disappeared. Worms, for instance, which are highly sensitive to most agrochemicals, are generally conspicuous by their absence; organic matter levels are lower than in the soils of organic farms. This has important implications for soil erosion.

On organic farms, animals are not subjected to the same degree of stress that conventionally reared livestock have to endure. Organic farmers rely on a preventive approach to health and therefore aim to keep their livestock in ways that are compatible with their innate behavioural needs.

There are other major disparities between organic and conventional livestock husbandry. Chief is the prophylactic use of antibiotics and other drugs, which is the norm on intensive livestock units because of the high risk of disease. By adopting lower stocking rates, which eliminates overcrowding, organic farmers reduce the incidence of diseases. Use of prophylactic medicines is, in any case, forbidden under organic standards, though veterinary treatment with modern drugs is allowed to treat serious illness and to alleviate genuine distress. Similarly, no growth-promoting substances may be added to organic livestock rations, though they are frequently used in conventional feeds.

There are, of course, other differences that shall be covered later, but I should not like to leave this section with the idea that the organic farm of today is a throwback to its pre-war antecedents. Whilst it is true that organic agriculture is rooted in the traditional wisdom of the past, the organic movement does not eschew modern scientific research. The past few years have seen rapid strides in the development of modern organic techniques, especially mechanical and thermal methods of weed control.

The structural problem of converting conventional, all-arable farms has led to experimentation with stockless systems in which green manures and nitrogen-fixing plants are used in fertility-building instead of animal manure. In this context, the move by local authorities to compost domestic rubbish rather than incinerate or dump it in landfill sites presents the opportunity for its re-use on the land. Sewage sludge, though currently not allowed to be used on organic farms, may also make a comeback once clean-up technologies that remove heavy metals and other contaminants are perfected.

Despite the near total neglect of research and development relating to organic agriculture during the past half century, the performance of many modern organic farms is in line with the average for farms as a whole—a fact that may surprise many.

1.3 ORGANIC FARMING TECHNIQUES

It is frequently assumed that organic farming involves little more than substituting manure for artificial fertilisers and "organic" sprays for potentially more toxic pesticides. This explanation is wide of the mark. As has been previously mentioned, unlike conventional agriculture—with its subdivision into specialist areas of pig production, broilers, arable, and the like—organic agriculture is a whole system approach in which a direct relationship among stock, land, feed, and manure is, as far as possible, maintained.

Whilst manure is important to organic farmers, its primary role lies in enhancing the soil, unlike artificial fertilisers that feed plants directly. Crop rotations are also an important key to fertility in organic systems. They provide a balance between feeding the soil on the one hand and exploitational cropping on the other. In the United Kingdom, this chiefly involves what is known as "ley farming".

At any one time, organic farmland will be split approximately into 60% grass and 40% crops. Most of the land that is growing grass will be of a temporary nature (known as a ley), though some will usually be classed as permanent (pastures that are never ploughed). During the period that the land is sown to grass, it will be grazed by cattle or sheep, and so enriched by the manure deposited as they feed. During the winter, when the livestock are taken into the farmyard and bedded on straw, piles of "muck" build up that are spread on the land the following year. Typically, the ley will be grazed for 3 to 5 years before being ploughed in. The accumulated fertility is then released for the benefit of the ensuing cash crops, usually wheat, barley, or other cereals. After 2 or 3 years of such cropping, this land is returned to grass.

Most leys have a mixture of perennial ryegrasses, meadow fescue, white clovers, and herbs to increase the diversity, production, and resistance to disease.

Some longer term leys involve the use of quite complex seed mixtures, the best known being the Clifton Park mix, which employs more than 30 plant species.

Though grass is the predominant component of a ley, nitrogen-fixing clovers and other legumes play a key role. Bacteria living in the vicinity of the roots of such plants enjoy a symbiotic relationship, converting atmospheric nitrogen that can be exploited by the plant in return for advantageous secretions from its roots. This is, in effect, free fertiliser and can be the equivalent of as much as 150–200 kg of nitrogen per hectare per year.

Legumes are also introduced into the cropping phase of a rotation in the guise of green manures—non-cropping plants grown specifically to enhance soil fertility. Though green manures may be grown at any time of the year to exploit a gap in cropping, they are usually sown in the autumn, immediately after harvest. They are allowed to grow through the winter before being dug in or incorporated into the soil prior to drilling the new year's crop. As they decay, nitrogen and other nutrients are released, which can then be utilised by the succeeding food plants. Another common technique is to sow the crop and clover at the same time (undersow). This gives the clover a chance to establish before the winter and also helps to suppress weeds.

Typical non-leguminous green manures are mustard and grazing rye. Nitrogen-fixing green manures include clovers, vetches, and field beans.

By such methods, the major field crops of wheat, barley, and oats can be grown almost indefinitely, without recourse to outside inputs. Though minerals are constantly being taken out of the land in the food that is destined for animal or human use, the biological activity of soil micro-organisms stimulated by applications of organic matter is continuously making nutrients available from the reservoir in the soil. There is evidence to support this from a long-running experiment on a Suffolk organic farm. It indicated that the topsoil alone contained sufficient quantities of the major plant nutrients potassium and phosphorus to allow for a thousand years of exploitation, not counting the reserves in the subsoil (Milton, 1975). Where deficiencies do exist, however, a remedy is sought by application of rock minerals such as rock phosphate and magnesian limestone. Other field crops, such as oilseed rape and linseed, which have been planted on a considerable scale in recent years in response to generous European Union (EU) subsidies, have as yet failed to find a market when grown organically.

The main factors that distinguish organic livestock husbandry from its conventional counterpart relate to stocking densities, animal welfare, dealing with diseases, and rations.

Just as artificial fertilisers are forbidden from use on field crops, they are disallowed on grass. Cattle and sheep farmers, therefore, have to rely on the natural productivity of the grass, of which the aforementioned clover plays a major role. This is the main reason why organic farms have lower stocking densities, typically

1.7 cows/hectare compared with 2, which is the norm on a conventional dairy farm. It ensures that organic farms do not suffer from the worst kinds of overgrazing that are commonplace on conventionally managed holdings.

Most organic dairy units try to provide as much of their own cattle feed as possible using conserved grass (silage). Up to 10 tonnes of dry matter per hectare per year can be expected from lowland leys cut for silage; in upland areas, the yield is usually less. All organic dairy farmers have to use grain or other concentrate feeds to lift production, but organic standards forbid this to exceed 40% of the daily dry matter intake. Wherever possible, this supplemental feed should come from organic sources. It might be worth mentioning that the use of processed slaughterhouse waste as an animal feed, believed to be the cause of Mad Cow Disease, has been banned by organic standards since 1973.

Veterinary medicines cannot be used routinely for prophylactic reasons, though they can be used on individual animals to treat specific illnesses. Mastitis, the bane of the dairy farmer's life, which produces inflammation of the udder amongst other things, is conventionally dealt with by the use of antibiotics. The organic approach is to maintain good standards of hygiene, reduce the causes of udder injury, keep stress to a minimum, and use homeopathic techniques. Similar emphasis on good husbandry practices (including rotational grazing of sheep to break the life cycles of parasitic worms) obviates the need for routine wormers and other medicines.

Organic vegetables, such as potatoes, carrots, and brassicas, are produced on a field scale by introducing them into the farming rotation as an alternative to cereals and by specialist horticulturists, operating from individual holdings. Clearly, the type of system appropriate for a farm of several hundred hectares is inappropriate for a plot of 10 hectares or less, for which the use of livestock would prove uneconomic. Under such circumstances, fertility will usually need to be brought in by way of imported animal manures. This, unfortunately, is usually from non-organic farms, and so it must be composted before use. However, manures are not allowed from conventional farming enterprises that involve battery egg, broiler, or other intensive livestock units.

With regard to pest control, many of the pesticides used on conventional crops are applied as a direct result of the adoption of monoculture practices, actively encouraging the build-up of specific diseases and weeds. Crop rotations break such cycles and thus are one of the key ways in which organic farms deal with pest problems. Weeds—such as wild oats and blackgrass, for example, which flourish in cereal crops—quickly succumb, once the field is turned over to grass and grazed by livestock. Within pastures, though, certain deep-rooting weeds such as docks can be troublesome, and a variety of techniques, such as hand pulling, topping, and selective grazing, must be adopted. Populations of soil-borne fungal organisms, such as clubroot disease of cabbages and other brassicas, will, if deprived of host plants during the rotational cycle, decline significantly.

Growing crops on a horticultural scale has traditionally relied on cultivation, involving a combination of hand weeding and hoeing, for the elimination of weeds. Though these methods are still employed, a range of labour-saving weeders has, in recent years, come onto the market. These include mechanical devices that pass between crops and tear out weeds and a set of revolving brushes that "sweep" them up. Thermal weeders, which pass over the land a precise number of days after the crop has been sown, apply an instantaneous burst of intense heat, exploding the cell walls of any weeds in their path. A day or two later, the crop germinates in a field virtually devoid of weed competition.

Insect pests are dealt with using a variety of strategies. A great deal of emphasis is placed on encouraging natural enemies—such as insect predators and parasites, insectivorous birds, frogs, toads, and other predating creatures—through the provision of suitable habitats (trees, hedgerows, ponds, "wild" areas) and the specific cultivation of beneficial insect-attracting plants. In horticulture, a range of porous synthetic fabrics has been developed to cover crops during seedling stages. Such "fleeces" provide highly effective physical barriers to many flying pests.

Biological pest control, involving the controlled and deliberate release of beneficial insects, has a long history of use in commercial glasshouses for the control of pests on tomatoes and cucumbers. Red spider mite and glasshouse whitefly are just two of the pests that are successfully controlled by such methods. However, the problems associated with pesticide resistance and the cost of developing new compounds have made biological control a more attractive option for all farmers and growers. In recent years, new biological control agents have been developed to deal with pests in the field, such as vine weevil of ornamentals, codling moth of apples, and field slugs. Similarly, the breeding of disease-resistant crop varieties offers an attractive, non-chemical solution to disease attack, for organic and non-organic growers alike.

One of the more successful biocontrols has been used against a number of caterpillar pests. *Bacillus thuringiensis*, or Bt as it is better known, only acts when the caterpillar has eaten the product, which is sprayed onto the crop. The bacillus produces toxic crystals that poison the caterpillars but do not harm any other organisms in the vicinity. Latterly, Bt has been targeted by agrochemical companies for use in their genetic engineering breeding work, and plants containing the "poison" gene are about to come onto the market. Such blanket use is bound to hasten the day when the insect pests become entirely resistant to Bt.

It would be untrue to say that organic growers have reliable alternatives to pesticides in all cases and, whilst sprays are rarely employed on cereal crops, they are used on vegetables and fruit. Such sprays are usually derived from plants and are of low toxicity and persistence; they are unlikely to cause any significant environmental problems, except perhaps in exceptional cases such as the continued use of copper fungicides on vines. The types of spray allowed to be used in organic systems are listed as part of the organic standards (Exhibit 1–1). No fungicides or other post-harvest and storage chemicals are permitted.

There are, even so, crops that are extremely difficult to grow organically because of pest and disease problems. Apples, pears, and other "top fruit" are a case in point. The cosmetic standards demanded by today's market require near-continuous prophylactic spraying of pesticides by conventional growers, which can amount to 30 or 40 applications in a season. Varieties, such as Cox's Orange Pippin, that are particularly susceptible to several intractable diseases are almost impossible to grow successfully organically. Though a number of effective biological controls are in use on the Continent, the expensive pesticide registration regulations that exist in the United Kingdom preclude home use. It is somewhat ironic that fruit growers must continue using far more deadly, but licensed, chemical pesticides on their crops when much safer biological alternatives exist. Given that top fruit is such a small market, however, unless there is a relaxation in the UK position to permit the use of products licensed elsewhere in the EU, this unsatisfactory situation will remain and organic fruit will continue to come in from abroad.

1.4 THE ENVIRONMENTAL IMPACT OF AGRICULTURE

Agriculture necessarily disrupts the environment, since it replaces indigenous plants with cultivated crops. Different types of agriculture have different impacts on the surrounding ecosystem, both within the confines of the farm and in the immediate countryside. Similarly, farming systems differ in their dependency on external inputs, which in turn may cause environmental damage in their manufacture and supply. Some aspects of farming, such as pesticide contamination, may even involve a risk to human health. The consideration of these issues will help us determine to what extent a farming system can be described as "sustainable".

Whereas in the past farms were largely self-sufficient, today's farmers are highly dependent on the purchase of fertilisers, pesticides, fuel, heavy machinery, animal feedstuffs, and other off-site-derived inputs. The benefit in terms of high yields is obvious, but at what cost to the environment?

1.4.1 Pesticides

Of all aspects of modern farming, the role of pesticides is the most controversial. Now almost universally accepted as being integral to crop production, their use has grown phenomenally. In 1950, there were just 15 types of insecticide and

Exhibit 1–1 Inputs Allowed by EEC Regulation No. 2092/91 for Use in Organic Agriculture

Allowable Fertilisers and Soil Conditioners

- Farmyard manure
- Dried farmyard manure and dehydrated poultry manure
- Composted animal excrements
- Slurry or urine
- Composted household waste
- Straw
- Peat
- Mushroom culture wastes
- Worm compost (vermicompost)
- Guano
- Composts from plant residues
- Processed animal products from slaughterhouses and fish industries
- Products and by-products of plant origin for fertilisers
- Seaweeds and seaweed products
- Sawdust and wood chips
- Composted bark
- Wood ash
- Soft ground rock phosphate
- Aluminum calcium phosphate
- Basic slag
- Crude potassium salt
- Potassium sulphate containing magnesium salt
- Stillage and stillage extract
- Calcium carbonate of natural origin
- Calcium chloride solution
- Epsom salt (magnesium sulphate)
- Gypsum (calcium sulphate)
- Industrial lime from sugar production
- Trace elements (boron, copper, iron, manganese, molybdenum, zinc)
- Elemental sulphur
- Sodium chloride
- Stone meal

Allowable Pest and Disease Control Products

- Azadirachtin extracted from *Azadirachta indico* (neem tree)
- Pyrethrins extracted from *Chrysanthemum cinerariaefolium*
- Rotenone extracted from *Derris* spp.
- Preparations from *Quassia amara*
- Metaldehyde traps
- Diammonium phosphate traps
- Sulphur
- Quartz sand
- Potassium soap (soft soap)
- Pheremone preparations
- *Bacillus thuringiensis* preparations
- Granulose virus preparations
- Mineral oils
- Paraffin oil
- Beeswax
- Gelatine
- Hydrolysed protein
- Lecithin
- Plant oils, e.g., mint oil
- Deltamethrin and lambdacyhalothrin traps
- Copper preparations
- Ethylene
- Potassium alum (kalinite)
- Lime sulphur (calcium polysulphide)
- Potassium permanganate

fungicide in common use. By 1960, world-wide sales had increased to 850 million dollars; over the next 3 decades, this figure increased by 3,100% to over 26,000 million dollars in 1990. Within Europe, sales rose from 4,800 million ECUs in 1993 to over 6,500 million ECUs in 1996 (European Crop Protection Association, 1996/97). However, the fastest growth is in the developing world, especially Asia and Latin America, with China, Brazil, and India all major users.

Rachel Carson, in her seminal work *Silent Spring* (1962), was the first to draw attention to the devastating effect that pesticides could have on wildlife. Organochlorine compounds such as DDT, Aldrin, Dieldrin, and others were discovered to be tenaciously persistent. They are stored in fatty tissue; hence, as one animal eats another, residues accumulate at ever higher concentrations. Species at the top of the food chain, such as birds of prey, are particularly at risk. During the 1960s, raptor numbers fell alarmingly, and populations only began to recover after use of these products was severely curtailed. Such is the persistence of organochlorine compounds, and their ability to travel the world on wind and water currents, that they even threaten creatures such as polar bears living far away from any agricultural activity. According to MAFF (1995), "persistent organochlorines are still a threat to wildlife despite being withdrawn in 1989".

One would have expected that the huge amount of pesticides applied worldwide would have had a demonstrable effect on the target species. However, between 1945 and 1989, whilst insecticide applications increased tenfold, total crop losses to insects nearly doubled, from 7% to 13% (Pimental et al., 1978). Although changes in farming practices, such as the widespread adoption of monoculture, explain part of the higher losses, increased resistance is a crucial factor.

Whenever pesticides are applied, beneficial insects and other creatures that are present in the crop usually suffer as well. Given the capacity of most pests to produce more offspring, faster, than their natural enemies, subsequent infestations are less likely to be checked. Furthermore, there are invariably individuals in any pest population that are able to withstand the effects of the spray to a greater or lesser degree. When susceptible individuals are wiped out through spraying, pests with resistance are able to breed in an environment of reduced competition and decreased natural predation—an ideal way of speeding up natural selection. The same pesticide, used time and again against a particular pest species, eventually leads to the evolution of a race that is entirely immune. This has now become a major problem with more than 900 species of fungi, insects, and weeds recorded as being resistant to pesticidal sprays. In some parts of Asia it is reported that farms have to apply pesticides at up to 18 times the recommended dose to ensure an effective kill (Brown et al., 1996).

Another unexpected side effect of pesticide use is the creation of secondary pests. Species that were originally minor problems can be catapulted into public enemy number one status when a spray programme eliminates the major pest species. This phenomenon has been most graphically demonstrated in the intensive cotton growing areas of the San Joaquin Valley in the United States, as one new pest species after another arose following an intensive spray programme.

Of course, it is not just the pests and the creatures feeding on them that are killed by pesticidal sprays. Other species that are in the path of the sprayer also suffer. Honeybees, for example, are particularly at risk in fields where crops such as oilseed rape are in flower. Though farmers do their best to minimise such damage by warning local beekeepers, wild bee populations inevitably suffer. Nematicides (compounds used against nematodes) are lethal to virtually all soil organisms (Edwards, 1984), and soil fumigants dramatically reduce earthworm numbers (Wright, 1979). It is impossible to estimate the direct impact of pesticide poisoning on wildlife, but it is salutary to note the comments of Pimental and Levitan (1986), who estimate that less than 1% of crop sprays reach their targets because of the small size of the pest, drift caused by wind, poor timing of applications, and other factors. As there is no official monitoring of wildlife losses, poisoning of smaller birds, mammals, amphibians, and reptiles of necessity goes unreported.

Apart from direct poisoning, there are indirect routes by which wildlife is adversely affected. Herbicides, which make up over half of the 23,000 tonnes of active ingredients of pesticides sprayed onto UK farms every year (British Agrochemicals Association, 1996), are frequently the chief culprits. The best documented example has been that of the decline of the grey partridge across much of southern Britain. This has been found, by the Game Conservancy, to be due to starvation of the newly hatched chicks (Potts, 1986). The young birds depend, during the first few weeks of their lives, on a plentiful supply of insects. Such insects in turn depend on there being weeds present in the crop. With the advent of cereal herbicides, which virtually ensure clean fields, there are no weeds and hence no insects, and the food chain collapses. Indeed, according to Potts (1986), the use of chemicals on cereals has seriously depleted or threatens 800 species of insects, 90 species of flowering plants, and 14 bird species.

Kickruth (1982) describes what happens when phosphonylsarcasine, the active ingredient of a commonly used herbicide, enters watercourses. The water flea relies on detecting sarcosine, a substance produced by decaying organic matter, to guide it to its food supply. The presence of the closely related herbicide confuses it so that it can no longer find its food. As the flea is at the base of the food chain, the entire chain is put at risk. This sort of effect is something that no amount of toxicity testing in advance of release into the environment could be expected to predict. Another unforeseen consequence of pesticide use has emerged in recent years; that of the potential threat to the ozone layer as a result of fumigation with the soil sterilant methyl bromide. It is estimated that it could be responsible for as much as 10% of observed atmosphere ozone loss (United Nations Environment Programme, 1992).

In recent years, the trend has been for agrochemical companies to produce very potent low-dose-per-treated-area pesticides. These are difficult to detect even by sensitive analytical equipment, and therefore it is much harder to assess potential dangers.

Pimental and colleagues (1992) are one of the few groups who have attempted to quantify the full costs of pesticide use in the United States. They come up with a figure of $8,000 million annually, of which $3,000 million is borne by the pesticide user and a further $5,000 million represents environmental and public health costs.

In the United Kingdom, an estimated £800 million to £1,000 million has been spent by water companies on treatment works for removing pesticides from drinking water. Every year, running costs amount to £80–100 million. Furthermore, the cost of monitoring pesticides in water supplies is thought to be in the region of £1 million per annum for each of the 21 water companies (Redman, 1996).

That countries can reduce the use of pesticides if the political will is there is shown by the example of Sweden. It cut consumption by half between 1986 and 1990, and by a further 64% between 1991 and 1996. Denmark, too, has an active pesticide reduction strategy.

Pollution by pesticides from organic holdings is relatively non-existent. Herbicides are not allowed under organic standards and those few insecticides that are permitted are unlikely to pose any long-term environmental hazard.

1.4.2 Fertilisers

One of the unwanted features of the intensification of agriculture is the pollution of the environment by fertilisers and manures. It is estimated that at least half of all nitrogen fertiliser applied to the soil is wasted (Cooke, 1979). Nitrate nitrogen dissolves easily in water. Once in the soil, any nitrogen not taken up by the crop finds its way into the field drainage system and hence either to surface waters or, by percolating down through the soil, eventually gets into deep underground aquifers. Any remaining nitrogen is bound up with soil particles or alternatively lost to the atmosphere, to return ultimately as acid rain.

But fertilisers are not the only sources of contamination of water courses. Ploughing up grassland releases nitrate nitrogen, and effluents from intensive livestock units also make a significant contribution, as do discharges from sewage treatment works. According to the Royal Commission on Environmental Pollution (RCEP) (1979), agriculture accounts for 50%–60% of the nitrate load of surface waters, and perhaps as much as 80% in certain cases (Moss, 1978).

One needs to distinguish here between "point source" pollution arising from a single, traceable source and "diffuse" pollution, which occurs when fertilisers routinely get washed off a field. In 1994, the National Rivers Authority (NRA) (now the Environment Agency) reported that 13% of all point source pollution, amounting to 3,329 incidents, was caused by farmers (NRA, 1995). Of these incidents, around a quarter were caused by leakages from slurry (liquid manure) stores, a similar amount from silage effluent, and approximately 20% from discharges of "dirty" water from cattleyards and milking parlours. The vast majority of point source pollution comes from industrial sources and sewage. However, according to Brown et al. (1996), leached and eroded nutrients help make agriculture the largest diffuse source of water pollution in the United States.

Once in water, such wastes are broken down by micro-organisms, resulting in a lowering of oxygen levels. Under such circumstances, fish and other aquatic species can quite literally drown. Alternatively, they may be poisoned by the ammonia present in nitrogen fertilisers. Salmon and related species are especially at risk in this respect.

But it is not just the nitrogen element of fertilisers and manures that causes damage. Excess phosphorus, by stimulating rapid algal and plant growth, can quickly lead to the choking of lakes and rivers. This process, known as eutrophication, also causes fish and other aquatic wildlife to die. In addition, the potentially carcinogenic toxins produced by some blooms of blue-green algae can pose acute health risks to humans and animals. Overall, in 1990, roughly as much phosphorus—33 million tonnes—was washed into the world's oceans as was applied to cropland (Brown et al., 1996).

On land, fertiliser application results in a number of adverse reactions. Soil can become acidified, resulting in the risk of a greater rate of leaching of other essential nutrients. Beneficial micro-organisms such as free-living and symbiotic nitrogen-fixing bacteria are suppressed, as are root-inhabiting mycorrhizal fungi that assist phosphate uptake. On permanent grassland, repeated use of nitrogen fertilisers and herbicides reduces the wildflower population, whilst newly seeded grass only contains those few species, such as rye grasses, that are fertiliser-responsive.

Because of the concerns to human health from excess nitrate in drinking water (see Section 1.6.1), the EU has fixed a maximum allowable concentration of 50 mg/litre. To comply with this legislation, many water companies have had to devise a range of strategies, including blending low and high nitrate supplies and building new treatment works. Redman (1996) estimates that the capital programme to date has cost £150–£200 million, with an annual running cost of some £10 million. With borehole supplies, the problem is likely to remain for many years, since nitrate-contaminated rainwater can take decades to percolate down through rock before reaching the aquifer.

A further cost arises in the United Kingdom, as MAFF compensates farmers who farm in areas where nitrate levels are high for crop reductions as a result of their using fewer inputs. The Nitrate Sensitive Area and Nitrate Vulnerable Zone schemes, together with associated programmes of research into nutrient losses, have cost the British taxpayer £74.9 million since 1989 and currently run at £13.4 million annually.

How do organic farms compare with their conventional counterparts in terms of pollution from fertility sources? There is, of course, no problem from synthetic fertilisers, as they are prohibited. However, there are other risks. Manure badly handled on an organic farm can be just as polluting as on a non-organic farm. Also, the ploughing in of a temporary grass/clover ley, especially if done in late autumn, can lead to a nitrate flush as the foliage decomposes. However, given the importance of such nutrients to organic farmers, great care is usually taken to minimise losses. Non-hardy winter green manures, for instance, may be sown with the following cereal crop immediately after turning in a ley in order to mop up nutrients that might otherwise leach away.

A number of researchers have tried to compare conventional and organic farms from the perspective of nutrient leaching. Fiege and Roethlingshoefer (1990) measured the nitrate concentration of drainage water and discovered that the peak was less from the organic field (180 vs. 100 mg/L nitrate) and the average total nitrogen loss was some 50%–60% less. Several other studies are reviewed by MAFF (1995), and the overall conclusion is that leaching from organic farms, taken over the entire crop rotation, is likely to be considerably less than from conventional farms. In this context it is illuminating to note that in 1998 Wessex Water introduced a grant worth £40/hectare to farmers in sensitive catchment areas to switch to organic farming.

1.4.3 Wildlife

Wildlife has undoubtedly been the loser in the post-war drive to intensify agriculture. A number of forces have been at work. We have already looked at the direct impact of pesticide and fertiliser use on flora and fauna. However, perhaps of even more importance has been the loss of habitat. The attractive prices paid to cereal producers throughout the 1970s and early 1980s made it economic to plough up grass and other land that would otherwise have been unsuitable for such crops. Grubbing up hedges, creating vast fields to suit the demands of large scale machinery, was commonplace. Woods were destroyed and ponds filled in.

The 1984 report from the erstwhile Nature Conservancy Council (NCC) charts the changes affecting specific habitats that occurred between 1945 and 1984 (Nature Conservancy Council, 1984). It makes grim reading:

- 95% of Britain's wildflower-rich meadows destroyed
- 30%–50% of ancient lowland woodlands destroyed
- 80% of lowland grasslands on chalk and limestone destroyed

- 50%–60% of lowland heaths destroyed
- 50% of lowland fens destroyed or damaged
- 50% of upland grasslands, heaths, and mires destroyed

Most noticeable of all has been the loss of hedgerows—some 140,000 miles in total, of which all but 20,000 miles is directly attributable to farming.

Needless to say, such a wholesale removal of unique habitats has taken its toll on wildlife numbers, and though extinctions are still, thankfully, rare, the status of a considerable number of animals and plants has been altered to rare or vulnerable. Table 1–1 is derived from figures supplied by the Department of the Environment and is contained in the report *Organic Farming and the Countryside,* edited by Mark Redman for British Organic Farms (1992).

Most striking of all has been the decline in once-common farmland birds. Every year the British Trust for Ornithology (BTO) publishes its Common Bird Census, charting the changes in bird populations. Its findings make grim reading. The latest census reported a decline of 41% in the number of linnets, 52% in song thrushes, 62% in bullfinches, and 33% in blackbirds over the period between 1972 and 1996 (Crick et al., 1998).

These trends are now being somewhat addressed by conventional farmers. Techniques such as integrated pest management (IPM), in which predator numbers are monitored and taken account of before spraying takes place, or the sowing of grass strips that act as "beetle banks" and habitats for other beneficial species, are beginning to be adopted. Nevertheless, for Britain to lose so many of its common farmland bird species is a shocking indictment of modern farming methods.

Are organic farms more beneficial to wildlife? As mentioned earlier, conservation features such as hedges, ponds, and woodland are an essential part of the organic strategy of combating pests and diseases by enlisting the help of natural enemies. However, in addition, desirable environmental and conservation require-

Table 1–1 Species Lost or Endangered

	Total Number of Species	*Number Extinct*	*Number Endangered, Rare, or Vulnerable*
Dragonfly, since 1953	43	3	12
Butterfly, since 1950	55	1	24
All insects, current	22,430	—	1,685
Plants, since 1930	1,423	10	149
Breeding birds, current	215	—	51
Mammals breeding on land, current	51	—	25

ments are incorporated into the Soil Association organic standards, prohibiting farmers from ploughing unimproved pastures, trimming hedges between the end of March and the beginning of September, ploughing species-rich grassland, or creating any new or improved drainage affecting wetlands of significant conservation value.

In recent years there have been quite a number of scientific investigations into the claims that organic farming confers benefits to wildlife. The Game Conservancy Trust sampled winter wheat fields on three organic and five conventionally managed farms to assess weed and insect populations in early summer, the crucial time for feeding game bird chicks. Grass weed species were no different, but there were significantly more species and greater cover of broad-leaved weeds in the organic fields. (Moreby and Southway, 1993).

Hald and Redderson (1990), comparing 15 paired fields in Denmark, showed weed number and biomass to be significantly greater in the organic fields. In Nebraska, bird population densities on fields managed organically were 6 times those on conventional farms (Ducey et al., 1980). This is backed up by a study in the Midwest by Gremaud and Dahlgren (1982), who found population increases of between six and eight times conventional levels on organic farms.

BTO has been looking at bird populations and species diversity on organic farms for a number of years, and in a variety of publications has concluded that birds appear to do better than on conventional holdings. Endangered species such as the skylark show a noticeable improvement.

At the instigation of the Prince of Wales, an intensive 3-year study, from 1994 through 1996, was carried out by an interdisciplinary team at his Gloucestershire farm. The interim results were reported at a seminar organised in March 1997. From the fauna study, it was found that there were more butterflies overall than on conventionally managed farms. Invertebrate numbers were also higher, though intriguingly the only pest species to be measured, aphids on winter wheat, were higher on the conventionally managed fields (Organic Farming Study Forum, 1997).

Summarising the results accumulated to date, the aforementioned MAFF review concluded that "sufficient data has been gathered to indicate that conventional farming practices and inputs pose far greater dangers to wildlife than organic practices and inputs".

The MAFF review was published in 1995, too early to assess the environmental impact of genetically engineered crops. Concern has been expressed, however, about the widespread introduction of herbicide-resistant crops. In early 1998, all four of the government's statutory nature advisors, including English Nature and the Joint Nature Conservation Committee, called for a 5-year moratorium on herbicide-tolerant crops such as oilseed rape and sugar beet, because their engineered tolerance to weedkillers means that farmers can now kill all the weeds that provide food for birds and host plants for insects in the spring. A further worry is that engineered resistance could spread to neighbouring wild weeds, which would then require extra herbicide applications to keep them under control. (Chevre et al., 1997).

Other preliminary data suggest that beneficial insects may also be at risk. Ladybirds feeding on aphids present on potatoes that had been engineered to include a gene from snowdrops lived only half as long as those fed on aphids from normal potatoes. ("Call for a Spindoctor", 1997) And the lacewing, another aphid predator, was shown to be at risk if it ate corn borers killed by genetically manipulated maize.

1.4.4 Soil Erosion

In any discussion of the deleterious effects of conventional farming methods, soil erosion rarely rates a mention, yet, as the global evidence shows, it is a serious environmental problem.

Reserves of topsoil, which take thousands of years to form, are being squandered at an alarming rate. World-wide, an estimated 25,000 million tonnes of topsoil are lost from croplands every year (Brown and Wolf, 1984). This is not confined to obviously susceptible areas, such as the Himalayas or the Ethiopian Highlands. The United States loses 1.7 billion tonnes of topsoil annually, and each year in the former Soviet Union, more than a million acres are abandoned because they are so severely eroded that they are no longer worth farming (Brown and Wolf, 1984). The Global Assessment of Soil Degradation, a 3-year study involving 250 scientists, found that in the 40 years since 1945, 679 million hectares had been lost to overgrazing, 579 million hectares to deforestation, and another 552 million hectares to agricultural mismanagement. Together with miscellaneous other losses, this constitutes a sixth of the planet's total vegetal land (Brown et al., 1996).

In the United Kingdom, the government's own research group reported that 44% of the country's arable land was at risk from erosion ("Soil Erosion in England and Wales", 1986), and on the Sussex Downs, erosion rates of 200 tonnes per hectare, representing 10% of the soil in the field, have been documented (Boardman, 1986). The Council for the Protection of Rural England, in May 1998, launched its "Charter for Soil" to draw attention to the threat to soil posed by modern farming methods.

The soil loss on the Downs typifies the land most at risk: slopes where the cultivation of winter cereals leaves the land exposed to the effects of rainfall that washes the soil downhill. This type of land should be permanently grassed, as indeed it was prior to the advance of the CAP regime, when the Downs were famous for their sheepwalks. Other aspects of conventional farming—hedgerow removal, soil compaction by heavy machinery leading to permanent tractor runnels (which are ideal for carrying away silt-laden water), and a general reduction in soil organic matter content—all contribute to the problem.

Nor should the problem of soil erosion be thought of purely in the context of degraded farmland. Streams and rivers also deteriorate as channels silt up, with a consequent loss of aquatic habitats. Wildlife may be further exposed to the additional pesticide load from contaminated soil particles.

To an extent, the declining fertility of soil is disguised by increased applications of artificial fertilisers and higher yielding plant varieties. However, it is imprudent, to say the least, to be moving towards a position where land is incapable of producing a crop without the addition of imported fertilisers.

By contrast, organic farms—with their 60:40 mixture of grass and arable cropping, greater density of hedgerows, enhanced soil organic matter (produced as a result of decomposing grass leys), and use of green manures and cover crops—are intrinsically less prone to soil erosion. This was borne out in a post-war study (Reganold, Elliot, & Unger, 1987) of two neighbouring farms—one conventional, one organic—which found that 5 cm of soil had been eroded from the organic fields, compared with 21 cm from the non-organic fields. In the case of the conventional farm, the soil loss represented one-third of the total topsoil present in 1948.

1.4.5 Energy and Non-Renewable Resources

Proponents of the farming status quo in the United Kingdom never tire of referring to the superior "efficiency" of conventional farming, over our much-derided "inefficient" continental partners in the EC. What is usually meant by efficiency, when thus expressed, is output per acre, per livestock unit or, indeed, per agricultural worker. However, if efficiency is instead defined by comparing the energy content, or calorific value, of the crop with that required to produce it, a very different picture emerges. More energy goes in than is produced, such is the energy demand of modern farming in terms of fuel to run and manufacture machinery, and to make fertilisers and pesticides.

By contrast, organic agriculture consistently uses up to 60% less fossil energy per unit of food produced than conventional agriculture (Lampkin, 1990). However, given that so much organic food currently sold in the United Kingdom is air freighted in from countries many miles distant, any localised energy consumption advantage is negated by energy used in transport. This whole issue of the transport costs of fragmented modern food production is well covered in a report by the SAFE Alliance (Paxton, 1996).

Opinion differs as to the availability of oil reserves, but as they become scarce, probably during the next century, such energy inefficiency might well be a determining factor in making organic farming more competitive. Other resource shortages will confront future generations. Current estimates put phosphate reserves at 400 years and potassium reserves somewhat less. As both are indispensable to conventional producers, it is clear that alternatives must ultimately be found. It is likely that recycling of waste products, especially household rubbish and human sewage, will feature significantly in the solution to this problem.

1.5 PHYSICAL AND FINANCIAL PERFORMANCE OF ORGANIC FARMS

As has been already noted, the market for organic food really took off in the mid-1990s in the United Kingdom and Western Europe. Food scares, such as the fear of pesticide residues, Mad Cow Disease and various bacterial contamination outbreaks, all contributed to a search by many consumers for a food source that was both safe and wholesome. Others found organic food a tasty alternative to the often bland fare produced by conventional methods. However, of the £260 million worth of retail sales recorded in the United Kingdom in 1997, the vast majority was imported. So why aren't UK farmers and growers taking advantage of the buoyant market and high premiums available to organic producers?

In some respects, the price premium is a disincentive to farmers considering conversion. Current premiums are so good that farmers reason they cannot last once the market moves into balance. This view is based on the belief that organic food will never be more than a niche market for the well-heeled. Though one encounters this opinion quite often, it flies in the face of repeated consumer surveys that indicate that a majority of consumers would choose to buy organically produced food if the price differential were lowered, typically from a 50%–100% premium to one nearer 10%–20% ("The Green Consumer", 1989).

Elsewhere in Europe, official support during the conversion period has persuaded farmers that, by adopting organic farming, they are moving in tune with government aspirations for more environmentally sustainable agriculture. This attitude has yet to manifest itself in the official pronouncements of the British government.

A more serious criticism of organic farming is that, because of lower outputs, a widespread switch would imperil food security. Indeed, a common claim emanating from supporters of the high-input approach to agriculture is that organic farming is the quickest way to mass starvation.

1.5.1 Crop Yields

In the past few years, the increasing number of organic farms has enabled more reliable data to be collated. Routine monitoring of yields and economic performance generally is conducted by MAFF and published in various forms.

The earliest study was conducted by Vine and Bateman in 1981. It indicated that organic yields average between 10% and 30% less than would have been expected if conventional methods had been used. However, they did find significant individual variations between farms; for winter wheat production, for example, organic yield varied from 3.0 to 5.3 tonnes/hectare, compared with conventional yields of 4.6–5.7 tonnes/hectare.

Much of the credit for present-day conventional high crop yields must go to plant breeders, who have produced a succession of high-yielding seed varieties. These have been bred to respond to high fertiliser inputs, and it cannot be assumed that such varieties will perform as well when grown under an organic regime, where nutrients are made available much more slowly and in a less concentrated form.

Another difficulty when comparing organic and conventional yields is the variability in farmland location. In the case of field vegetables, for example, where much of the organic production is concentrated in West Wales, the yields cannot be expected to equal those of crops grown in the rich fen soils of the east of England.

Because organic farmers rely to a large extent on fertility derived from the rotational system of farming, the yield from any particular crop depends to a large extent on its position in the rotation. A winter wheat crop, exploiting the abundance of nutrients following the ploughing in of a temporary ley, would reasonably be expected to outperform a similar crop sown before the field is returned to grass. It follows that any proper comparison of organic and conventional yields needs to be aggregated over the entire rotational period of 7 years or more.

More recent assessments, based on yields achieved during the 1996 season, confirm cereal yields of 50%–75% of that of conventional farms. Some farmers do better than that—in 1996, some harvests of 9 tonnes/hectare were reported, though the norm was 4–5 tonnes.

In 1996, fresh fruit and vegetables accounted for 44% of all retail sales. Of this, 70%–80% was imported. The organic horticulture industry in the United Kingdom is still very fragmented, with a lot of small producers (see Section 1.5.5). Statistical data for this sector are, therefore, very patchy, though it would appear that organic growers are able to achieve yields of up to 75% of their chemical-using counterparts. Results from the MAFF experimental research station at Terrington in Norfolk showed that potato yields in the organic plots were equal to those supplied with chemical fertilisers for 3 years out of 5. (Ministry of Agriculture, Fisheries and Food, 1998).

The markets for crops such as sugar beet, linseed, and oilseed rape are undeveloped, so what little organic acreage is grown does not yield sufficient data to permit informed comment.

Overall, the evidence to date indicates that, whilst yields would decline under an organic regime, the declines would not be catastrophic; indeed, in the context of a European over-supply of cereals and other agricultural commodities, it could be seen as a positive advantage. It should also be remembered that, for the past 40 years, organic agriculture has been effectively starved of research and development; with an increasing effort in this area, considerable productivity gains are possible.

1.5.2 Livestock Outputs

The organic livestock sector is still poorly developed, with zero penetration in many parts of the country. Whereas most supermarkets now stock organic fruits and vegetables, organic meat and milk are still hard to find. With the establishment of marketing initiatives in the late 1990s, the situation is expected to improve, but the figures for retail sales in 1996—7% for dairy sales and 13% for meat—tell their own story.

Comparisons of relative outputs of livestock under organic and conventional systems are difficult to evaluate because of their very different natures. For instance, organic livestock is managed extensively, with stocking rates generally 20% or so less than those reared conventionally. Feedstuffs are, in the main, home produced, relying heavily on forage from the grass/clover ley element of the rotation plus cereal legumes such as field peas and beans. This compares with conventional production, which is centred around the feeding of high-protein concentrates, much of which, as in the case of soya beans, is imported.

For example, an organic dairy unit stocked with black and white Holstein/Friesian cows could be expected to produce between 5,000 and 6,000 litres of milk per cow based on a stocking density of 1.7 cows per hectare. A comparable conventionally managed herd would produce more or less the same amount per cow, say 5,500 litres, but the overall productivity of the farm would be higher because of the higher stocking rate of 2 cows per hectare. From the point of view of national self-sufficiency and food security, it is worth reflecting on the annual consumption of cereals for livestock feed, which accounts for around 45% of all cereals grown in the United Kingdom.

It is, perhaps, surprising that there is still very little organic pork produced in the United Kingdom, since a quarter of the nation's pigs are reared mostly outdoors. Perhaps it is the requirement to provide organically derived feed, to practise shorter rotations, and to finish piglets outdoors rather than under cover that puts off non-organic pig farmers.

Poultry, too, despite being the most popular meat with consumers, has yet to find a volume market. Of all livestock, chickens are the most intensively reared: Broilers are routinely slaughtered at 6 weeks, compared with the 10 to 11 weeks required for organic birds. Figures for productivity of both layers and table poultry raised under organic regimes are not yet readily available because of the relatively small market.

1.5.3 Prices

Any discussion of prices must take account of the prevailing circumstances at the time of writing; in a rapidly growing market such as the current one, figures may quickly become outdated. Nevertheless, some general points can be made.

A price premium has almost always existed for organic produce, reflecting the farmers' higher costs of production and the customers' belief that they are buying a superior product. Furthermore, the strong demand for organic produce has been associated with an undersupply in the marketplace, producing an inevitable upward pressure on price.

During the early 1990s, demand for organic food levelled off in much of Europe and, whilst recessionary conditions persisted, the premium market almost disappeared. However, as the new millennium approaches, premiums for most commodities are holding strong. For example, for horticultural produce, price premiums of 50% or more are the norm. Milling wheat for bread making was fetching an extra £70–100 a tonne over non-organic wheat in 1997. In the same year, demand for organic feed wheat to support the rapidly growing organic livestock industry was generating an additional return of around £40/tonne.

It is difficult to generalise about meat and livestock products because of the fragmentation of the market. However, in 1997, many organic dairy farmers were getting a 30% premium, with milk fetching 29p/litre compared with the 21p–23p being paid to conventional farmers. A more realistic premium, though, would be around 3.5p/litre, or 15%. Beef and sheep meat typically attracts a 10%–20% premium, and pigmeat perhaps as much as 60%. In many cases, similar premiums can be obtained for other quality meat schemes (see Section 1.6.5). The strong consumer demand for organic table birds as an alternative to the insipid broiler product has allowed producers to get away with up to double the price, whilst organic eggs usually sell at a margin of at least 30% above conventional prices.

1.5.4 Costs

As with prices, it is hard to be specific about costs, other than in general terms. On organic farms, there is clearly a significant reduction in outgoings as a result of farmers' not having to buy in artificial fertilisers and pesticides. However, this is offset to an extent by the increased number of mechanical cultivations that may be required, particularly for weed control. In certain cases, seed costs are also higher, because of the higher density sowing rates required to crowd out weeds.

Contrary to popular belief, organic farming is not intrinsically more labour-intensive than non-organic farming, except in specialised circumstances such as the intensive growing of vegetables, which may require a considerable amount of hand weeding. However, if an all-arable conventional farm converts to an organic mixed farm, the additional livestock enterprise (which must be developed for fertility-building reasons) will probably require additional labour.

The cost of conversion is the main stumbling block for many farmers contemplating change. Capital costs can be enormous, particularly if livestock must be purchased, buildings and stockproof fencing erected, new machinery obtained, and so on. As part of its Organic Aid Scheme Review, the UK government investigated conversion costs prevailing during 1997. It found that, depending on the type of enterprise, costs ranged from £50 to £600 per hectare, spread over the complete conversion period (Ministry of Agriculture, Fisheries and Food, 1998). The ability or inability of government-funded conversion schemes to cover these losses is a key determinant in the take-up of organic agriculture by the farming community.

1.5.5 Profitability

Given the combination of generally lower yields and variable costs, higher labour costs, and a price premium over conventional produce, what is the overall effect on profitability?

Vine and Bateman (1981) found that the price premiums obtained were not sufficient to compensate for lower yields and that the majority of farmers in their survey could have obtained a better return had they farmed conventionally. The main weakness of the study was that it dealt with only 30 farms and 1 year's figures, and that, at the time, commercial organic farming was in its infancy.

A later study from West Germany compared 57 organic farms with 223 conventional farms, though again over only a single year, 1987–88 (West Germany Federal Ministry of Agriculture [BMELF] Annual Report, 1989). The study demonstrated the same profitability for both groups, even though the output of the organic farms averaged 18% less. This was mainly because costs were less for the organic farms, by 23% on average. Another report (Lampkin, 1993a), covering the years 1985–86 to 1990–91, reinforced these results, with the organic farms, if anything, outperforming the conventional farms.

A similar conclusion was reached in a detailed study of 14 paired organic and conventional farms in the US cornbelt for the period 1974–1976. This was repeated in 1977–78 with 20 farms, again with the finding that net returns for both sets of farms was more or less equal, mainly because costs incurred by organic farms were 25%–50% less than the average for the conventional farms (Lockeritz et al., 1984).

Also in the United States, the National Research Council (NRC), reporting in 1989 on a four-year study of "alternative agriculture" that included organic farms and other low-input systems, concluded that well-managed alternative farms "use less synthetic chemical fertilizer, pesticides and antibiotics without necessarily decreasing per-acre crop yields and productivity of livestock systems. Wider adoption of proven alternative systems would result in even greater economic benefits to farmers and environmental gains for the nation".

However, a trial conducted by Cambridge University (Murphy, 1992), surveying the results from 238 organic farms and smallholdings for the period 1989–90, painted a much more dismal picture. The conclusion—that more than half of the farms were losing money and one in eight was technically bankrupt—led to numerous banner headings in the press at the time, dismissing organic farming as a viable option for farmers.

What the study failed to take into account was that, during the year in question, many conventional farms also failed to make money. Indeed, the data for 38 farms in Wales were subsequently re-analysed and compared with published acreage figures for conventional Welsh farms (Lampkin and Bateman, 1993). The authors conclude that "if farms are distinguished into full and part-time types, the organic mixed and dairy farms perform as well, or better, than their conventional counterparts, despite smaller than average farm size".

The most recent figures for profitability of organic farming in the United Kingdom are shown in Table 1–2 (Lampkin, 1997). Apart from hill livestock, all sectors of agriculture currently show that it is more profitable to farm organically.

These findings were also backed up in the case of the Prince of Wales' farm study, which considered the economics during the latter part of the conversion. It

Table 1–2 Financial Returns, Including Fixed-Cost Changes, under Conventional and Organic Management and During the Transition Period (£/ha, 1997 prices)

Farm Type	*Conventional*	*Transition**	*Organic*
Specialist dairy	1,489	1,566	1,860
Mainly dairy	1,077	1,137	1,356
Stockless arable	799	848	942
Mainly arable	692	627	654
Lowland livestock	655	589	595
Upland livestock	450	459	482
Hill livestock	338	326	318

*5-year average

found that the farm as a whole was more profitable by 15%–20% on gross margins compared with non-organic agriculture with current EU subsidies in place, and by 25%–30% with all subsidies removed (Organic Farming Study Forum, 1997).

1.6 STANDARDS FOR ORGANIC FOOD AND FARMING

1.6.1 Human Health Considerations

Though an increasing number of consumers indicate in surveys that environmental considerations play a part in their food purchasing decisions, it would appear that health reasons are the predominant motive for buyers of organic products, especially concern over the presence of pesticide residues and other contaminants, which routinely occur in most conventionally produced food.

As a species at the top of a food chain, humans are at the most potential risk from poisoning by pesticides. The fear is not that of acute poisoning (though, regrettably, this does occur, especially in the developing world) but more of the possible long-term adverse effects brought about by continually eating food contaminated with low levels of pesticide residues. Many pesticides are known to cause cancer, genetic damage, and other disorders in laboratory animals. Of the pesticides evaluated by the US Environmental Protection Agency (EPA), for example, 92 are listed as possible, probable, or known carcinogens (EPA, 1992).

It is thought that one route by which cancer might be induced by pesticides is through a suppression of the body's immune system. Research by the World Resources Institute ("Pesticides and the Immune System", 1996) with animals subjected to high pesticide exposure demonstrated such a deterioration. Numerous studies also show a higher risk of cancer in people, such as farmworkers, who are occupationally exposed to pesticides. Typical cancers include those of the lung, kidney, and testicles, and various leukaemias.

In recent years, a great deal of attention has been focused on farmers suffering illnesses as a result of dipping sheep with organophosphate pesticides. The same type of pesticides, known as o-p compounds, have also been implicated in Gulf War Syndrome. Symptoms include nausea, headaches, chronic fatigue, cardiac problems, and psychiatric and nervous disorders. The generalised nature of many of these symptoms, which could readily be attributed to other causes, is one of the main reasons why doctors frequently overlook possible pesticide poisoning when making their diagnoses.

The agrochemical industry dismisses such fears on the basis that any residues present in food are in such minute quantities as to be harmless. However, it is worth considering the practice of homeopathic medicine, whereby beneficial effects on human health are produced by taking potions diluted to concentrations several orders of magnitude less than those of most pesticide residues. It is also the case that the individual effect of a particular compound may be amplified a hundredfold or more when it reacts synergistically with other unrelated compounds in the body. As yet (in part because of the difficulty of devising effective epidemiological trials), there is no substantive body of scientific opinion proving the case one way or the other. In effect, most of the population are unwitting guinea pigs in a long-term experiment.

Most governments carry out routine monitoring of food for pesticide residues. In the case of the United Kingdom, residues are typically found in up to a third of all samples, though at levels said to be "harmless". In 1995, though, MAFF issued a health warning on carrots because of the exceptionally high o-p residues present. Carrots were not to be eaten raw unless first peeled, "topped and tailed".

Paradoxically, people who profess to be "healthy eaters", consuming raw and whole foods and including many fruit and vegetables in their diet, may be putting themselves more at risk of consuming excess pesticides, which are present on the outer surfaces of raw food products. Young children also may be exposed to two or three times the equivalent adult dose.

Turning now to the problem of the acute toxicity of pesticides: According to the World Health Organisation, each year more than 3 million people world-wide suffer acute pesticide poisoning, of whom 20,000 may die (Pesticides Trust, 1992). Many of these poisonings occur in the developing world, which is the fastest growing market for pesticides (Jeyaratnam, 1990). Such tragedies frequently arise when agricultural workers, many of whom are illiterate, fail to take any precautions or use any protective clothing when spraying chemicals, many of which are banned in the West. Much of this food is grown for export, however, and so the pesticides return in the form of toxic residues—the so-called "Circle of Poison".

In November 1992, Britain was convicted in the European Court of Justice for breaching the legal EC nitrate limit of 50 mg/L for drinking water, as set out in directive 80/778 issued in July 1985. Monitoring by the European Environment Agency in 1995 indicated that 85% of the agricultural land in Europe has nitrate levels above the guide level of 25 mg/L, and 20% of this land exceeds the 50 mg/L permitted maximum.

What are the health hazards from nitrate-polluted drinking water? At high concentrations, it can produce methaemoglobinaemia or "blue baby syndrome" in new-born infants, though this is unlikely to occur at levels below 100 mg/L. There is also a theoretical hazard, whereby nitrates that are changed by enzyme action into nitrites after being cooked or simply ingested react with secondary and tertiary amines to form nitrosamines, which are known cancer-producing agents. Although each of these links has been demonstrated, and stomach cancers have been produced in laboratory animals, there is no firm evidence that this is yet happening in humans. Indeed, paradoxically, the incidence of stomach cancer appears to be less in areas that have high nitrate levels in the water supply.

Though most public attention has focused on nitrate contamination of drinking water, an even greater public health threat exists in the form of nitrate levels in food. Four-fifths of the nitrate entering our bodies comes from food rather than from water. Of this, the nitrate in vegetables exceeds the total for all other food sources ("The nitrogen cycle of the United Kingdom," 1983). Salad vegetables such as lettuce and spinach are most susceptible. Interestingly, analytical comparisons between organic and conventionally grown lettuces show significantly less nitrate accumulation in the organic crop, except during winter months (Vogtmann, 1984).

A further complication occurs when crops have been treated with any of the fungicidal dithiocarbamate (DTC) sprays. Nitrites are capable of reacting with the breakdown products of these sprays to form powerful cancer-producing chemicals, further illustrating the difficulties of assessing potential hazards of pesticides under laboratory conditions. This so-called "cocktail effect", and the synergy or enhancement of toxicity that frequently accompanies it, is extremely worrying, as it can occur without our realising it, with totally unknown consequences.

A third area of concern for consumers is the use of antibiotics and hormones in livestock husbandry. In the case of the former, the greater threat to human health comes not so much from the presence of residues in meat but from the potential impact in the treatment of human diseases.

Antibiotics are routinely included in the diets of many farm animals, as growth promoters and as a way of dealing with ill health caused by the intensive conditions in which they're reared. Until banned by the EU in 1997, avoparcin, the most commonly used antibiotic, was fed on a daily basis to 80% of all broiler hens, 30%–40% of all pigs, and 30% of dairy cows in the United Kingdom (National Consumer Council, 1998). The reason for its banning was concern that pathogenic bacteria in livestock were becoming resistant and that this resistance could pass, in a phenomenon known as "cross resistance", to organisms causing some of the most dangerous human infections; in this case, a strain of resistant enterococci that could overcome the antibiotic Vancomycin, which derives from the same bacteria as avoparcin.

Several authorities have agreed that widespread use of livestock feed antibiotics is responsible for the wholesale contamination of food by resistant strains of salmonella and virulent strains of *E. coli* (Harvey and Mason, 1998). Evidence for this theory comes from Sweden, where all farm-feed antibiotics have been banned since 1986, and the problem of resistant pathogens is very much decreased.

In the past few years, many antibiotics that have been the mainstays of modern medicine have been overcome by resistant bacteria. Vancomycin is a drug of last resort in the treatment of life-threatening conditions such as those caused by methicillin-resistant *Staphylococcus aureus*, a hospital superbug that infects 20,000 people annually in the United Kingdom, killing 5,000 of them (Harvey and Mason, 1998).

Various types of hormones—both "natural", like progesterone and testosterone, and synthetic, like zeranol and trenbolone—as well as other growth promoters such as corticosteroids and beta-agonists (typically Clembuterol) are used by farmers to enable livestock to put on weight more quickly and to improve the ratio of lean to fat flesh. Debate has raged over the safety of such substances ever since evidence was found in the 1970s of young children developing breasts and enlarged genitals following the consumption of hormone-contaminated meat. In December 1985, the EU finally passed a regulation forbidding all hormone use, though this decision had more to do with political concerns over the size of the "beef mountain" than with health. Even so, illegal use of hormones by farmers would appear to be an intractable problem, though more so in some countries than others. A survey in Belgium in 1995 found that 22% of beef liver was contaminated with Clembuterol; the figure for Spain was 35%. In 1990, 135 people in Spain and 22 in France were hospitalised after eating Clembuterol-contaminated liver (National Consumer Council, 1998). The United States, along with Canada, New Zealand, and Australia, still allows the use of natural hormones, plus zeranol and trenbolone, for fattening animals.

No section on human health considerations could avoid mentioning bovine spongiform encephalopathy (BSE), which, since it was first recognised in the mid-1980s, has affected cattle on 35,000 farms in the United Kingdom (National Consumer Council, 1998). Scientists are still not sure precisely what caused it, though most believe it was due to the inclusion in cattle rations of improperly treated slaughterhouse waste containing offal contaminated by organisms that cause the brain disease scrapie in sheep. In March 1996, the UK government announced the discovery of a new variant of the human brain disease Creutzfeldt-Jakob Disease (CJD), expressing the view that a likely explanation for its occurrence was a direct link to BSE.

There is not space to go into the ramifications of the BSE epidemic here. A simple observation is that, had modern farming not set off down the intensification route and instead adopted the approach taken by organic farmers, which eschews the use of processed slaughterhouse feed and all animal growth promoters, the BSE crisis could have been avoided. Nor would we be worrying about the problems posed by antibiotic resistance or hormone residues in our food.

1.6.2 Animal Welfare

In 1995, the Co-operative Wholesale Society, in a survey of 30,000 UK shoppers, discovered that over a third had boycotted a shop or product because of concerns over issues such as animal rights and the environment. Nearly double that number, 60%, would do so in the future ("Organic and Ethical Foods", 1997).

The transport of live animals and calves kept in veal crates were just two of the animal welfare issues that achieved widespread prominence during the 1990s. Whilst vociferous protesters took to the streets, other, less demonstrative, consumers voted with their purses and switched to organic products, which, as we saw in Section 1.3, are produced using the highest welfare standards.

The conditions by which most conventionally produced pigs and poultry are raised causes most concern. In the United Kingdom, 30 million hens live five to a cage with no more than 450 cm^2 (the size of an A4 sheet of paper) of space each and a sloping wire mesh for a floor. Though many pigs are raised outdoors, most still lead lives of stultifying boredom in cramped conditions indoors, with their tails cut off to prevent mindless chewing.

Dairy cows suffer a less obvious form of cruelty. Bred to produce more and more milk, modern breeds have huge pendulous udders that are prone to infection from mastitis (for which they are prophylactically treated with antibiotics) and are responsible for lameness. After four lactations (pregnancies), most dairy cows have to be slaughtered; under more natural conditions, a dairy cow could easily produce 10 or more calves. Even so, the agrochemical industry has been pressing hard to lift the European ban on the use of bovine somatatrophin (BST), a genetically engineered hormone that is claimed to raise milk yields by 10%–20%. Apart from adding to the glut of milk, this can only add to the suffering of dairy cattle.

In June 1996, the EU Amsterdam Summit recognised farm animals as being "sentient" beings, a move that animal welfarists believe will be a useful lever in raising standards. Discussions are under way to increase space in battery cages to 800 cm^2, and the veal crate has been banned in the EU, though not before 2007. This rate of progress is too slow for many people who have reacted by becoming vegetarians. Mintel ("Organic and Ethical Foods", 1997) reports that there were 3.1 million vegetarians in the United Kingdom in 1997, up a million from the start of the decade, and that 5,000 people a week stop eating meat.

However, from an organic perspective, the prospect of people turning away from the products of animal husbandry, which is a mainstay of organic agriculture, is alarming. Richard Young, a leading organic proponent, believes that "the organic movement and the vegetarian movement have some aims in common but overall they are on a collision course" (Young, 1998).

1.6.3 Organic Standards

Organic food is perceived by many people as a safe alternative. Yet how valid is this assumption and what guarantees are there that organic produce is, in fact, genuine?

Standards for organic production have been in place for many years, having been first developed in the 1960s by the Soil Association. This organisation retains a pre-eminent position today, responsible for approximately 70% of UK organic acreage. Farmers belonging to the Soil Association's symbol scheme must undergo an initial two-year conversion period, to rid the soil of pesticide residues and to begin to re-establish the soil micro and macro flora and fauna. During this period, and subsequently, they must adhere to a well-defined set of requirements. These requirements are backed up by an inspection procedure that includes at least one annual visit, scrutiny of the farm's production returns and access to all of the farm's financial records. Farmers who successfully comply with these standards are entitled to use the Soil Association logo on their produce. During the 1970s, another organic symbol scheme was set by Organic Farmers and Growers, Ltd, a marketing co-operative that catered primarily to cereal producers in the east of England.

There was nothing in law during this period to compel anyone to join either of these schemes; thus, produce could be sold as organic without any check being made on its validity (subject always, of course, to the possibility of action by Trading Standards Officers [TSOs]). All of this changed on 1 January, 1993, with the implementation of the EC Regulation on Organic Food (No. 2092/91). Since then it has been illegal to market crops and plant-derived foods unless they have been properly inspected and certified by an approved certification body. This covers not only the growing of the crops but also all processing, packing, wholesaling, and, in some cases, retailing—in the latter cases to guard against unwanted contamination or the substitution of conventionally grown food. Meat and animal products do not currently fall under this legislation, but they are expected to be incorporated into the standards by early in the next century.

Each member state has the obligation of policing the regulations, which in the United Kingdom is shared between the TSOs and the United Kingdom Register of Organic Food Standards (UKROFS)—an organisation set up in 1987 under the auspices of "Food from Britain" and that produced unified standards in May 1989. Though farmers and food processors may register directly with UKROFS, the preferred route is via one of the approved certifying bodies, which as of 1998 were:

- Soil Association Certification
- Organic Farmers and Growers, Ltd
- Organic Food Federation
- Scottish Organic Producers Association
- Irish Organic Farmers and Growers
- BioDynamic Agricultural Association

The EC regulation enshrined the principle, already followed by the Soil Association and the International Federation of Organic Agricultural Movements, that organic food should be defined in terms of the production process and not by reference to an analysis of the end product. The latter approach was felt to be unworkable, not least the problem of defining acceptable residues of pesticides in organic food, because, despite its absence during cultivation, stray contamination via air and water pollution makes any claim of freedom from pesticide residues virtually impossible to justify.

This has not prevented TSOs from bringing actions against companies selling organic food considered to contain unacceptably high pesticide residues. But except in the case of post-harvest pesticides, which are much less likely to arise through accidental contamination, these actions have all failed. Processed organic food has a much more subscribed list of non-agricultural ingredients that may be added than its conventional counterpart (Exhibit 1–2). Post-harvest chemicals, ethylene gas for ripening, and chemical and synthetic agents used for preparation, presentation, and preservation are all banned, as are irradiated products. Successful prosecutions based on finding such products using analytical methods should not be ruled out.

The rules governing organic production throughout the EC also apply to imported produce from third countries. Beginning in January 1995, to qualify for admission into EC countries, importers or certifying bodies overseas had to satisfy the commission that their standards and inspection and certification schemes complied with the EC regulation.

1.6.4 Labelling

The EC regulation prescribes exactly how organic food may be defined. Only food that has been produced to authorised organic standards may be described as such, though in the case of products containing a number of ingredients, up to 5% of the total may be of non-organic origin, provided they are on an approved list of ingredients (Exhibit 1–2). Products that contain 70% or more organic ingredients may refer to their organic origin only as part of the ingredients list and not especially highlighted in any way. Where organic ingredients make up less than 70% of the total, the fact that they are organically produced must not be referred to anywhere on the label.

All organic products must carry the logo of one of the certifying organisations or its code number, consisting of the prefix "UK" followed by a number from 1 to 7. Each European certifying body has its own code number.

To avoid any contamination by or substitution with non-organic produce, most organic food, especially that sold in supermarkets, is pre-packed.

Exhibit 1–2 Permitted Substances Used in Organic Food Processing—Annex VI to EC Regulation No. 2092/91 as of 29 January, 1993

Section A: Ingredients of Non-agricultural Origin

E170 Calcium carbonates
E220 Sulphur dioxide
E270 Lactic acid
E290 Carbon dioxide
E296 Malic acid
E300 Ascorbic acid
E306 Tocopherol-rich extract
E322 Lecithins
E330 Citric acid
E333 Calcium citrates
E334 Tartaric acid
E335 Sodium tartrate
E336 Potassium tartrate
E341(i) Monocalciumphosphate
E400 Alginic acid
E401 Sodium alginate
E402 Potassium alginate
E406 Agar
E410 Locust bean gum
E412 Guar gum
E413 Tragacanth gum
E414 Arabic gum
E415 Xanthan gum
E416 Karaga gum
E440(i) Pectin
E500 Sodium carbonates
E501 Potassium carbonates
E503 Ammonium carbonates
E504 Magnesium carbonates
E516 Calcium sulphate
E524 Sodium hydroxide
E938 Argon
E941 Nitrogen
E948 Oxygen
Natural flavourings
Drinking water
Salt
Microorganism preparations
Minerals, vitamins, amino acids, and nitrogen compounds where legally required

Section B: Processing Aids and Other Products That May Be Used for Processing of Organically Produced Ingredients of Agricultural Origin

Water
Calcium chloride
Calcium carbonate
Calcium hydroxide
Calcium sulphate
Magnesium chloride
Potassium carbonate
Sodium carbonate
Citric acid
Sodium hydroxide
Sulphuric acid
Carbon dioxide
Nitrogen
Ethanol
Tannic acid
Egg white albumen
Casein
Gelatin
Isinglass
Vegetable oils
Silicon dioxide gel
Activated carbon
Talc
Bentonite
Kaolin
Diatomaceous earth
Perlite
Hazelnut shells
Rice meal
Beeswax
Carnauba wax
Saltpetre (potassium nitrate)
Preparations of micro-organisms and enzymes

continues

Exhibit 1–2 continued

Section C: Ingredients of Agricultural Origin That Have Not Been Produced Organically

Unprocessed vegetable products	Dried banana powder
Coconuts	Gooseberries
Brazil nuts	Dried strawberry powder
Cashew nuts	Dried raspberry powder
Dates	Dried redcurrants
Pineapples	Lesser galanga (*Alpina officinarum*)
Mangoes	Allspice (*Pimenta dioica*)
Papayas	Wild rice (*Zizania plauspra*)
Sloes	Sesame seed
Cocoa	Algae, including seaweed
Maracujas (passion fruit)	Olive oil
Colanuts	Sunflower oil
Peanuts	Cane and beet sugar
Rosehips	Starches produced from cereals and tubers
Swallowthorns	Rice paper
Blueberries	Gluten
Maple syrup	Fructose
Quinoa	Honey
Amaranth	Gelatin
Horseradish seed	Buttermilk powder
Pine kernels	Edible aquatic organisms not originating from aquaculture
Radish seeds	Lactose
Acorns	
Fenugreek	
Acerola	

1.6.5 Other Schemes

The 1980s witnessed the development of a whole range of "green" or environmentally friendly foodstuffs, representing a "halfway house" between organic and conventional food.

The most well-established organisation is the Guild of Conservation Food Producers, whose Conservation Grade evolved from the secondary conversion standard originally developed by Organic Farmers and Growers, Ltd. This scheme allows a restricted range of artificial fertilisers and pesticides that are claimed to have less environmental impact than other conventional inputs. Examples include the herbicides glyphosphate and asulam, and fertilisers such as Chilean nitrate and superphosphate. A published set of standards is available from the Guild, and all farms are independently inspected. There is, as yet, no market for fruit and vegetables and other food produced during the 2-year or more period when a farm is undergoing conversion.

The Conservation Grade also includes production standards for livestock husbandry, which cover aspects such as welfare, the prohibition of routine and prophylactic use of drugs, and restrictions on the use of feed additives.

Livestock schemes comprise all the other halfway house labels on the market. Most address the issue of animal welfare (in particular the abuses of intensive livestock systems) and have been introduced as own-brand initiatives by supermarkets. Comforting names such as "traditional" and "naturally reared" are used, but little information is currently offered to consumers as to exactly how the animals are reared.

Freedom Foods has been introduced by the Royal Society for the Prevention of Cruelty to Animals (RSPCA) to denote livestock reared so as to be free of "thirst and hunger", "pain, injury and disease", "discomfort", and "fear and distress", plus the positive freedom "to express normal behaviour". However, it is hard to see how pigs reared indoors under this label are much better off, as farrowing crates, tail docking, and tooth clipping still apply. The only benefit over intensive pig raising would appear to be shorter journey times to the slaughterhouse and a little straw for bedding; hardly anything to shout about. With poultry, Freedom Foods bans the use of battery cages but permits "barn and perchery" systems, which are almost as intensive, and allows beak clipping to prevent feather pecking—a sure sign that birds are unnaturally bored.

Neither Freedom Foods nor any of the other so-called "welfare systems" has anything to say about the use of prophylactic antibiotics and other growth promoters.

Before leaving this section, it is worth referring to the products of biodynamic farming marketed bearing the Demeter trademark or, for farms in conversion, the Biodyn symbol. Though not widely practised in the United Kingdom, biodynamic farming is more common in mainland Europe. It does not differ significantly from organic farming in that the basic premises are the same. However, it goes further in attempting to work more closely with subtle forces such as lunar and planetary influences acting on the whole farm environment. This leads to practices such as undertaking cultivation by reference to an astronomical calendar (Thun, 1993) and the use of sprays, derived from animals and plants and diluted to homeopathic levels, in crop production and on compost and manure heaps. For a full account of biodynamic farming, see *Biodynamic Farming Practice* by Fritz Sattler and Eckard von Wistinghausen, 1993.

1.7 FUTURE TRENDS

To devise and implement systems of financial support that encourage maximum production was, for 40 years after the end of World War II, official agricultural strategy. This policy proved successful without doubt. Indeed, it can be said to have been too successful: For much of the period, Europe produced more food, over a range of staple commodities, than it needed to feed its own population. Surplus food was stored, destroyed, or sold onto world markets at prices less than the cost of production.

At the same time, agricultural exporters elsewhere in the developing world, producing food at prices below European levels, were simultaneously denied access to the EU market and forced to compete against heavily subsidised European exports. Blatant "dumping" of produce on third world countries de-stabilised and destroyed fragile local markets of farmers and hastened their exodus to the cities.

Much more vociferous, though, in their condemnation of the Common Agriculture Policy, were the United States and the food exporters of Canada and Australasia, the so-called Cairns group. Crunch time came with the Uruguay round of the GATT world trade negotiations of the early 1990s. Agricultural support was the key sticking point for many months, but eventually a deal was reached that obliged Brussels to reduce commodity prices and curb production. The instrument to achieve the latter was compulsory "set-aside", whereby farmers were paid to leave 15% of their land idle each year. Quotas were set for dairy products, and the market prices for cereals were progressively reduced each year to bring them closer to world levels. By way of compensation, farmers' incomes were protected by a new flat rate payment based on the acreage of their land. Stock farmers were given a subsidy (headage payments) based on the number of beef cattle or sheep they kept.

This policy had the desired effect of reducing surpluses, though at considerable cost to the European taxpayer. In 1998, it has been calculated that European farmers collected subsidies worth around £27.2 billion, nearly half the entire EU budget. The British contribution, by way of taxation, was £4.3 billion.

One positive outcome of the CAP reform was the introduction, for the first time, of environmental payments. The agri-environment programme was designed to encourage farmers to adopt environmentally friendlier practices, including conversion support for organic farmers. In comparison with price support and arable area payments, however, the amounts offered are tiny.

In July 1997, the European Commission published its proposals for the Union's financial framework beyond the year 2000. The package, known as Agenda 2000, includes recommendations for further reform of the CAP. This has been devised in a climate of further liberalisation of world trade and reduced agricultural support, and in the context of the next round of multilateral trade negotiations at the World Trade Organisation in 1999.

Environmental protection and rural development are firmly embedded in the proposals. Member states must establish environmental requirements that are a pre-requisite for any direct payments. Farms that fail to respect these conditions, known as cross compliance, will have their aid redirected or cancelled. However, set-aside is no longer compulsory, though voluntary set-aside will still be allowed.

Nevertheless, the core proposals still hinge on the twin pillars of commodity support prices (admittedly at lower levels) and compensatory area payments. It is the view of most environmental organisations that Agenda 2000 does not go anything like far enough in tackling environmental problems. Furthermore, it is hard to see these proposals being acceptable to the United States.

1.8 TOWARD A SUSTAINABLE AGRICULTURE

Farmers have traditionally been looked on by city dwellers as custodians of the countryside. It is only during the last decade or so that this view has changed, in response to the major landscape alteration, environmental pollution, and destruction of wildlife that have accompanied latter-day food production policies. The farming community tries to improve its image through good public relations, but it will take radical change in farm practices to alter people's poor perception.

The conclusions of the Rio Summit of 1992 include a clear commitment by the participating governments to encourage more sustainable forms of agriculture. Conventional high-input agriculture—with its over-dependency on fossil fuels and agrochemicals and deleterious environmental consequences, such as increased soil erosion and pollution of air, land, and watercourses by pesticides and nitrates—could hardly be further from this ideal. It is true that work is under way to modify existing systems to devise low-input models for agriculture, but this research is in its infancy, and there is no guarantee of success. It is also clear from a number of surveys of farmers' attitudes that many are unhappy with present directions in farming and would prefer to reduce their reliance on agrochemicals.

Agriculture requires a new way forward, one that seeks to minimise the market distortions inherent in price support whilst rewarding farmers for producing food in ways that meet environmental criteria. To an extent, this already happens, but the approach is piecemeal, with various schemes currently in operation in the United Kingdom—Environmentally Sensitive Areas, Farm and Conservation Grant Scheme, Nitrate Sensitive Areas, Countryside Premium Scheme, Countryside Stewardship Scheme, Sites of Special Scientific Interest, Farm Woodland Premium Scheme, and the Organic Aid Scheme for converting farms—receiving less than 5% of that spent on market support.

The Agenda 2000 proposals go somewhat further in this direction but are not sufficiently radical. Many believe that farm support must be decoupled from production, with market subsidies eliminated, leaving only payments for genuine environmental improvement. A CAP reorganised along these lines would also allow for the progressive enlargement of the European Union by the admittance of countries from Central and Eastern Europe in the early part of the next century.

A further incentive towards creating a sustainable agriculture could come via taxation on fertilisers and pesticides. The Netherlands government plans to put a levy on agrochemicals by increasing the value-added tax (VAT) from 6% to 17.5%, raising an extra £50 million to be spent on environmentally friendly farming. In the spring of 1998, the Environment Ministry of the British government floated a similar idea, which was immediately attacked by the farming industry.

Such a regime of "stick and carrot" could work wonders for the organic sector. Organic farming, is, by contrast, sustainable, both by nature and by definition. Because of the more limited use and restricted range of fertilising materials, plant protection products and other inputs, a more widespread adoption of organic farming would confer significant environmental benefits as discussed earlier. There would also be cost savings attaching to a reduced need for activities such as nitrate and pesticide clean-up of water supplies—costs that are not currently borne by agriculture. The reduction in crop yields would provide further savings on storage and disposal costs.

Currently, the organic sector, though growing fast, is still tiny compared with conventional agriculture. In countries where governments have recognised the potential of organic farming to deliver a combination of environmental improvements, rural development, and food the consumer trusts—and have tailored their agricultural policies accordingly—growth has been rapid.

Assuming that current growth rates persist, organic farming will account for 10% and possibly 15% of all European agriculture by the year 2005. This implies more than half a million organic farms on 15 million hectares and a retail market in the EU worth more than £30 billion (Lampkin, 1997).

Such a transformation would bring credibility to the sector, and economies of scale, further reducing the price differential in the shops. With environmental payments replacing traditional agricultural support, the way would be clear for a significant further expansion of organic farming in the first decades of the new millennium.

1.9 FURTHER READING

For a comprehensive, fully referenced textbook of organic agriculture, the outstanding work is *Organic Farming* by Nicolas Lampkin, Farming Press, 1990. *The Organic Farm Management Handbook*, published biennially by Elm Farm Research Centre (Lampkin et al.) gives up-to-date statistics of the industry. A more personal and philosophical account has been written by HRH the Prince of Wales in conjunction with Charles Clover: *Highgrove, Portrait of an Estate*, Chapman, 1993. *The Shopper's Guide to Organic Food* by Lynda Brown (Fourth Estate, 1998) gives a comprehensive account of many organic products on sale and lists outlets in the United Kingdom.

REFERENCES

Balfour, E.B. (1975). *The living soil and the Haughley experiment.* London: Faber and Faber.

BMELF (West German Federal Ministry of Agriculture). (1989). *Annual report.* Bonn: Landwirtschaft und Forster.

Boardman, J. (1986). The context of soil erosion. *SEESOIL, 3,* 2–13.

British Agrochemical Association. (1996). *Annual report.* Peterborough, United Kingdom.

Brown, L., Abramovitz, J., Bright, C., Flavin, C., et al. (1996). *State of the world, 1996.* London: Earthscan.

Brown, L.R., & Wolf, E.C. (1984). *Soil erosion: Quiet crisis in the world economy* (Worldwatch Paper 60). Washington, DC: Worldwatch Institute.

Call for a spindoctor. (1997). *New Scientist,* November 1.

Carson, R. (1962). *Silent spring.* London: Penguin Books.

Chevre, A.M., et al. (1997). Gene flow from transgenic crops. *Nature, 389,* 924.

Cooke, G.W. (1979). Some priorities for British soil science. *Journal of Soil Science, 30,* 187–213.

Crick, H., et al. (July 1998). *Breeding birds in the wider countryside: Their conservation status (1972–1996).* London: British Trust for Conservation.

Ducey, J., et al. (1980). *A biological comparison of organic and chemical farming.* Lincoln, NE: University of Nebraska.

EC (Commission of the European Communities). (1991). *The agricultural situation in the Community: 1990 report.* Brussels.

Edwards, C.A. (1984). Changes in agricultural practice and their impact on soil organisms. In D. Jenkins (Ed.), *Agriculture and the environment, Proceedings of the Institute of Terrestrial Ecology Symposium* (pp. 55–65). Zurich: Institute of Terrestrial Ecology.

Environmental Protection Agency (EPA). (1992). *List of chemicals evaluated for carcinogenic potential.* Washington, DC: EPA, Office of Pesticide Programs.

European Crop Protection Association. (1996/97). *Annual report 1996/97.* Brussels, Belgium.

Fiege, W., & Roethlingshoefer, R. (1990). Leaching of nitrate on two fields with different farming methods. *Zeitschrift fuer Kulturtechnik und Landentwicklung, 31,* 89–95.

Georghiou, G. (1986). *Pesticide resistance: Strategies and tactics for management.* Washington, DC: National Academy Press.

The green consumer. (1989). London: Mintel.

Gremaud, J.K., & Dahlgren, R.B. (1982). Biological farming: Impacts on wildlife. In R.H. Dahlgren (Ed.), *Workshop in midwest agricultural interfaces with fish and wildlife resources* (pp. 38–39). Ames: Iowa State University.

Hald, A.B., & Reddersen, J. (1990). *Food items of birds in cereal fields: Insects and wild flora.* Miljoprojekt No. 125, Miljostyrelsen, Copenhagen.

Harvey, J., & Mason, L. (1998). *The use and misuse of antibiotics in UK agriculture.* Bristol: Soil Association.

Howard, A. (1940). *An agricultural testament.* London: Oxford University Press.

Jeyaratnam, J. (1990). Acute pesticide poisoning: A major global health problem. *World Health Statistics Quarterly, 43,* 139–144.

Kickruth, R. (1982). Die okologische landwirtschaft. *Alternative Konzept, 40,* 89–99. Karlsruhe, Germany: C.F. Muller.

Lampkin, N. (1990). *Organic farming.* Ipswich: Farming Press Books.

Lampkin, N. (1993a). The economic implications of conversion from conventional to organic farming systems. Unpublished doctoral dissertation, University of Wales, Aberystwyth.

Lampkin, N. (1997, November 4). Opportunities for profit from organic farming. *Organic Farming: Science into Practice.* Stoneleigh Park, Warks: National Agriculture Centre.

Lampkin, N., & Bateman, D. (1993). *The economies of organic farming in Wales, 1989.* Department of Economics and Agricultural Economics, University of Wales, Aberystwyth.

Lampkin, N., & Midmore, P. (1988). *Organic production as an alternative to set aside: A response to the Agricultural Department's Consultative Document on Extensification.* Department of Economics and Agricultural Economics, University of Wales, Aberystwyth.

Lockeritz, W., et al. (1984). Comparison of conventional and organic farming in the corn belt. In D. Bezdicek, et al. (Eds.), *Organic farming: Current technology and its role in a sustainable agriculture* (Special Publication No. 46). Madison, WI: American Society of Agronomy.

Milton, R.F. (1975). The Haughley experiment: Some unsolved problems, their significance and suggested methods for investigation. In E.B. Balfour (Ed.), *The living soil and the Haughley experiment* (pp. 274–281). London: Faber and Faber.

Ministry of Agriculture, Fisheries and Food. (1988). *Output and utilisation of farm produce in the United Kingdom, 1981–1987.* London: HMSO.

Ministry of Agriculture, Fisheries and Food. (1995). *The effect of organic farming systems on aspects of the environment.* London: HMSO.

Ministry of Agriculture, Fisheries and Food. (1998). *Review of the organic aid scheme.* London: HMSO.

Moreby, S., & Southway, S. (1993). How "green" are your organic cereals? *Game Conservancy Review*, 75–76.

Moss, B. (1978). The ecological history of the medieval man-made lake Hickling Broad, Norfolk. *Hydrobiologia*, *60* (1), 23–32.

Murphy, M. (1992). *Organic farming as a business in Great Britain*. Cambridge: University of Cambridge, Agricultural Economics Unit.

National Consumer Council. (1998). *Farm policies and our food: The need for change*. London.

National Research Council. (1989). *Alternative agriculture*. Washington, DC: Academy Press.

National Rivers Authority (NRA). (1995). *Water pollution incidents in England and Wales 1994* (NRA Water Quality Series No. 25). London.

Nature Conservancy Council. (1984). *Nature conservation review of Great Britain*. London.

The nitrogen cycle of the United Kingdom. (1983). London: The Royal Society.

Organic and ethical foods. (1997). *Market intelligence*. London: Mintel.

Organic Farming Study Forum. (1997, March 26). *Proceedings of the Joint Research Council, Organic Farming Study Forum*. Cirencester: Royal Agricultural College.

Paxton, A. (1996). *The food miles report: The dangers of long distance food transport*. London: SAFE Alliance.

Pesticides and the immune system: The public health risks. (1996). Washington, DC: World Resources Institute.

Pesticides Trust report. (1992). London: The Pesticides Trust.

Pimental, D., Krummel, J., Gallahan, D., Hough, J., et al. (1978). Benefits and costs of pesticide use in US food production. *Bioscience*, *28* (12), 772–784.

Pimental, D., & Levitan, L. (1986). Pesticides: Amounts applied and amounts reaching pests. *Bioscience, 36* (2), 86–91.

Pimental, D., et al. (1992). Environmental and economic costs of pesticide use. *Bioscience, 42* (10), 750–760.

Potts, G.R. (1986). *The partridge*. London: Collins.

Redman, M. (Ed.). (1992). *Organic farming and the countryside*. Bristol: Soil Association.

Redman, M. (1996). *Counting the cost*. Bristol: Soil Association.

Reganold, J.P., Elliott, L.F., & Unger, Y.L. (1987). Long-term effects of organic and conventional farming on soil erosion. *Nature, 330* (6146), 370–372.

Royal Commission on Environmental Pollution. (1979). *Agriculture and pollution*. London: HMSO.

Sattler, F., & von Wistinghausen, E. (1993). *Biodynamic farming practice*. Stourbridge, West Midlands: Biodynamic Agricultural Association.

Soil erosion in England and Wales. (1986). Soil Survey of England and Wales. London: HMSO.

Thun, M. (1993). *Working with the stars: A biodynamic sowing and planting calendar*. East Grinstead: Lanthorn Press.

United Nations Environment Programme. (1992). *Synthesis report of the methyl bromide interim scientific assessment and methyl bromide interim technology and economic assessment*. Nairobi.

Vine, A., & Bateman, D.I. (1981). *Organic farming systems in England and Wales: Practice, performance and implications*. Department of Agricultural Economics, University of Wales, Aberystwyth.

Vogtmann, H., et al. (1984). Accumulation of nitrates in leafy vegetables grown under contrasting agricultural systems. *Biological Agriculture and Horticulture, 2,* 51–68.

Wright, M.A. (1979). Effects of benomyl and some other systematic fungicides on earthworms. *Annals of Applied Biology, 87,* 520–524.

Young, R. (1998, Jan-Mar). Livestock, an essential part of the organic cycle. *Living Earth*, 10–12.

Chapter 2

Environmental Implications of Genetic Engineering in the Food Industry

Sibel Roller

2.1 INTRODUCTION

Since the end of World War II, insect and pest control in agriculture has been heavily dependent on the use of chemical insecticides and pesticides. Several of these synthetic compounds accumulated in the environment, the food chain, and human adipose tissue, and had acute or chronic side-effects in non-target organisms. Furthermore, pesticide resistance developed in some populations, creating a need for alternative pest control strategies. The high cost of early biotechnological developments, combined with the potential for very high profitability for the industry, ensured that the first wave of research in food-related biotechnology was supported by the large agrochemical companies. This has resulted in a preponderance of novel plant, animal, and microbial strains genetically engineered for the benefit of the agrochemical industry, farmers, and food processors. It is only very recently that new genetically modified food plants with quality improvements designed to appeal to the consumer have started to emerge from the laboratory; the large majority of these have yet to reach the supermarket shelves.

Currently, terms such as "transformed", "transgenic", "recombinant", "genetically engineered", and "genetically modified" are all used to describe plants, animals, or micro-organisms that have had DNA introduced into them by means other than the combination of a sperm and an egg. The collective term "genetically modified organisms" (GMOs) is used frequently in regulatory documents and in the scientific literature. In this chapter, as in most scientific and popular literature, all of these terms are used interchangeably.

2.2 FEEDING AN INCREASING WORLD POPULATION

In terms of food quantity, it is worth remembering that famine is still common in some parts of the world. During the last decade, the levels of world food stocks have gravitated downwards from the Food and Agriculture Organization (FAO) safety level of 60 days (Dunwell, 1995; Dunwell and Paul, 1990). Crop production remains vulnerable to climatic hazards, and this situation may be exacerbated in the future as a result of global warming. As has been pointed out by Dunwell and Paul (1990), in a country like the UK, where only 2.1% of the population is employed in agriculture and where population growth in the 1980s of 1.2% was accompanied by an increase of 11.4% in cereal production, it is easy to neglect developing countries such as Rwanda, where, before the recent civil war, 92% of the population was employed in agriculture and a population growth of 30.8% was accompanied by a rise in cereal production of only 3%.

Although the world's population has not reached the critical mass predicted in the 1960s, it is nevertheless expected to increase from the current 5.8 billion to 10 billion by the year 2050 (James, 1997). It has been suggested that without the improvements in yields made possible by the application of the new biotechnology, shortfalls in food supply will become increasingly prevalent in many countries. The problem could be further exacerbated if global food demand grows at a faster rate than that of population growth, as has been forecast by some commentators (James, 1997). Clearly, some of the problems of food supply can only be remedied by political solutions, but gene technology offers an additional tool by which adequate food quantity could be ensured for the entire world population.

2.3 FOOD QUALITY

While the genetic modification of agronomic traits in crops may help to improve the supply and quantity of food available in the near future, improvements in food quality brought about by the genetic route are likely to develop more slowly for both technical and regulatory reasons. Nevertheless, the first examples of quality-improved GM foods and food ingredients are already available commercially, as shown in Table 2–1. A well-known example of a whole food genetically modified for improved quality has been the tomato (Schuch, 1994). The gene for the softening enzyme endopolygalacturonase has been effectively "switched off" so that the GM tomato can be left to ripen on the vine and develop its full flavour without losing the firmness required for modern food distribution (Giovannoni et al., 1989; Kramer et al., 1990; Kramer et al., 1992). In 1994, the FlavrSavr tomato received clearance from the US Food and Drug Administration (FDA), and by the end of May 1995, Calgene was supplying it to more than 1,700 retail outlets in the Midwest, New England, and the West Coast of America (Calgene, 1995). Similarly, in the UK, Zeneca's processing tomato received regulatory clearance in 1995 and the resultant paste was sold in selected supermarkets in 1996 (Redenbaugh et al., 1994; Roller and Harlander, 1998).

Table 2–1 Examples of Foods or Food Ingredients Modified by Gene Technology To Improve Quality

Food, Food Ingredient, or Processing Aid	*Application*	*Manufacturer*	*Status*
Chymosin	Coagulant used in cheese-making; genetically modified variants replace calf rennet, supplies of which are inadequate	Gist-brocades, Genencor International, Pfizer (Chr. Hansen's)	Permitted in Europe and United States since early 1990s; market penetration extensive in United States
Glucose isomerase	In high fructose corn syrup production, improvements in enzyme stability led to cost savings for processor	Genencor International	Not known but possibly in use since early 1990s
Baker's yeast	Faster carbon dioxide production, hence improved dough characteristics	Gist-brocades	United Kingdom approved in 1990 but not currently used commercially
Brewer's yeast	Able to degrade starch; hence useful for making reduced-calorie beer	Brewing Research International	United Kingdom approved in 1994; used on pilot scale
Rapeseed oil (canola)	Increased content of lauric acid; hence higher solid fat content to replace chemically hydrogenated vegetable oil	Calgene/Monsanto	Contract grown since 1994 for personal care products; food applications under development
Soybean oil	High oleic acid content	DuPont Agricultural Products	Field testing approved in May 1997
Fresh tomato	Firmer texture; hence can be left to ripen on vine and so develop better flavour	Calgene/Monsanto	First sold in United States in 1994; United Kingdom approved in 1996
Processing tomato	Higher solids; hence cheaper to produce	Zeneca	First sold in United Kingdom in 1996

Other examples of quality improvements include modifications of lipid composition in oilseed-bearing crops (Table 2–1). For example, rapeseed oil (canola) has been modified genetically to contain up to 40% lauric acid, which gives the oil a higher solid fat content (Kinney and Knowlton, 1998). The recombinant rapeseed was contract-grown in Canada in 1996 with primary commercial targets in the soap and cosmetics industries. Potential food applications for high-laurate oil as a cocoa butter substitute have also been suggested. Another modified plant oil approved by the FDA in May 1997 and now very close to commercialisation is the high oleic acid soybean (Kinney and Knowlton, 1998). It is conceivable that in the near future, oils high in mono- and polyunsaturated fatty acids may be developed for use in "healthy" food applications while high stearate oils may replace hydrogenated fats in processed foods.

Many plants produce anti-nutritional factors that may cause poisoning or even death if ingested in large quantities. For example, many legumes contain lectins, which, if not removed by soaking and cooking, can cause severe gastrointestinal symptoms. Genetic modification has been used to produce plants low in these anti-nutritional factors; however, because these compounds play an important role in the plant's resistance to pests and pathogenic micro-organisms, the GM variants frequently suffer from increased sensitivity to plant diseases and consequently yield poorer harvests. Clearly, further advances in both fundamental and applied knowledge will be necessary before the practical products of these efforts can enter the markets.

2.4 THE BASIC TECHNIQUES OF GENETIC MODIFICATION

The manipulation of genetic traits of plants and animals is not a new concept. Selective breeding has been practised by farmers for centuries, well before the molecular basis of the procedures was understood. All classical breeding relies upon and is limited by the inherent characteristics of a particular species. Thus, a black tulip cannot be produced by classical crossing because the gene coding for this characteristic is not inherent to this species of flower. The new recombinant DNA technology is different from traditional breeding methods in three principal ways: (1) it reduces the random nature of classical breeding; (2) it achieves the desired results much more quickly and predictably, and (3) most important, it makes it possible to cross the species barrier.

The transfer of genes from one cell to another can be likened to the old-fashioned editing process of "cutting and pasting". The genetic engineer uses very specific bacterial restriction enzymes that recognise, cut, and ligate (join) DNA as molecular "scissors and tape" to isolate the section of DNA coding for a specific trait (e.g., the ability to ferment a sugar). Because DNA does not readily move asexually from one organism to another, "vehicles" known as plasmids are required to transfer the isolated gene from one organism into another. Plasmids are usually small rings of DNA that contain a limited number of genes. The new code is then inserted into the genome of the target organism so that it forms part of the normal cell machinery and is expressed together with all the other proteins that the organism normally produces.

In plant biotechnology, a very specific plasmid from the plant pathogen *Agrobacterium tumefaciens* can be used, provided the tumour-inducing ability of the specific strain used has been eliminated. The bacterium dissolves the plant cell wall and inserts DNA into the chromosome of the target cell. The entire plant can then be re-grown from the single modified cell. For more details on the *Agrobacterium*-mediated transfer of DNA in plants, see the reviews by Zambryski (1992) and Hooykaas and Schilperoort (1992).

Not all cells subjected to genetic modification techniques are successfully modified. Consequently, it is necessary to identify the modified cells (whether from plants, animals, or micro-organisms) using "marker genes" closely linked to the genetic material to be transferred. This is often accomplished by using antibiotic resistance marker genes as, for example, in the case of the kanamycin-resistant marker gene used in the development of the genetically modified tomato. Such antibiotic resistance can be detected rapidly in the laboratory without the need to regenerate the whole plant or animal. In cereals, two frequently used markers are the phosphinothricin acetyl transferase *(pat)* gene or the closely related bialaphos resistance *(bar)* gene. These two genes originate from species of *Streptomyces* and enable the transformed organism to inactivate the normally toxic compounds phosphinothricin (an active ingredient of AgrEvo's herbicides Basta, Challenge, Ignite, Finale, and Harvest) and bialaphos (an active ingredient of Meiji Seika's Herbiace), respectively (Dunwell, 1995; Christou, 1998).

2.5 REGULATIONS CONTROLLING THE RELEASE OF TRANSGENIC ORGANISMS INTO THE ENVIRONMENT

Regulatory controls over the release of transgenic products have, of necessity, been developed on an ad hoc basis as a result of the virtual absence in the 1980s of quantitative data on the ability of GM organisms to survive in the environment. However, during the past two decades, much evidence has accumulated demonstrating the general safety (for humans as well as for the environment) of GM products, so that regulations and guidelines can now be developed on a somewhat more rational basis (Crawley et al., 1993; Engel et al., 1995; Kok et al., 1993; OECD, 1993; Tomlinson, 1998; World Health Organisation [WHO], 1991). These guidelines and regulations will need continuous revision and updating as new data on GMOs become available.

In most countries, the release and use of GMOs are strictly controlled, and permission has to be obtained from at least one (and very often more than one) government authority (for details, see Tomlinson, 1998). Unfortunately, the ease with which different countries issue permits is variable across the world. Up until 1997, permits for field releases in France were obtainable within 1 month at minimal cost; in the UK, the process could take up to 6 months and cost over $3,600; and in Germany, the general political attitudes towards biotechnology have been so hostile that permits have been difficult to obtain. Not surprisingly, the difficulties in obtaining permits are reflected in the share these countries have in the total number of releases in Europe: 30%, 17%, and 9% for France, the UK, and Germany, respectively, during the period 1991–1995 (Lloyd-Evans and Barfoot, 1996). However, extensive efforts are being made to harmonise regulations world-wide, and this trend is likely to continue until global consensus on the best way to ensure the safety of genetically modified foods is eventually reached (Tomlinson, 1998).

2.6 FIELD TRIALS

New crops modified for agronomic traits such as herbicide tolerance and insect resistance have constituted the first wave of development in food-related biotechnology for a variety of technical and socio-political reasons. Undoubtedly, the potential profitability of the market was an important driving force in the development of plant biotechnology. In 1996, the global market for agrochemicals for the top five crops grown in the United States and Europe (maize, cotton, wheat, oilseed rape, and sugar beet) was estimated at over $9.5 billion (Lloyd-Evans and Barfoot, 1996). In addition, many agronomic traits in plants are relatively simple technical targets for genetic modification because they are determined by a single biochemical pathway that is, in turn, controlled by a single gene. For example, sensitivity to the herbicide glyphosate (the active ingredient in Monsanto's broad-spectrum herbicide Roundup) is controlled by the single enzyme 5-enolpyruvyl-shikimate-3-phosphate (EPSP) synthase. Because conventional crops as well as weeds are killed by this herbicide, the use of glyphosate in conjunction with a tolerant GM variant (e.g., the GM soybean Roundup Ready) allows for reduced levels of application, thereby potentially conferring benefits for the environment.

By the late 1990s, more than 80 crops have been genetically modified for one or more traits. Transgenic derivatives of at least 48 of these species have been field-tested in the United States alone (Animal and Plant Health Inspection Service of the US Department of Agriculture [APHIS-USDA], 1998). Some examples of the numbers and types of crops most frequently subjected to genetic modification are shown in Exhibit 2–1 and Figure 2–1. As shown in Figure 2–1, the vast majority of field trials carried out in the United States to date have been on traits of agronomic significance. The picture looks very similar in other countries. Furthermore, the rate of field-testing has increased rapidly, especially in the United States, where the number of trials has doubled each year since 1987. In addition, the first-ever field trials of transgenic life forms other than plants such as arthropods and nematodes were approved in 1998 (APHIS-USDA).

In terms of field releases, the European Union lags well behind North America. For example, there were more than 500 field releases in the EU between 1991 and 1995, but Canada had the same number of releases in 1995 alone (Lloyd-Evans and Barfoot, 1996). It has been estimated that approximately 25,000 transgenic field trials were carried out between 1986 and 1997 in 45 countries (James, 1997). More than 70% of these trials were conducted in the United States and Canada, followed in descending order by Europe, Latin America, and Asia, with very few trials conducted in Africa (James, 1997). The large number of field trials of GM plants conducted to date without any adverse environmental effects represents considerable accumulated experience of benefit to developers and regulators alike.

Exhibit 2–1 Frequency of Worldwide Food Crop Field Trials From 1987 to 1998

200 to > 1,000	*10 to 200*	*< 10*	
• Corn (maize)	• Cucumber	• Apple	• Mustard
• Melon and squash	• Rice	• Asparagus	• Papaya
• Potato	• Sugar beet	• Aubergine (eggplant)	• Pea
• Rapeseed	• Sunflower	• Barley	• Peanut
• Soybean	• Wheat	• Blueberry	• Pepper
• Tomato		• Broccoli	• Plum
		• Cabbage	• Radicchio rosso (chicory)
		• Carrot	• Raspberry
		• Cauliflower	• Strawberry
		• Cranberry	• Sugarcane
		• Grape	• Sweet potato
		• Kiwi	• Walnut
		• Lettuce	• Watermelon

Source: Data from Animal and Plant Health Inspection Service of the US Department of Agriculture, 1998; and C. James, *Global Status of Transgenic Crops in 1997*, Brief No. 5, © 1997, ISAAA.

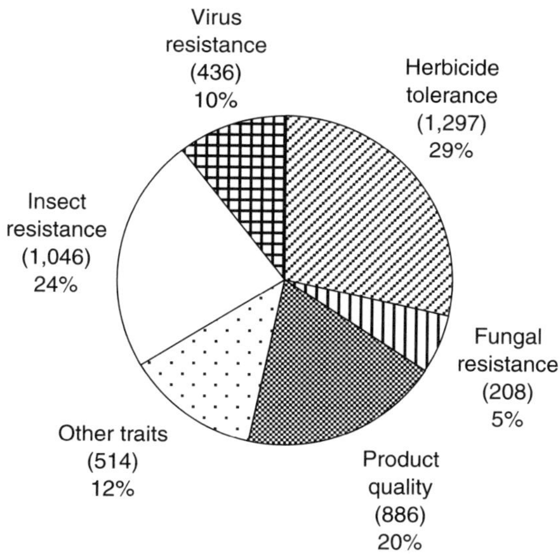

Figure 2–1 Number of field releases in the United States from 1987 to May 1998 and their most frequent categories. *Source:* Data from Animal and Plant Health Inspection Service of the US Department of Agriculture, 1998.

2.7 LARGE-SCALE CULTIVATION OF GM CROPS

Towards the end of the 1990s, more than 20 GM variants of several food crops, including maize, rapeseed, and soybean, had received regulatory approval for large-scale sowing and use in foods in the United States and Canada (Christou, 1998; Tomlinson, 1998). It has been estimated that 12.8 million hectares (31.5 acres) have been planted world-wide with transgenic crops (Table 2–2). Notably, this represented a 4.5-fold increase from the 2.8 million hectares (7 million acres) planted globally in 1996, and this rate of growth is expected to increase further in the next few years (James, 1997). As shown in Table 2–2, by far the largest acreage planted with GM crops has been in North America, although plantings in China and Argentina have also been significant. Transgenic soybean and corn accounted for 40% and 25% of the global acreage planted in 1997, respectively, with the rest made up of tobacco, rapeseed, cotton, and tomato (James, 1997).

Now that substantial acreage is being planted with transgenic crops globally, it is possible to make initial assessments of the benefits associated with this new technology. For example, corn modified to be resistant to the European corn borer

Table 2–2 Global Area of Land Planted with Transgenic Crops in 1997

	Area of Land Planted with Transgenic Crops		
Country	*Millions of Hectares*	*Millions of Acres*	*% of Global Total*
United States	8.1	20.1	64
China	1.8	4.5	14
Argentina	1.4	3.5	11
Canada	1.3	3.3	10
Australia	0.1	0.1	< 1
Mexico	< 0.1	< 0.1	< 1
World total	12.8	31.5	100

Source: Data from C. James, *Global Status of Transgenic Crops in 1997*, Brief No. 5, © 1997, ISAAA.

has been reported to produce yield increases of 9% (valued at $190 million) in 1997 alone. It has been reported that herbicide-tolerant soybeans in the United States in 1996 resulted in 10% to 40% less herbicide requirement, improved yield, no carry-over of herbicide residues, and better control of weeds. The use of herbicide-tolerant rapeseed in Canada has been estimated to bring benefits valued at $6 million (James, 1997).

2.8 ENVIRONMENTAL CONCERNS

The development of genetically modified crops has provoked much controversy, particularly in Europe. Proponents are predicting higher crop yields, better food quality, and reduced reliance on agrochemicals, with subsequent benefits for the environment, whilst critics are warning against the potential for reduced biodiversity and "genetic pollution", with traits such as herbicide tolerance spreading to weeds (Karp, 1993; Rissler and Mellon, 1994; van Wagner, 1993). As might be expected, the true picture is rather complex.

Current agronomic practices involve the preventive application of several selective herbicides before weed infestation has started (Dunwell, 1995). Given the uncertain effectiveness of some older herbicides, the farmer may apply these pre-emergent herbicides in excess of requirements, with potential detrimental consequences for the environment. It has been argued that the new genetically modified crops will allow farmers to use newer, more ecologically friendly broad-spectrum herbicides at lower but more effective concentrations and only when weed pressure actually demands it. The availability of the new herbicides, coupled with "matching" GM crops, could change agronomic practices profoundly, with the environment as the ultimate beneficiary.

However, one of the concerns frequently expressed about herbicide-resistant plants is their potential for becoming weeds that can stifle useful crops and wild plants. The characteristics of weeds are usually controlled by many complex genes, and it is highly unlikely that the genetic manipulation of a single gene trait in a crop would increase its tendency to become a weed (Taylor and Thomas, 1994). Of greater concern is the risk of the herbicide-resistant gene spreading through pollen to plants that are already weeds. Field studies have shown that pollen from GM plants is rarely carried over very long distances. For example, the percentage of cross-pollination in a field of GM rapeseed has been determined experimentally as 0.1%, 0.0003%, and zero at distances from the test site of 6, 47, and 70 metres, respectively. For GM potato, cross-pollination was 0.02% at 10 metres and zero at 20 metres from the site (Dale, 1998). One problem that remains is in predicting the effects of extremely rare hybridisation events that may not be evident in relatively small experimental plots of a hectare or two but may occur when GM crops are grown on thousands of hectares (Dale, 1998). There are historical examples of genes moving from one organism to another by unknown mechanisms over several million years, and such events may eventually result in the transfer of a genetically modified trait to a weed. However, these occurrences are so rare that it would be impossible to monitor them in practice (Taylor and Thomas, 1994).

When considering the potential threat of a GM plant to the environment, several properties of the crop must be taken into account. For example, maize plants have no seed dispersal mechanism, as the seeds are wrapped in cobs and adhere tightly to each other (Taylor and Thomas, 1994). Therefore, gene transfer can be expected to be limited, as there are no wild relatives of maize in the United States, where much of the global crop is planted. However, in Mexico and Central America, where maize originated and where wild relatives exist, careful monitoring may be necessary (Morrow, 1996). Similarly, wheat has no wild relatives in the United States. However, other plants may cause more concern. For example, the cultivated oat and sorghum cross readily with their wild relatives, many of which are serious agricultural weeds (Morrow, 1996; Taylor and Thomas, 1994); therefore, genetic modification for multiple herbicide resistance in these crops may well pose a threat to the environment. Clearly, each genetic modification and application to a specific crop needs to be scrutinised on a case-by-case basis.

It is now widely recognised by responsible authorities that the introduction of one or more novel genes into an organism requires careful risk assessment of the possible environmental impacts (Dale, 1998). These and other environmental concerns have been reflected in the regulations (Food and Drug Administration, 1992; Olempska-Beer et al., 1993; Tomlinson, 1998).

2.9 CONCLUSION AND FUTURE PROSPECTS

The new biotechnology has the potential to ensure environmentally sustainable supplies of safe, nutritious, affordable, and, above all, enjoyable food (Fraley, 1992). Ultimately, it is conceivable that crops with a combination of recombinant genes could be tailored to suit a variety of specific localities or climatic conditions to give optimum agronomic performance. However, the development of such "designer crops" may well be some way in the future; in the meantime, many commentators have said that the extent to which the potential of gene technology as applied to foods will be realised will depend entirely on public opinion. With the recent issue of bovine spongiform encephalopathy (BSE) in Europe, public trust in the food chain has been eroded, which may be one of the reasons why there is much more concern in Europe than in the United States about gene technology. It will be essential for the acceptance of food products developed by genetic modification that they offer benefits over conventional products not only for the suppliers (plant breeders, farmers, agrochemical companies, processors, and retailers) but also for the consumer. It will be vital to communicate the personal and tangible benefits of biotechnology to an uncertain public.

Currently, the benefits of improved agricultural efficiency brought about by biotechnology may not be equally available. However, it is entirely conceivable that plant genetic modifications possible without excessive capital investment in expensive equipment will bring about benefits for the developing world in the future.

REFERENCES

Animal and Plant Health Inspection Service of the US Department of Agriculture (APHIS-USDA). (1998). http://www.aphis.usda.gov/biotech/

Calgene updates FlavrSavr introduction. (1995, June). *Genetic Engineering News*, 26.

Christou, P. (1998). Cereals. In S. Roller & S. Harlander (Eds.), *Genetic modification in the food industry* (pp. 228–249). London: Chapman & Hall.

Crawley, M.J., et al. (1993). Ecology of transgenic oilseed rape in natural habitats. *Nature, 363,* 620–623.

Dale, P. (1998, January). Genetically modified crops: Impact and behaviour of transferred genes. *Biotechnology and Biological Sciences Research Council Annual Report,* Swindon, United Kingdom.

Dunwell, J.M. (1995, September). Transgenic cereal crops. *Chemistry & Industry*, 730–733.

Dunwell, J.M., & Paul, E.M. (1990). Impact of genetically modified crops in agriculture. *Outlook Agriculture, 19* (2), 103–109.

Engel, K., Takeoka, G.R., and Teranishi, R. (1995). *Genetically Modified Foods. Safety Issues.* ACS Symposium Series 6. Washington, DC: American Chemical Society.

Food and Drug Administration. (1992). Statement of policy: Foods derived from new plant varieties. *Federal Register, 57,* C.F.R. 229843005.

Fraley, R. (1992). Sustaining the food supply. *Bio/Technology, 11,* 656–659.

Giovannoni, J.J., et al. (1989). Expressions of a chimeric polygalacturonase gene in transgenic rin (ripening inhibitor) tomato fruit results in polyuronide degradation but not fruit softening. *Plant Cell, 1,* 53.

Hobson, G. (1993, January). The transgenic tomato: The story so far. *Grower,* 14.

Hooykaas, P.J.J., & Schilperoort, R.A. (1992). *Agrobacteritum* and plant genetic engineering. *Plant Molecular Biology, 19,* 15–38.

James, C. (1997). *Global Status of Transgenic Crops in 1997.* Ithaca, NY: International Service for the Acquisition of Agri-Biotech Applications.

Karp, A. (1993, May/June). Are your plants normal? Genetic instability in regenerated and transgenic plants. *Agro-Food Industry Hi-Tech,* 742.

Kinney, A.J., & Knowlton, S. (1998). Designer oils: The high oleic acid soybean. In S. Roller & S. Harlander (Eds.), *Genetic modification in the food industry* (pp. 193–213). London: Chapman & Hall.

Kok, E.J., Reynaerts, A., & Kuiper, H.A. (1993). Novel food products from genetically modified plants: Do they need additional food safety regulation? *Trends in Food Science and Technology, 4,* 42–48.

Kramer, M., Sanders, R.A., Bolkan, H., Waters, C., et al. (1990). Field evaluation of tomatoes with reduced polygalacturonase by anti-sense RNA. In A.B. Bennett & S.D. O'Neil (Eds.), *Horticultural Biotechnology* (pp. 347–355). New York: Wiley, Inc.

Kramer, M., Sanders, R.A., Bolkan, H., Waters, C., et al. (1992). Postharvest evaluation of transgenic tomatoes with reduced levels of polygalacturonase: Processing, firmness and disease resistance. *Postharvest Biology and Technology, 1,* 241–255.

Lloyd-Evans, M., & Barfoot, P. (1996, July). EU boasts good science base and economic prospects for crop biotechnology. *Genetic Engineering News,* 16.

Miflin, B.J. (1993, January-February). Bringing plant biotechnology to the market: The next steps. *Agro-Food Industry Hi-Tech,* 3–5.

Morrow, K.J. (1996, September 12). Debate on the environmental transfer of engineered plant genes resurfaces. *Genetic Engineering News*, 37.

Organisation for Economic Cooperation and Development (OECD). (1993). *Safety evaluation of foods derived by modern biotechnology.* Paris.

Olempska-Beer, Z.S., et al. (1993, December). Plant biotechnology and food safety. *Food Technology,* 64–72.

Redenbaugh, K., Hiatt, W., Martineau, B., Kramer, M., et al. (1994). Regulatory assessment of the FlavrSavr tomato. *Trends in Food Science and Technology, 5,* 105–110.

Rissler, J., & Mellon, M. (1994, February). No commercial gene-altered crop approvals until Fed Government assesses the ecological risks. *Genetic Engineering News*, 4, 12.

Roller, S., & Harlander, S. (Eds.). 1998. *Genetic Modification in the Food Industry.* London: Chapman & Hall.

Schuch, W. (1994, November). Improving tomato quality through biotechnology. *Food Technology,* 78–83.

Taylor, P., & Thomas, J. (1994, July/August). Genetically modified plants: Benefits, concerns and control. *Lipid Technology*, 91–92.

Tomlinson, N. (1998). Worldwide regulatory issues: Legislation and labelling. In S. Roller & S. Harlander (Eds.), *Genetic Modification in the Food Industry* (pp. 61–88). London: Chapman & Hall.

van Wagner, L. (1993, December). Jurassic Park meets the food industry. *Food Processing*, 79–80.

World Health Organisation (WHO). (1991). Strategies for assessing the safety of foods produced by biotechnology. Geneva.

Zambryski, P.C. (1992). Chronicles from the *Agrobacterium*-plant cell DNA transfer story. *Annual Reviews of Plant Physiology and Plant Molecular Biology, 43,* 465–490.

CHAPTER 3

Energy Efficiency and Improved Profitability

Jeremy Tait

3.1 INTRODUCTION

This chapter will help you to make your company more profitable: Every £1,000 saved from your energy bill means £1,000 more profit, year on year. In achieving this, you will also cut your company's impact on global warming. Whichever of these motivates you, this chapter aims to:

1. highlight the possible scale of energy wastage at your plant
2. suggest many practical steps you should consider taking
3. help you to prioritise action for your own company

Prior knowledge of the engineering of factory services is not assumed—the aim is to give basic guidance so you can ask the right questions to achieve improvement. This chapter can only be a starting point and, as it's best to start small, the focus is on zero-cost and low-cost steps on four fundamental factory services.

The first section shows that energy deserves a place on the board agenda and sets out an approach to bringing your energy bills under control. Section 3.3, Essentials of Plant Efficiency, introduces and prioritises the key actions for boilers and steam plant, motors and drives, refrigeration and compressed air. The reference section at the end of the chapter suggests ways to pursue more details on your priority areas.

The material in this chapter is compiled from the publications of the UK Energy Efficiency Best Practice Programme, managed by the Energy Technology Support Unit on behalf of the government's Department of the Environment, Transport and the Regions. Crown copyright is reproduced with the permission of the Controller of Her Majesty's Stationery Office.

Energy usage in food factories: The predominant energy usage is for hot water and steam raising at around 49% of energy usage. Electric motors consume some 24% (including refrigeration and air compressors). The use of ovens (16% of total energy usage) is the largest single use of fuels for direct heating. Figure 3–1 shows an analysis of energy in the UK food and drinks industry by factory service.

The global warming impact of energy use: It is now proven beyond doubt that the concentration of carbon dioxide (CO_2) is increasing in our atmosphere. Scientists are increasingly confident that this will mean higher temperatures and higher sea levels. It is likely that this will also give more extreme weather conditions locally. The environmental impacts of nuclear and even hydro-electric schemes still impose a responsibility on us to minimise demand.

In the United Kingdom during 1997, for every kWh of electricity used, about 0.55 kg of CO_2 was released into the atmosphere, along with 0.002 kg NO_x and 0.007 kg SO_x (based on the average emissions for all generators). For comparison, natural gas emits around 0.19 kg CO_2/kWh, coal 0.32, and oil 0.28 (all UK average data—this varies around the world according to fuel purities and efficiency of distribution networks).

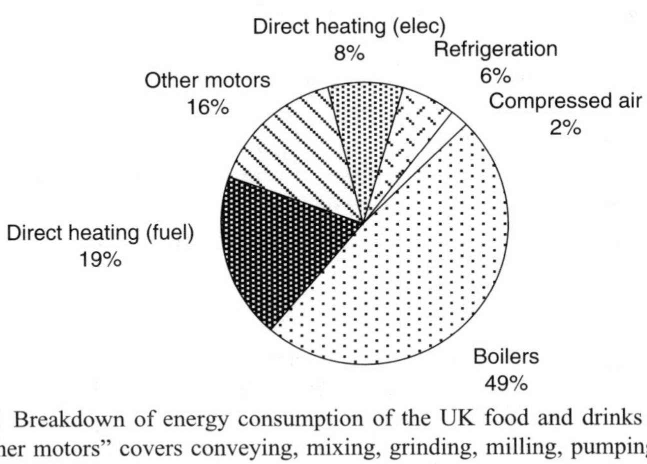

Figure 3–1 Breakdown of energy consumption of the UK food and drinks industry by utility. "Other motors" covers conveying, mixing, grinding, milling, pumping, and other motive-power.

3.2 GETTING STARTED ON ENERGY

The first priority is to make sure that you are buying the energy at the best price; the second is to minimise the amount that you use. But first, a few words on why energy won't be neglected if it is looked at in the right light.

3.2.1 Persuading Management That Energy Is Important

"Men do not realise how great an income thrift is" —Cicero, 48 B.C.

It is useful to realise why energy consumption is often neglected: There are three fundamental blocks on progress, none of which stands up to scrutiny:

1. "Energy is just an overhead, a fixed cost. There's nothing I can do about it".
2. "Energy? My other costs are larger and more important".
3. "Energy efficiency means expensive new plant and complex technical gadgets".

The following may help to persuade you, your management, and your staff otherwise:

1. "Energy is just an overhead, a fixed cost. There's nothing I can do about it": There is insurmountable evidence from almost every branch of industry, and especially from the food sector, that energy is a controllable cost. Most companies could make substantial savings if they were to focus on the task.

- **Question:** Do you waste 10, or even 5, times as much of your raw materials as your direct competitors?
- **Answer:** Not if you want to stay in business.

Dozens of surveys of industrial sectors by the UK Government's Energy Efficiency Best Practice Programme show that many companies are way ahead of their competitors in energy efficiency. Why are so few companies looking at energy costs in the same light as their other raw materials?

2. "Energy? My other costs are larger and more important": Look again! Your energy costs probably amount to 2% to 4% of your production costs if you are a typical food processor; for a few sub-sectors it is higher: for example, distillers with 8%. However, if you are looking to improve competitiveness, then "important" costs are those you can reduce. Try these eye-openers:

- List all your production costs, then delete those that are already at a practical minimum. The truth is that energy will account for some 20% of your controllable costs.

- Calculate the typical energy savings indicated in this chapter as a percentage of your profits; the proportion may surprise you.
- Find out your profitability per £ sales and calculate the extra sales that would be needed to generate the same amount of extra profit as you could get from energy savings. Then ask yourself how much would have to be spent on marketing, extra plant, etc., to achieve those sales. Is energy efficiency an effective way to improve profitability?

You will find that your other costs may be larger, but you may have overlooked some very important profits.

3. "Energy efficiency means expensive new plant and complex technical gadgets": Only think about making big investments once you have the zero-cost and low-cost basics under control. Basic housekeeping issues are very effective and will save you a lot of money. To launch into investment in the latest heat recovery technology may be foolish if, with some lateral thinking, you can eliminate or cut down the heat-creating process. This chapter should provide plenty of low- and zero-cost actions to get a campaign going.

3.2.2 Getting the Best Price

Often it is not until companies start to look at their energy consumption that they realise that they are on high or inappropriate tariffs, which may have seemed good 2 years ago. Energy markets are developing fast in most countries: deregulation, supplier competition, value-added services, complex tariff structures, maximum demand charges, demand-side management options, and much more. Look for competing suppliers and request information on tariff structures; consider seeking independent advice.

3.2.3 Using Less Energy

The majority of companies could cut their energy consumption by 20% using zero- and low-cost measures and modest investment, which would pay for itself within 2 years. The following sections give an introduction to some of the principles of energy management.

3.2.3.1 Pick the Cherries First

In looking for energy-saving options for the first time at a company, there will usually be several major "cherries" to be picked at zero or very low cost. This chapter will help you spot such opportunities.

> *When did you last check your bills?:* A cannery manager was sufficiently persistent to extract the site's electricity bills from his accounts

department. He was appalled to find that the cheap night rate charge was nil in every case, despite their three-shift work pattern. He soon found that the electricity meter clock had failed and successfully reclaimed two years worth of night rates.

The next step is to have a more thorough review of your energy consumption, and plan a campaign to escalate the savings.

3.2.3.2 Find a Champion

Experience shows that the most important factor in successful energy-saving campaigns is an enthusiastic champion—someone to lead the day-to-day campaign with enthusiasm, drive, and persistence. Such champions need to be nurtured and their achievements recognised.

3.2.3.3 Set a Budget and Start Small

But even champions need to be paid, and will expect some investment of time and money from others: Apply resources in proportion to the likely cash-saving benefits. Many campaigns start over-enthusiastically with too-high expectations. If they fail to deliver, much good will is lost, and from there it is difficult to recover. The best approach is to start small and build steadily. Some energy managers have persuaded management to ring-fence any savings for re-investment in energy efficiency the following year; in this way they have developed substantial campaigns slowly but surely.

3.2.3.4 What Gets Measured Gets Done

A useful management adage says, "What gets measured gets done". This is particularly true for energy, as most people have little clue as to how much they use and why. Understand your energy consumption patterns—where and when the energy is used, both in the course of the day for electricity, and across the months of the year for all fuels. This will help you set priorities by focusing on the biggest energy users first. Work out how your consumption pattern relates to the timing of tariff bands, agreed maximum demand levels, etc. Data on your consumption patterns are invaluable when negotiating energy supply contracts. The more you know about how much and when you use energy, the stronger your negotiating position.

3.2.3.5 Publicise Your Specific Energy Consumption

Understand how energy usage relates to your production or throughput to help identify the important factors that affect consumption. You can compensate for the effect of ambient temperatures on space heating energy use by means of "degree days" to expose other, more useful, influencing factors such as use of particular machinery, shift patterns, type of product being made, faulty valves, etc. (Degree days are a measure of the variation of outside temperatures to help building designers calculate heating demands—ask your building services trade association.)

Calculate and track your specific energy consumption (SEC); for example, your kWh per tonne of product or therms per tonne of raw material processed. This will provide a yardstick by which to measure the impact of your campaign. Choose the units carefully to ensure that they will be easily understood by management and the workforce. Wherever possible, present information graphically (graphs, pie charts, etc.) and keep it simple: No one can digest reams of data every week.

Note: SEC is a good start, but it will not reveal all the trends and so on that an energy manager can use to be most effective. There are other fairly straightforward but powerful analysis methods, such as CUSUM (cumulative sum). This is worth looking up in energy management texts.

> *Students can work wonders:* Summer placement students have time and again proven invaluable, and very cost-effective, in getting energy campaigns under way. The projects can be extremely rewarding for the students and also give them an excellent overview of company activities.

3.2.3.6 Set Targets and Track Progress

Agree on targets for improvement, or at least maximum acceptable levels for SECs for the factory, production line, or even for a machine if data can be gathered and the savings look good enough. Track progress towards the targets and revise them as required. In particular, record energy patterns before and after any changes or investments and make sure everyone knows about your successes. This is crucial to prove to management and staff that the investment of time and/or money was worthwhile. There's nothing like a good track record for winning backing for the next enterprise.

> *Monitoring energy brings maintenance benefits:* An over-stretched engineering manager had to keep 14 separate production lines going with inadequate maintenance resources. However, his enthusiasm for energy efficiency led to improvements in reliability: Energy sub-meters on key plant were read weekly and recorded on graphs on his office wall. The graphs proved excellent indicators of the general health of the systems, with unusual consumption patterns flagging maintenance needs long before failure.

3.3 ESSENTIALS OF PLANT EFFICIENCY

Most processes in food factories rely upon a few basic utilities, and this chapter concentrates on minimising the energy consumption of four of them. These four have been selected because they cover a large proportion of the total energy consumption in typical food factories, and they also have plenty of scope for simple, high-yield savings at relatively little or no cost:

1. steam and hot water production and distribution (for sterilisation, cooking, plant washing, etc.)
2. motors and drives (for mixing, mashing, pumping, conveying, etc.)
3. refrigeration (for freezing, storage, tempering, etc.)
4. compressed air (for packaging, handling, conveying, etc.)

3.3.1 Steam and Hot Water Production and Distribution

Figure 3–2 summarises the key action areas for steam and hot water systems. When you are looking for savings in steam and hot water systems, control wastage before improving generation efficiency. Start at the point of use, then look at the distribution system and finally at the boiler.

3.3.1.1 Point of Use

Ask yourself the following questions:

- Can the supply pressure of steam be reduced? Why is the pressure as high as it is?
- If a high pressure is required for only a small area of the plant, consider installing a separate system for that and lowering the main circuit pressure. Pressure selection is always a compromise between capital costs and efficiency: Higher pressure means smaller (but thicker walled) pipes and fittings, less insulation volume, but higher radiation losses and more leakage. For example, raising from 5 bar to 10 bar increases heat losses per square meter by 15% and doubles leakage losses.
- Is the temperature of the plant using the steam correct and effectively controlled? Could it be reduced?
- Is the plant insulated as far as is practicable?
- Is air vented from the plant and are steam traps checked regularly?
- Are the heat transfer surfaces clean? Oil or contamination films forming on the surfaces will cut the heat transfer coefficient.

3.3.1.2 Distribution System

A survey of more than 100 industrial steam systems in the United Kingdom showed that an average of 18% of the energy absorbed in generating the steam was lost during distribution. Losses ranged from 5% to 30%. ("Steam Distribution Costs", 1997).This is in distribution alone—boiler and point-of-use losses are not included.)

Figure 3–2 Improving energy efficiency: what to do first.

Systems have often evolved over many years, being added to as demand has increased and the factory has grown. It therefore makes sense to have a periodic review and rationalisation: Once every 5 years is recommended. Eliminate or isolate redundant pipework and check that pipe sizes are still appropriate for the demand: Too big a pipe means more heat losses, too small means problems meeting demand and high pressure drops.

When were the steam traps last checked? Traps drain water from the lines and often stick. If they are stuck open, steam is vented continuously; if they are stuck shut, the steam line becomes wet or even waterlogged. Wet steam causes erosion, especially at bends and valves. Routine preventive maintenance of traps is usually cost-effective if suspect parts are replaced regularly; if you have high repair labour costs, selection of easily repairable or replaceable traps is essential. A single continuously leaking 12.5 mm steam trap can waste £1,000 per year.

Leaks should be fixed. A 0.8 mm hole in a 7 bar pipe is costing you £125 per year; a 1.5 m steam plume could cost over £1,000 per year.

Do you make use of your condensate? Around 20% of the heat from the boiler ends up as hot condensate; this should be used or at least piped back to the boiler as feed water if it is economic to do so. Returning clean condensate reduces water treatment costs and blow-down losses.

Is all pipework well insulated? Insulating unlagged sections of pipework will pay back in less than a year. A survey of UK sites found that over three-quarters of the 100 sites could benefit from improved insulation. ("Steam Distribution Costs", 1997). An unlagged 100 mm steam pipe will lose around 800 W/metre; compare that to between 40 and 60 W/metre for a well-insulated pipe. An uninsulated flange wastes as much heat as 0.6 metres of uninsulated pipe.

3.3.1.3 Boiler Plant

There are three main heat losses from a (non-electric) boiler: Around 75% of heat lost goes up the flue, with the remainder split equally between radiation losses from exposed surfaces and blow-down. Logging fuel use in each boiler will help you to track performance and prompt maintenance well before failure: Trends or unusual declines in performance will indicate the build up of fouling or blocked burners, etc. When operated correctly, all modern steam boilers are capable of converting fuel energy to steam with efficiencies in the region of 80%. Even higher efficiencies are possible if an economiser is fitted (flue gas heat recovery).

Surveys of UK boiler plant showed that the average thermal efficiency was 74%, ranging from 51% to 82% ("Steam Generation Costs", 1997). This corresponds to an average steam generation cost of 0,81 p/kWh, ranging from 0,73 to 1,18 p/kWh. Average savings of 7% are possible by improving the efficiency of steam generation alone.

Some low-cost measures are described below:

- Controls: Consider operating the minimum number of boilers at optimum efficiency to meet the demand through sequential switching and eliminating boilers on hot standby at full pressure. This is particularly effective where low load conditions persist for extended periods—-it may also be worth investigating alternative means of providing heat for these periods, for example, immersion heaters. Make sure that boilers are fired up as late as possible and switched off as early as possible.

- When did you last check the combustion efficiency and fuel:air ratio? A UK survey recommended this in over one-third of the 100 sites ("Steam Generation Costs", 1997). This is done by measuring the temperature, oxygen, and carbon dioxide content of exhaust gases and consulting standard tables (ask your supplier), which indicate the combustion efficiency. Too much air means fuel wastage as excess air is heated and exhausted. Too little air also means fuel wastage as a result of incomplete combustion and can be dangerous because of the production of carbon monoxide. Getting this ratio right is the main method of reducing heat losses to the flue, but it needs to be done by a competent, qualified person: Consult your burner supplier for detailed advice. A 2% reduction in flue gas oxygen will give a fuel saving of 1.2%.
- Heat exchanger surfaces can get fouled with soot on the flue gas side or scale inside, which leads to high heat losses up the flue. The cost of cleaning can be well worthwhile. Check with your supplier how often this should be necessary.
- As part of the 5-year review, check whether the boiler type and pressure is still appropriate to your demand.

3.3.2 Motors and Drives

The low cost of buying a motor is deceptive. A modest-sized 11 kW induction motor costs as little as £300 to buy, but it could accumulate running costs of over £30,000 over its 10-year life. The electricity bill for a motor for just 1 month can be more than its purchase price. This means that it is particularly important to think carefully when buying or replacing motors and drives.

It is not always easy to decide on the best course of action for cutting motor energy bills, but it is a significant area to tackle. There may be a very large number of motors distributed around the site, and collecting data may be very time-consuming; even then, the best technical and economic option may be difficult to establish. However, following the advice in this section will make this task easier.

The first task is to estimate the running costs of the motors and focus on the most costly ones. For a first order of magnitude estimate, multiply the rated power of the motor, the number of hours it runs per year, and the cost of electricity in £/kWh. It is surprising how this adds up: For example, an 11 kW motor running for 5,000 hours per year at 4.5 p/kWh costs £2,475 per year to run. This is only a rough estimate, but it gives a reasonable starting point without having to take measurements. Remember that the more energy a motor uses, the greater the potential to save.

Once a costly motor is selected for attention, work through the following checklist:

1. Is the equipment still needed? This is not a flippant question! An energy manager at a food packaging plant realised that because the natural ventilation was so good, a 22 kW extraction fan was not actually necessary. Turning off the fan saved £10,000 per year.
2. Switch the motor off.
   - Set up a timer to switch the motor according to a fixed programme or schedule.
   - Use system condition detectors such as a high or low temperature to switch off the motor when it is not needed.
   - Sense the motor load so that the motor is switched off when it is idling.
3. Reduce the load on the motor. Optimising the motor and drive could be a waste of time and money if what the motor is driving is fundamentally inefficient or unnecessary.
   - Is the system doing a useful and necessary job?
   - Is the transmission between motor and driven equipment efficient?
   - Is the driven machine efficient?
   - Are the machine maintenance programmes adequate?
   - Have losses due to, for example, pipework, ducting, or insulation been minimised?
   - Is the control system effective?
4. Minimise motor losses.
   - Always specify higher-efficiency motors, since they are now available at the same price as standard motors and save around 3% compared with standard motors.
   - When a motor fails, minimise subsequent motor losses by always using a good quality motor repair shop.
   - Avoid using greatly over-sized motors.
   - Consider permanent reconnection in "star" wiring configuration (as opposed to "Delta") as a no-cost way of reducing losses in a lightly loaded motor.
   - Check that voltage imbalance, low or high supply voltages, harmonic distortion, or a poor power factor are not causing excess losses.

Treat "energy optimisers" or "motor controllers", or "power factor controllers" with caution: Whilst often portrayed as miracle energy-saving devices, they will not make useful energy savings unless the motor is running most of the time at less than one-third full load. Replacing an over-sized motor with a smaller one may be better. Before investing in optimisers for your motors, persuade the supplier to lend one or more for trials, and be sure to monitor true kWh consumption to check their performance.

5. Slow down the load. In pump or fan application where the cube law applies (i.e., energy consumed is proportional to the speed cubed), just a small reduction in speed can give large energy savings.
   - Use variable speed drives (VSDs) where several discrete speeds or an infinite number of speeds are required. Although often the most expensive option, the many benefits and large energy savings from VSDs make them the usual choice for speed control.
   - Use multiple speed motors where two (and up to four) distinct speed requirements exist.
   - For belt drives only, a low-cost option is to change the pulley ratio.

A rice mill realised that the blower for their pneumatic conveying system was costing them a fortune to run and was over-sized for the job. They cut the blower duty by changing the pulley size on the belt drive—the lowest cost way to cut the speed, at £60. They are now saving nearly £5,000 per year, every year. (Get specialist advice before trying this; pneumatic conveyors need careful balancing).

3.3.3 Refrigeration

The running costs of most refrigeration installations can be reduced by 20% to 30% by using simple, cost-effective measures to improve efficiency.

Refrigeration plant is usually essential to food businesses, and as long as it is working, no-one pays it much attention. This complacency may be eating up your profits: Refrigeration plant is fairly robust and usually has bags of spare capacity. This means that its performance can deteriorate and become very inefficient before you even notice that something is wrong. The measures recommended here are particularly useful as most also improve the reliability of your plant. The advice applies equally to blast freezers, cold stores, or water chillers—indeed to most types of plant, large or small.

Big savings are possible with refrigeration: A frozen food distributor had ageing refrigeration plant that struggled to cope in summer and ran on chloroflurocarbons (CFCs). Wanting also to improve his energy efficiency, the manager took professional advice and decommissioned the old plant, selected mainly secondhand replacements, and now pays less than 50% of the previous running costs. His marginal investment in energy efficiency (over and above the basic replacement cost) paid for itself within 10 months (independently assessed Good Practice Case Study).

There is a close analogy between weight lifting and refrigeration:

$$\text{Energy to raise the weight} = \text{lift} \times \text{load}$$

In a refrigeration system, the lift is the difference between the "evaporating temperature" at the cold end where heat is absorbed and the "condensing temperature" at the warm end of the plant where the heat is dumped (usually to outside air). To reduce the lift, lower the condensing temperature and/or raise the evaporating temperature.

The load is the heat that is absorbed from the process or product. To reduce the load, eliminate or reduce as many of the non-essential or parasitic heat sources as you can. Cutting the lift and load are therefore the first issues to examine, followed by refrigerant leakage and general maintenance.

3.3.3.1 Cut the Temperature Lift

A 1-degree Celsius reduction in temperature lift saves 2% to 4% on energy cost. It may be difficult to estimate what the best evaporating and condensing temperatures are for your plant, and this is where the services of a technician or consultant can be helpful. A number of reasons for higher than necessary temperature lift are summarised in Table 3–1. Note that condensing temperature is also referred to as "head pressure".

3.3.3.2 Cut the Heat Load

The most significant load whenever product is stored is the air change load, entering through open doors. When a door is open, cold air that you paid to chill is escaping and being replaced by warm air from outside. This brings an additional problem: The warm air will carry in more moisture, which will condense on the evaporator as frost. Frost cuts the heat transfer efficiency, pushing the evaporating temperature down and driving costs further up. Reducing the air change load will give good rewards.

Any electrical items operating in the store—such as lights, fan motors, fork lifts, etc.—will be giving out heat. You are paying twice for that energy: once for the electricity to power the item and a second time to run the refrigeration plant to remove the heat that the item generated. Cut these parasitic loads to the minimum by turning off whenever possible, and select low-energy models when you replace them. This also applies, for example, to the chilled water pumps used in process

Table 3–1 Some Causes and Solutions for Excessive Temperature Lift

Condition	*Cause*	*Remedy*
Condensing temperature too high	Blocked or corroded condenser*	Clean/replace
	Condensing temperature controlled high	Reset or remove control
	Air onto the condenser too warm	Relocate, shade, or isolate the condenser
	Air in the condenser*	Purge the air
	Condenser too small for its load	Replace it
Evaporating temperature too low	System short of refrigerant because of leakage*	Find, repair leaks, recharge system
	Evaporator dirty or fouled*	Clean it
	Evaporator frosted up unnecessarily*	Reset the defrost system control
	Contamination (moisture, oil) in refrigerant*	Drain and clean system
	Process/storage temperature too low	Reset thermostat
	Evaporator too small for its load	Replace it

*Result of poor maintenance.

cooling: The energy driving the pumps will often end up as heat in the chilled water.

Other ways of reducing the heat load are shown in Table 3–2.

> A freezer store door left open costs £6 per hour; A chill store door left open costs £3 per hour: If a frozen food cold store door is left open for an entire shift, 5 days a week, this will cost £12,000 per year. A large number of cold stores have leaking seals and/or damaged door strips. On these stores the additional heat load is adding to running costs.

3.3.3.3 Refrigerant Leakage Costs Much More Than Just the Refrigerant

Leakage is a major problem in commercial and industrial refrigeration plants and accounts for an average 11% increase in the running costs of these systems. When the plant does not have enough refrigerant, its cooling capacity drops, which means that it has to run longer to achieve the same cooling effect. Even modest leakage can result in the plant's costing twice as much as necessary to run. Leakage also makes the plant more likely to fail at times of high demand, such as in hot weather. For an ammonia plant, leakage can be inwards, as the plant may operate at a slight vacuum. Any air that gets in must be purged, and automatic plant purgers can be very cost-effective.

> Leakage means increased refrigerant costs, increased running costs, and reduced reliability.

The most common causes of leakage are flare joints on copper pipe (reportedly responsible for over half of all leakage), shaft seals, other mechanical joints, small bore (signal) lines, valves and glands, and vibration (e.g., from pipes connected to compressors). A responsible management approach to leakage includes:

- ensuring that a service and maintenance log book is kept for your plant
- regularly logging the refrigerant charge, leakage rates, and running costs of your plant
- following a comprehensive maintenance programme, including leak testing
- ensuring that all technicians working on the plant are appropriately trained, qualified, and equipped
- eliminating flared joints on copper pipe wherever possible
- ensuring that gas-tight valve caps are used

Table 3–2 Methods of Reducing the Heat Loads on a Cold Store

Heat Load	*Methods of Reduction*
Gain through insulation	Use good quality insulation of appropriate thickness
	Ensure joins in insulation panels are sealed*
	Ensure store lining is not damaged, allowing moisture to penetrate insulation*
Air changes through access	Ensure good door/lid seal*
	Keep doors closed whenever possible**
	Use good quality, undamaged strip curtains, etc.*
Occupancy	Ensure loading/unloading is as quick and efficient as possible**
Lighting	Use low-energy lighting
	Switch on lights only when needed**
Fans	Use low-power fans
Defrost	Ensure defrost is not more frequent or on for longer than is necessary*

*Achieved by good maintenance.

**Achieved by zero cost "good housekeeping" measures.

Consider using one of the proprietary ultra-violet (UV) dyes for leak detection. The dye is added to the refrigerant and carried around the system. At any leak site, the dye is left behind and fluoresces brightly when in a UV beam. Even intermittent leaks are then easy to find.

3.3.3.4 Maintenance Is Money Well Spent

Regular maintenance saves money by reducing running costs, reducing or eliminating service costs (both labour and replacement parts), and reducing or eliminating the costs associated with plant failure, including lost product, inconvenience, and management time.

> A cheese maturation plant had three refrigeration plants on the roof. A very long overdue maintenance call found that one of the three had been running flat-out for months with virtually no refrigerant at all. Worse, for most of the year, the third plant was not even needed because of the very low cooling loads.

The notes below are intended to help end users draw up a comprehensive maintenance schedule. For more detailed independent help, contact a refrigeration consultant.

1. Technicians who work on your plant should be trained as competent refrigerant handlers.
2. Consumables such as refrigerant and oil should be included in the maintenance contract.
3. Contractors should be aware of and work to all relevant Standards and Codes of Practice. In the United Kingdom, relevant standards might include:
   - Code of Practice for the Commissioning and Maintenance of Refrigerating Systems Using Chloroflurocarbons, Parts 1 and 2, Institute of Refrigeration
   - BS4434:1995 Specification for Safety and Environmental Aspects in the Design, Construction and Installation of Refrigeration Appliances and Systems, BSI
   - Code of Practice for the Minimisation of Refrigerant Emissions from Refrigerating Systems, Institute of Refrigeration
4. The following should be included:
   - Check temperature of product/process/room.
   - Check compressor oil level.
   - Check compressor drive belts/couplings and guards if necessary.
   - Carry out maintenance of compressor (only necessary for industrial compressors).

- Clean evaporator if necessary.
- Check evaporator defrost and drain.
- Check operation of evaporator and condenser fans as necessary.
- Clean condenser and check condition.
- Check fan controller operation.
- Check condenser cooling water treatment if necessary.
- Check for blockage of liquid line filter.
- Check operation of system and safety controls.
- Check pipework vibration and insulation.
- Note suction and discharge pressure and temperature of system.
- Check refrigerant charge.
- Leak-test entire system.
- Check integrity of pressure vessel and operation of pressure relief valves.
- Check condition of insulation and door seals and strip curtains where appropriate.

The frequency of each action depends on the application, size, and environment of the installation. For example, some condensers need cleaning every 2 weeks in areas where there is severe air pollution, but usually every 3 months is sufficient.

3.3.4 Compressed Air

Compressed air is expensive—over 50 p/kWh at point of use (assuming an electricity price of 5 p/kWh). However, most people using compressed air systems are not aware of this and treat it like a free and convenient resource.

When you are looking for savings, compressed air systems usually offer cheap and simple ways to cut costs, with typical savings of 30% from cost-effective measures. For example, an 85 kW compressor delivering 250 litres per second, working at 75% load in a 48-hour week, will cost £7,800 per year to run (at 4.5 p/kWh); savings from simple measures could amount to £2,300 per year.

Air compressors are typically 10% efficient, and after treatment and distribution losses, only 5% of the energy bought for the compressor is converted and available at the point of use. Start your consideration of compressed air systems by asking two fundamental questions about the system:

1. *Could you cut the operating pressure?* The higher the operating pressure, the more work done by the compressor and the more expensive the air. Also, lower pressure means less loss from leakage. Find out what the pressure requirement is for the machinery being run, and cut as close as possible to that. If you have to set the supply pressure at more than 10 psi over that minimum pressure, the pipework may need upgrading—get expert advice

on pipe sizing. If all but one or two machines could use a much lower pressure, consider installing a separate, smaller volume high-pressure supply to meet that demand. It may also be possible to reduce pressure during some periods if high-pressure demand is not constant.

By cutting the supply pressure from 7 bar to 6 bar for a 6 mm nozzle, the cost saving is £260 per year for this nozzle alone.

2. *Could you run fewer compressors, or a smaller compressor, and still meet demand?* If you run several compressors, is demand sufficient to require all of them? A simple option is to try shutting one down, yielding significant savings at no cost. Alternatively, various control systems are available for optimising compressor use; modern control systems can automatically sequence compressors, shutting down machines that are off-load to save 5% to 20% of total generation costs.

Is compressed air used to sweep dry waste from surfaces where a brush would do? If it has to be air, a low-pressure system or blower may do.

Air knives are invaluable for drying glassware and so on but may be a misuse of compressed air: Dedicated high-pressure blowers are far less costly and can be linked to the conveyor to switch off when not needed.

3.3.4.1 Leakage

Attention should be given to leakage and the air treatment equipment. Cutting air leakage from the system is one of the simplest ways of reducing costs. Regular systematic surveys are the best method, using proprietary bubble fluids. Many leaks are repaired easily with spanners. Ultrasonic leak detectors are particularly useful for identifying leaks where bubble fluids are impractical and there is significant background noise. Removing redundant pipework eliminates many potential leak points and means fewer pipes to check. The compressed air system should be re-tested for leakage at frequent intervals, for example, every 2 to 3 months.

A survey of compressed air systems at 47 UK industrial sites showed an *average* leakage rate of 39%, and some were losing over 90% of the air ("Compressing Air Costs", 1994). After corrective action, the average leakage over these sites had dropped to 18%.

A 3-mm hole discharging from a pressure of 7 bar wastes about 11 litre/sec of air. If the system is on-line for 120 hrs/week, this corresponds to an annual cost of £1,050.

To understand the scale of the problem, it is useful to find out the leakage rate, either by using a measuring device or by carrying out a no-load test. The best method of measuring the leakage rate is to install a flow meter and pressure gauge in the main supplying the distribution system. Monitoring the output from the flow meter during a representative period will allow you to establish the leakage rate.

An alternative method is applicable to compressors that are operated in on-/off-load; that is, when the compressor is on-load it produces a known amount of air:

1. Close down all equipment operated by compressed air.
2. Start the compressor and operate it to full line pressure, when it will off-load. Air leaks will cause the pressure to fall and the compressor will come on-load again.
3. Record the average on-load time (T) and average off-load time (t) over a number of cycles.
4. Calculate total leakage using the following equation: leakage (1/s) = Q x T/(T + t), where Q = air capacity of the compressor (litre/sec).

Hold a "Leak Week": One soft drink company runs a very successful annual Leak Week: The week before shut-down, staff are issued rolls of yellow tape, which they use to mark each leak they find in the course of their work. During shut-down, maintenance staff work quickly round the system, guided by the tape. Leak Week also demonstrates clearly to staff that wastage of air is a serious cost to the company.

3.3.4.2 Check the Treatment System

Treating air is costly because of the pressure drop across filters and the energy consumed by dryers. The following should be part of your regular maintenance checks:

- Check the pressure drop across the pre- and after-filters. This should not exceed 0.4 bar with the system on full demand.
- Measure the dryer pressure drop. This should not exceed 0.3 bar with the system on full demand.
- Measure the dryer inlet temperature. This should not exceed 35° C with the compressors on full load.
- Measure the temperature of the dryer room. This should be within 5° C of the outside ambient temperature.
- Check that the condensate collection system is working properly and that there is no constant bleed of air from the condensate drain traps.
- Consider treating the bulk air supply to the minimum acceptable level and upgrading the quality, if necessary, at the point of use.
- Don't forget to check the compressor inlet filter!

REFERENCES

Compressing air costs: Leakage. (1994). *Energy Consumption Guide, 41*. Department of the Environment, Transport & the Regions.

Steam distribution costs. (1997). *Energy Consumption Guide, 67*. Department of the Environment, Transport & the Regions.

Steam generation costs. (1997). *Energy Consumption Guide, 66*. Department of the Environment, Transport & the Regions.

ADDITIONAL RESOURCES

GLOBAL WARMING

Houghton J. (1995). *Global Warming, The Complete Briefing*. Oxford: Lion.

CADDET

CADDET (Centre for the Analysis and Dissemination of Demonstrated Energy Technologies) is an International Energy Agency organisation that disseminates case studies and technical booklets about energy efficiency in 15 countries around the world, including the United Kingdom, United States, New Zealand, Australia, and Canada. See their Web site at http://www.caddet-ee.org/. Alternatively, contact the CADDET Energy Efficiency Centre to find out about your national CADDET agency: CADDET Energy Efficiency, P.O. Box 17, 6130 AA Sittard, The Netherlands, Tel.: +31–46–420–2224, Fax: +31–46–451–0389, e-mail: caddet@caddet-ee.org

UNITED KINGDOM ENERGY EFFICIENCY BEST PRACTICE PROGRAMME

An information programme funded by the UK Government's Department of the Environment, Transport and the Regions. It publishes independently prepared advice on energy efficiency free of charge to all UK industry and commerce. Find out more at the programme Web site: http://www.etsu.com/eebpp/home.htm. You can contact the enquiries bureau at Energy Efficiency Enquiries Bureau, ETSU, Harwell, Oxfordshire OX11 0RA, UK, e-mail etsuenq@aeat.co.uk

Selected publications are available free of charge to UK companies from the Energy Efficiency Enquiries Bureau. There is a charge of a few pounds per copy for orders from outside the United Kingdom. Well over 500 publications and case studies are available—ask for publications lists for the topics in which you are interested.

–*Energy management:* Good Practice Guide 186, *Developing an Effective Energy Policy*

–*Boilers:* Good Practice Guide 30, *Energy Efficient Operation of Industrial Boilers*

–*Motors and drives:* Good Practice Guide 2, *Energy Efficient Operation of Electric Motors and Drives*

–*Refrigeration:* Refrigeration publications list, *The Essentials of Refrigeration*

–*Compressed air:* Good Practice Guide 126, *Compressing Air Costs*

ENERGY IN FOOD PROCESSING

Crawford, A.G., & Elson, C.R. (1982). *Energy use in the food manufacturing industry*. Leatherhead, UK: Leatherhead Food Research Association, Scientific and Technical Survey No. 128.

Singh, R.P. (1986). *Energy in food processing*. Amsterdam: Elsevier.

Casper, M.E. (1977). *Energy saving techniques for the food industry*. Park Ridge, NJ: Noyes Data Corporation.

CHAPTER 4

Noise and Air Pollution in the Food Industry: Sources, Control, and Cost Implications

PART 1: NOISE POLLUTION

Philip Walsh

4.1 SOURCES

Noise is defined as undesired sound, and sound is a sensation produced in the ear caused by small variations of air pressure. There are many sources of generation of the air pressure changes; the following is a list of the main ones:

- air excited by the vibration of a surface
- air turbulence generated by a moving surface, such as a fan
- air turbulence generated by mixing of gases, such as a flue or exhaust port
- the flow of air over a discontinuity, causing turbulence; for example, a whistle

All the above sources of noise exist in the food industry in one form or another. The cause of vibration of a surface can be obscure, but some simple examples are:

- material being dropped onto a plate
- impacts between parts in a machine
- out-of-balance forces from a motor
- electromagnetic excitation
- turbulent fluid flow within a pipe

4.2 EFFECTS ON HEALTH

There are two aspects to noise pollution: (1) the effects of relatively high noise levels on people's hearing and (2) the disturbance to people's concentration, enjoyment, or sleep from lower levels of noise.

In most cases, the effects on people's hearing take place in the workplace, fairly close to the source of noise. On initial exposure, people may notice a slight dullness in hearing or a faint ringing in the ear when they are away from the noise, rather as experienced from a cold or flu. These effects are usually short term after first exposure, lasting perhaps only minutes or hours. With prolonged exposure, the dullness of hearing and in some cases the ringing last longer and longer, until eventually it is there until the next day, when the process begins again. As the effect is very slow to build up, it is not readily noticed by most people, and they often claim that they have "got used to it." It is not until after many years of exposure that the reduced ability to understand speech becomes clearly apparent.

The short-term effect is called temporary threshold shift (TTS) and the long-term effect is permanent threshold shift (PTS). It is called threshold shift because this is the effect that can be measured; the threshold being the lowest level of sound that can be heard. As the names imply, TTS is a reversible phenomenon, whereas PTS is the condition of permanent, irreversible impairment.

As is the case for other exposure-generated effects on the body, high noise levels do not produce equal results on everyone. Some workers seem to have the ability to be exposed to high noise levels for their entire working life and suffer little if no reduction in hearing ability, whereas others suffer PTS after comparatively short-term exposure. This difference in susceptibility cannot be measured before exposure, and thus everyone needs to be considered at risk from high noise levels and prolonged exposure.

Very high noise levels (in excess of 140 dB re 20 μPa) can cause instantaneous damage to the ear. Fortunately, such levels do not occur too often in industry; they are usually generated from explosive devices such as explosive cartridge-operated tools or very heavy impacts. Noise-induced hearing loss mostly occurs after prolonged exposure and can be assessed in terms of the exposure to a certain energy per day. Because the ear can usually recover from short-term exposure, if day-long exposure is limited, PTS can be avoided.

Noise levels that are low enough not to be considered hazardous can have other effects on workers that are difficult to quantify. Noise can induce fatigue and irritability that can result in reduced efficiency. Conversely, extremely quiet environments can also lead to reduced efficiency, because any sound not produced as a consequence of your own actions becomes a disturbance.

Noise emitted from a factory to residential areas often leads to complaints. Whilst this noise is rarely directly hazardous, it can make the receiver extremely annoyed or lead to sleep disturbance and thus produce adverse medical effects.

4.3 LEGISLATION

4.3.1 Noise at Work

In the United Kingdom since 1975, employers have had a general obligation under the Health and Safety at Work (HSW) Act 1974 to safeguard workers' health. Since January 1990, the law also has required specific steps under the Noise at Work Regulations 1989. The regulations are based on a European Community (EC) directive requiring similar basic laws throughout the Community for reducing the risk of hearing damage as far as is reasonably practicable. Thus, in general, the limits on noise levels and actions required by employers that will be outlined apply to most of Western Europe, with detailed differences from one country to another. In the United States, there is some difference in the way noise exposure is assessed, but the principles are the same.

The Noise at Work Regulations 1989 deal only with people at work and with risks to hearing, not other aspects of health, safety, and welfare. The HSW Act is more general and can require employers to take action if noise creates risks other than hearing damage or creates risks to persons other than workers. An example of such action is noise that reduces the audibility of a warning sound; in this case, the warning sound may need to be made louder or the noise level reduced.

Three action levels of noise are given in the Noise at Work regulations:

- First action level: a daily personal noise exposure of 85 dB(A)
- Second action level: a daily personal noise exposure of 90 dB(A)
- Peak action level: a peak sound pressure of 200 Pa (140 dB re 20 μPa)

Every employer must, when any employees are likely to be exposed to the first action level or above or to the peak action level or above, ensure that a competent person makes a noise assessment. This assessment must cover the following:

- identify the employees who are likely to be exposed
- provide sufficient information to enable appropriate action to be taken

If you are not sure if an assessment is necessary, a simple guide is that it will be required wherever people have to shout or have difficulty being heard clearly by someone about 2 m away.

It is not usually necessary to measure the noise level at the position of every worker in a factory; the noise can often be assessed in an area where there are a number of people by taking a typical case. Measurements can often be made for a relatively short sample period and then calculated for a full working day. Where exposure is to widely varying noise levels over a working day, special equipment can be used to assess the day-long level.

Reassessments should be undertaken at periods of not more than 2 years or whenever any change has taken place that may affect noise levels, such as the installation of new machinery, or after a programme of noise reduction has been undertaken.

The employer should ensure that the person undertaking the assessment is competent to do so. This may mean sending someone from the company on a suitable course to learn how to undertake noise measurements, or employing an outside consultant who has the necessary equipment and knowledge to undertake the work.

All assessments should be kept at least until a new one is taken, but most companies are wise to keep all records so that long-term trends can be monitored, or in case there is some dispute in the future concerning individual noise exposure levels.

The assessment of noise levels should be seen only as a start, as the regulations require employers to reduce the risk of damage to the hearing of their employees from exposure to noise to as low a level as possible. Although the first action level is set at 85 dB(A), there is a small risk to some people from noise less than this. Thus, if it is reasonably practical, noise levels should be reduced even lower.

Where it is found from the assessment that employees are exposed to noise levels greater than the second action level, 90 dB(A), or to the peak action level or above, the employer must take action to reduce noise levels, so far as is reasonably practical (by means other than the supply of ear protectors).

There are clearly many noises that there are currently no practical means of reducing to a safe level. In these cases, until means are found, ear protection must be used. At the first action level or above (but below the section action level), every employer must provide employees, if they request, with suitable ear protectors. At or above the second action level or the peak action level, every employer must ensure that ear protectors not only are provided but that they are worn.

As many factories have some noisy and some quiet areas, the regulations require that ear protection zones be clearly marked where workers are likely to be exposed to the second action level or above or to the peak action level or above.

Any equipment installed to reduce noise levels or ear protection devices provided to workers must be kept in good order by employers. The regulations also require workers to cooperate in all noise assessments and to use all noise control measures provided.

Employees have the right to know what levels of noise they are likely to be exposed to and should be provided with adequate information concerning the risks and consequences of noise exposure.

Factory inspectors may insist on an assessment of likely noise exposure levels. If they find particularly high noise levels, they can issue an enforcement notice requiring an employer to undertake such work as is necessary to reduce noise to a safe level.

If employers are ignorant of the consequences of high noise levels or ignore the legislation, they may find at some time in the future that current or past employees make claims against the company for compensation for loss of hearing. The costs of contesting such claims or payment of compensation can be several thousands of pounds per employee. To prevent this eventuality, good employers should undertake all the requirements of the Noise at Work Regulations. In addition, to protect themselves and identify people with existing hearing defects, the use of screening audiology tests for new workers is sensible.

4.3.2 Neighbourhood Noise

Complaints of noise nuisance from a factory are usually investigated by a local council agency; in the United Kingdom this would be the Environmental Health Department. A noise may be determined as a nuisance according to the professional opinion of an environmental health officer. This is necessarily a somewhat subjective appraisal, and the officer relies on his or her experience with similar problems; the concept of "public nuisance" as formally held; and a British Standard, BS 4142: 1997—Method of Rating Industrial Noise Affecting Mixed Residential and Industrial Areas. This latter document is used as a guide to decide whether a given noise will excite complaint (i.e., as a predictive technique), and is not meant as a technical substitute if complaints have actually been received. However, to determine the "reasonableness" of a complaint, some recourse is necessary for assessing a problem against average noise susceptibility and a standardised situation. Thus, despite the above comments, BS 4142: 1997 is widely used as a criterion for rating an existing situation.

There is no absolute threshold on noise nuisance; rather, the sound complained of is compared with the background level in the absence of the noise and, depending on the level difference, the tonal character of the noise, or its impulsive nature. The probability of complaint may be determined and the complaint deemed reasonable or otherwise. If it is not possible to measure the background sound level in the absence of the offending sound, a "notional" level may be deduced. If the noise measured from the factory, together with any additions due to the tonal or impulsive nature of the noise, exceeds the background noise by more than 10 dB(A), it is generally accepted that there is a justified nuisance.

The powers of a local authority in these matters are potentially very strong. At first, a notice can be issued giving the owners of a factory a certain period to undertake such work as is necessary to abate the nuisance. If this is not adhered to, fines can be imposed, and ultimately a factory could be closed down. However, if the owners of a factory can show that they have undertaken all practical means of abating the nuisance, this can be a defence: "best practical means."

4.4 METHODS OF MEASUREMENT

As previously described, sound is as a result of fluctuations in the pressure of air. To quantify this, one can measure the pressure fluctuations. In addition, the rate of change of fluctuations—commonly known as the frequency—can be determined.

The frequencies of interest for the effects on humans are those to which the human ear responds, from about 30 cycles per second or Hertz (Hz) to 20,000 Hz, although only young people can hear the very high frequencies. Frequencies above 10,000 Hz are not usually a problem and are usually ignored.

The range of sound pressures to which the ear responds is very large, although the pressures themselves are very small. Pressures are measured in terms of the Pascal (Pa) which is 1 Newton/m^2.

At the frequencies at which the ear is most sensitive (about 2,000–4,000 Hz), the threshold of hearing occurs at a sound pressure level of about 0.00002 to 0.00003 Pa, whilst at the other end of the scale, at the pressure at which the ear begins to hurt, the pressure is about a million times greater, 20–30 Pa.

In comparison, the atmospheric pressure is about 100,000 Pa. It can be seen that if sound pressure were described in terms of Pascals it would be a large and awkward range of numbers to deal with. An added complication is that subjective changes in loudness are not equal for uniform changes in sound pressure. The ear would hear the same relative difference in sound levels between 0.0002 Pa and 0.002 Pa as between 2 Pa and 20 Pa, although the actual changes in sound pressure are widely different. Thus, a scale is required that is based on ratios, which is the basis of the decibel scale.

It should be noted that the decibel is not an absolute measure but always a relative one; it is not only used for measurement of sound, it gives the ratio of one thing to another. In the case of sound, it is the ratio of two pressures and is defined as 20 times the logarithm (to the base 10) of the ratio of the two pressures, that is,

$$\text{decibel ratio} = 20 \log(p_1/p_2)$$

where p_1 and p_2 are the two pressures being compared. For example, if the pressure is doubled, the decibel ratio is 20 log(2/1), which is 6 decibels (dB).

To enable everyone to work to the same standard, it is useful to have a reference pressure; this has been set internationally as 0.00002 Pa (which corresponds approximately to the average person's threshold of hearing at the ear's most sensitive frequencies). If the sound pressure level produced by, say, a pneumatic drill is 2 Pa, this is 100,000 times the reference pressure. The noise can then be described as 100 dB greater (20 log 100,000) than the reference level, usually written as 100 dB re 20 μPa.

The ear does not hear all frequencies equally well; it is developed to hear and understand speech and is thus optimised to work best at those frequencies containing the intelligibility content of speech, about 1,000–4,000 Hz. Thus, the perceived loudness of a sound depends on both its pressure and frequency.

The basic equipment used to measure sound consists of a microphone, which converts the fluctuating air pressure into a voltage; an amplifier; and an indicating device. These are usually combined into one piece of equipment: the sound level meter.

There is a wide range of sound level meters on the market today, offering different combinations of signal processing. All sound level meters should have a "weighting network" (an electrical circuit arranged to attenuate different frequencies by different amounts) built in, which causes the meter to mimic the frequency response of the human ear. This is called "A" weighting, and decibel readings on a sound level meter using this weighting are called dB(A). Use of the (A) after the reading also implies that the standard decibel reference pressure level of 20 μPa has been used.

To measure noise levels for the purpose of assessing the noise in a factory, many sound level meters incorporate an integrating function. This continuously adds the energy incident on the microphone and divides it by the time during which the measurements have been taken. The reading is called the "equivalent continuous level," abbreviated L_{eq}. Where it is not practical to assess the noise level over a sufficiently long period of time because a worker is moving from one position to another or the noise is greatly varying, the worker can be fitted with a dose meter that will measure the equivalent continuous noise level over a complete working day.

To help with the investigation of noise problems, more sophisticated equipment can be used that will measure the levels in various frequency bands or will show the noise level variation with time. To monitor the background noise level in cases of neighbourhood noise, it is usual to use a statistical meter that gives the level of noise exceeded for 90% of the time (the L_{90}).

Measurement and analysis of vibration levels on machinery can sometimes be used to help solve noise problems. Most sound level meters can be adapted for this purpose.

As stated previously, the decibel scale is a relative one and thus it is essential that all noise-measuring equipment is calibrated regularly. A simple calibration can be made before each measurement, using a calibrator coupled to the microphone.

4.5 METHODS OF REDUCTION

4.5.1 Ear Defenders

Ear protection devices come in two basic forms: those that fit in the outer ear passage (often called ear plugs) and those that cover the outside of the ear (ear muffs). Either of these should only be used as a protection against high noise levels if no other means of reducing noise levels are possible, or until other means are undertaken. In many sections of the food industry, there are particular problems in wearing ear protection devices because of the high temperatures and humidities involved. Sweating inside an ear muff can lead to ear infections. Ear plugs can be a problem where there is a danger of their falling into food to be processed, although there are some on the market that are linked to a cord that can be attached to clothing or to a flexible rod that can be placed around the head.

If ear protection must be used, advice should be sought on the suitability of the particular types, to ensure that they afford sufficient protection to the noise levels present, suit the working environment, and fit the wearer.

Dispensers for the supply of disposable ear plugs can be placed at the entrances to noise hazardous areas.

It is a good idea to insist that supervisory and management staff wear ear protection when they enter noise hazardous areas, even if it is only for a short period, to encourage other workers to wear them.

4.5.2 Reduction of Noise at Source

This is the best but the most difficult way of solving noise problems. It can usually only be done at the design stage of a new machine, but users can encourage this process by insisting on lower noise levels when they are purchasing machinery. In the food industry, there are particular problems with the use of certain materials because of hygiene requirements. This often leads to machines' being covered in large panels of stainless steel or aluminium, which act as good radiators of sound.

Noise is typically a waste of energy, so a noisy gearbox or squeaking conveyor belt should be seen as something wrong. A common source of noise in many food processing factories is leaking compressed air. Compressed air seems to be treated by some people as free, although it is expensive as well as being a source of unnecessary noise. Noise from air exhaust ports can be reduced by the fitting of a pneumatic silencer costing less than £3.

The operation of plant can be changed in some cases to reduce noise levels. For example, the impacts between cans on a conveyor line could be reduced if the line speed were continuously changed to match the supply of cans to the line, rather than running at fixed speed.

4.5.3 Noise Enclosures

With existing machines, acoustic enclosures can sometimes be used, consisting usually of a metal box around the machine with controls mounted externally. Careful design of any necessary openings in the enclosures are called for. It is usual for the inside of an enclosure to be covered in a sound-absorbing material (mineral fibre or cellular foam plastic); for many applications in the food industry this is not possible for hygiene reasons, and thus the enclosure needs to be built especially well. Costs of enclosures range from £1,000 to £10,000 per machine and make noise reduction exercises very expensive. A full noise survey of a factory should include measurements of the noise from individual machines, thus enabling the noisiest machines to be tackled first. In many cases, however, satisfactory noise levels are not achieved until all noisy machines are enclosed.

In some cases, it may be possible to enclose the operator in a refuge rather than enclose the machine.

4.5.4 Reduction in Reverberation

Food factories are often very reverberant as a result of the use of tiled or hard plastered surfaces for hygiene requirements. This leads to higher noise levels than might otherwise be present. In some cases, a reduction in reverberation is worthwhile, and there are now available sound-absorbing materials that can be used on the ceilings and upper walls of food processing plants or hung in panels. These materials can be subjected to the normal washing processes used in good processing plants and will not support the growth of bacteria. Typical material costs are on the order of £50 per square metre for flat ceilings. Before embarking on the installation of acoustic treatment, one should be sure that it will be worthwhile. In most instances, the level of noise that workers are subjected to is a result of direct noise from machinery rather than reflected noise from the building.

4.5.5 Reduction of the Time Exposure

Not only are the levels of noise important, the length of time exposed to them is as well. One way of reducing the total exposure of individuals to noise is to reduce the time of exposure. This must be done on a per day basis; it is not acceptable to be exposed one full day to noise and then not on another day. The legislation assumes an 8-h working day; if exposure to noise is reduced to 4 h per day, the limits can rise by 3 dB(A); reduced to 2 h per day, the limits can rise by 6 dB(A); and so on, each halving of time allowing a 3 dB(A) increase.

4.5.6 Reduction of Noise to Neighbors

If this is a problem one should first undertake a detailed study of the noise sources. Where possible, noise measurements should be made with individual items of plant running at the time of day or night that complaints have been made. This should be done near the complainants' houses, or it may be necessary to do it closer to the factory, so that the measurements are clearly discernible above the background noise level.

Reduction of noise at source or enclosure of machines will reduce the noise emitted from a factory, but this is often not practical or cost-effective. In some cases, the sound insulation of the building needs to be improved. This may mean filling in holes in the walls, bricking up or double-glazing windows, building lobbies to doors, and so on. If all such work does not sufficiently reduce the level of noise emitted, it may be necessary to build a false wall inside or outside the existing walls.

In many instances, the source of noise nuisance from food processing factories is ventilation or cooling fans. Attenuators can usually be fitted to these to reduce noise to an acceptable level.

Screening a factory from houses by building an earth bund or wall can be a useful noise-reduction measure. It is necessary to screen the line of sight from windows in the affected houses and any potential sources of noise. The screen is most effective if it is close to the source of the noise or the potential receivers, and will give a 10 dB(A) reduction if line of sight is just broken—more if the screen is considerably higher than the original line of sight.

PART 2: AIR POLLUTION

Stephen Toogood and Martin Key

In assessing the food industry, or indeed any industry, with respect to air contamination, it is important to distinguish between pollution, where emissions from a process will or could cause environmental damage, and nuisance, where the effects are noticeable to people outside the factory, but where there is no environmental harm. The most obvious example of nuisance in this context is odour, which is treated separately in Section 4.10, because the requirements for monitoring, and to some extent for abatement techniques, are rather different. The other main causes of nuisance are noise, which is discussed in Part 1, and dust, which, because it may also be regarded as a form of pollution and because pollution control techniques will also control dust, is included in this section.

4.6 AIR POLLUTION SOURCES AND EFFECTS

4.6.1 Introduction

In general, food processing operations are not major sources of air pollution compared with other industries such as metal or chemical production. There are nonetheless several important emission sources. The load and composition of these are dependent upon the process materials and operations. They typically come from processes such as handling dusty raw materials, cooking, and grinding/milling, and we can characterise emissions from sources like these. In addition, there may be emissions from utilities such as heat or power production at the site. The main types and sources of emissions into the air can be summarised as follows:

- particulate matter and gases (carbon dioxide, sulphur dioxide and nitrogen oxides) from combustion processes
- particulate matter from process operations such as size reduction and raw materials handling, and from cooking and heating processes
- volatile organic compounds (VOCs) and chemical emissions (e.g., ammonia, hydrogen chloride, sulphides) from processes involving the use of organic solvents or potentially volatile chemicals
- VOC and chemical emissions from processes involving the heating or cooking of food, or from fermentation processes

- halogenated VOC and other chemical emissions leaking from refrigeration systems
- VOC emissions from packaging printing and manufacture

4.6.2 General Effects of Emissions

Emissions into the air can cause a wide range of environmental effects. Before we can consider these, we need to define the term pollution. Though many different shades of meaning have been used, the UK Environmental Protection Act 1990 defines pollution in terms of the capacity of a substance to cause harm to man or other living organisms supported by the environment. The term "harm" can be defined as:

> harm to the health of living organisms, interference with ecological systems of which they form part, or offence caused to any sense of man or harm to his property

The range of effects of substances emitted include:

1. chronic and acute toxicity to humans, flora, or fauna
2. impairment of health or dysfunction of humans, flora, or fauna
3. dry deposition leading to soiling of buildings and other surfaces, and possible toxic effects caused by ingestion of deposited materials
4. wet deposition of soluble materials leading to damage to building materials and effects in water such as acidification
5. induced changes in environmental systems as a result of atmospheric chemistry, such as ozone depletion, climate change, and photochemical ozone production

These effects are considered in greater detail in the following sections, which also describe the potential emission control and minimisation techniques available for the various pollutants.

4.7 LEGISLATION GOVERNING EMISSIONS INTO THE AIR

4.7.1 Introduction

Although legislation is specific to a particular country, similar principles underlie most of the international air pollution control legislation. We will consider here the general principles, but must also refer to the relevant legislation in the country where the process operates.

4.7.2 Legal Framework

In most countries, the environmental protection legislation has developed considerably since the 1970s, the intention being to produce a system that is:

- precautionary
- preventive
- pre-emptive

This results in a system that:

- relies on scientific evidence but does not require absolute proof that an emission produces environmental damage (precautionary)
- prefers preventive measures to clean-up (preventive)
- involves the issue of licences, permits, or authorisations (pre-emptive)

In addition, there is an important international approach to pollution control that is central to most legal systems: the "polluter pays" principle. The central idea is that because the capacity of the environment to absorb waste products is limited, any misuse of that capacity ought to be reflected in the cost of production and use of the goods producing the waste. Put another way, it is intended to ensure that irresponsibility is never the cheapest option. Environmental resources are limited and, as their use is not given a direct value, the misuse of such resources is not reflected in the market. It therefore seems reasonable to expect that the cost of pollution prevention and control measures should be reflected in the cost of the goods and services that cause the pollution either in production or consumption, so that the rational and careful use of environmental resources is encouraged.

4.7.3 Responsibilities for Enforcement

Most countries have a two-tier system for controlling air pollution from industry. The first tier includes certain listed processes that fall under the control of state or regional authorities, requiring the issue of a licence specifying operating conditions. The second tier concerns non-listed processes, which are usually controlled by a reactive regulation system that responds to nuisance or environmental effects after they have occurred. We will look in more detail at this approach in Section 4.11.

In addition, air pollution control is usually divided between state/regional authorities and national government agencies. The responsibilities of these bodies vary from country to country; for example, in the United States, the federal (national) Environmental Protection Agency (EPA) sets standards of control and oversees the enforcement responsibilities of the state governments, although the states are responsible for setting the conditions in the licences issued. In Australia, the state authorities specify the industries that require licensing and set their own emission limits. This can lead to processes requiring licensing in some states but not in others, and the industries may also be subject to differing emission limits. It is usual for national government agencies to establish standards and guidelines for the operation of processes subject to licensing by the state or regional authorities.

In most cases, food processing operations are not included in national lists of processes requiring licensing or authorisation, although there may be some local variations in state regulations. The main operations that may be carried on at food processing sites and could be listed for permit-type control are power or heat production by combustion processes, waste incineration, product drying, and certain processes involving the use of organic solvents or hazardous chemicals.

4.7.4 Air Pollution Control Standards

Most countries control emissions from processes by including conditions on licences that require the installation of pollution control equipment. The level of control is often determined by considering the costs and environmental benefits of the equipment; the standard will require the installation of the most effective technology whose cost is deemed reasonable. The term used varies from country to country, but the basis of the approach is similar. Section 4.7.5 outlines the principles behind the United Kingdom's approach of best available techniques not entailing excessive cost (BATNEEC).

There are basically three methods of establishing the performance standards that may be applied to processes and used to establish conditions on licences, authorisations, and permits:

1. stipulate technologies that must be fitted to process plants to control emissions
2. set emission concentration limits for discharges that effectively set the level of control required
3. set ambient air quality standards that are then used to determine the effect of emissions on air quality

In many cases, the strategy used is a combination of these three. The use of emission limits and quality standards is discussed further in Section 4.7.6.

4.7.5 Best Available Techniques Not Entailing Excessive Cost

BATNEEC is the name given to the approach in the United Kingdom. The European Communities directive on the combating of air pollution from industrial plants (84/360/EEC), the so-called "Framework Directive", required member states to take all necessary measures:

1. to require operators of certain categories of industrial processes to seek prior authorisation
2. to use all appropriate preventive measures against air pollution, including the application of BATNEEC

In the United Kingdom the requirement was implemented slightly differently: Processes use BATNEEC for preventing the release of substances to any environmental medium or, where it is not practicable by such means, for reducing the release of such substances to a minimum and for rendering harmless any substances that are released.

The accepted definition of the term BATNEEC is that:

- *Best* should mean the most effective in preventing, minimising, and rendering harmless polluting emissions. There may be more than one set of techniques that can be termed best.
- *Available* should mean procurable by any operator of the class of process in question. It should not imply that the technique is in general use, but it does require general accessibility.
- *Techniques* include the process itself and how it is operated, including the concept and design of the process, staff numbers, supervision, training, and working methods.
- *Not entailing excessive cost* needs to be taken in two contexts, depending on whether it is applied to a new or existing process. The presumption is that the best available techniques will be used, although this can be modified by economic considerations where the cost of applying the best available techniques would be excessive in relation to the environmental protection achieved.

The principle in the application of BATNEEC is therefore the use of the most efficient pollution control technique having regard to a balance between economic costs and environmental costs. This approach is common to most countries, though it may have different names.

4.7.6 Establishing the Level of Control Required

It is in practice quite difficult to determine directly the limit of BATNEEC. It is more usual for governments to set either emission limits or ambient air quality standards to establish the level of control they require. To determine what particular control strategy will fulfil the objective, a systematic and analytical approach must be taken. This might be:

1. Identify the pollutants produced by the process.
2. List the control techniques available.

3. Assess each of these as to how effective it is at controlling the pollutants, given how they are produced in the process.
4. Determine the capital and operating costs of each technique.
5. Compare how the efficiency of pollution control and costs relate to one another.

A judgement then has to be reached as to where the balance lies between the cost and the environmental protection achieved. This is the single most difficult part of the procedure, but again there are criteria that can make the process more systematic:

- any air quality guidelines that exist for the pollutants
- the ecotoxicity of the pollutants
- any potential the pollutants have to cause nuisance, for example odour
- any additional environmental problems and supplementary costs incurred by particular techniques, for example, the production of liquid waste or solid waste that will then require treatment and disposal

Clearly, the effect of the pollutants on air quality is important, and most countries now have statutory air quality standards. These can be used to measure the effectiveness of the controls specific to the process (an approach used by the EPA in the United States) or can be combined with emission limits to establish dual controls on emissions (an approach used in Australia).

In the European Community, the directives issued indicate a similar dual approach; there are directives that set statutory air quality standards (for example, 80/779/EEC on sulphur dioxide and suspended particulates, 822/882/EEC on lead in air, and 85/203/EEC on nitrogen dioxide in air), while others set control standards for types of industrial process (for example, 88/609/EEC on large combustion plant).

4.7.7 Best Practicable Environmental Option (BPEO)

As mentioned in Section 4.7.6, air pollution control plant often converts the pollution to liquid or solid waste for disposal. Recognising this, international cooperation has developed an approach integrating the control of emissions to air and water and of solid waste generation, to ensure that pollution problems are not simply transferred from one environmental medium to another.

This has resulted in the best practicable environmental option (BPEO) concept. The UK Royal Commission on Environmental Pollution twelfth report (1988), entitled Best Practicable Environmental Option, explained the term:

> A BPEO is the outcome of a systematic consultative and decision making procedure which emphasises the protection and conservation of the

environment across land, air and water. The BPEO procedure establishes, for a given set of objectives, the option that provides the most benefit or least damage to the environment as a whole, at acceptable cost, in the long term as well as in the short term.

The aim is to consider the environment as a total system: Any proposals to reduce pollution to one part of that system should not adversely affect another part of that system.

4.7.8 Development Planning Control

One common theme among all pollution control activities is the importance of planning. In terms of development control, when a new development is proposed, assessment should include consideration of environmental hazards. To do this, environmental impact assessment is now required for larger proposals, particularly industrial processes with significant potential to pollute.

Environmental impact assessment is a systematic procedure for collecting, analysing, and presenting environmental data to ensure that the likely effects on the environment of a new development are fully understood and taken into account before the development begins. Though requirements vary between countries, as well as developments where assessment is mandatory, for some other types of scheme in the United Kingdom it may be required at the discretion of the planning authority, and for this reason it is now unlikely that any significant development by the food industry would be allowed without an Impact Statement.

4.8 EMISSIONS, EFFECTS, AND CONTROLS

4.8.1 Introduction

As we discussed in Section 4.7, most air pollution control legislation assumes that emissions from the process can be prevented. This can usually only be achieved by changing the process operations or the raw materials used. For example, it is possible to prevent the emission of volatile organic compounds (VOCs) from processes by ceasing to use organic solvents. However, in most cases, the options for process change are limited by the process technology, so that "end-of-pipe" pollution abatement equipment has to be used to minimise emissions.

4.8.2 Combustion Process Emissions

There are two main methods of using energy in food processing:

1. indirect use, where steam or hot gases from combustion are used to heat the process through heat exchange, not involving contact with the product

2. direct use, where the steam or hot gases are passed directly into the process equipment and contact the product

Most energy use in food processing is indirect, and activities such as steam raising account for the biggest of combustion process emissions. Nevertheless, some drying and cooking processes involve the direct injection of steam or direct use of combustion gases. The combustion process releases a range of pollutants, the most important being particulate matter and smoke, carbon dioxide, sulphur dioxide, nitrogen oxides, and organic compounds. The emissions produced from combustion processes are largely dependent on the fuel used. The particular characteristics are listed by fuel in Table 4–1.

4.8.2.1 Health Effects

By "particulate matter" we mean to include both relatively large particles such as grit, through dust, to very small sizes found in smoke and fume. Particulate matter has its principal effects in the respiratory system, and the severity of the effects is related to the particle size range. Particles may also carry other toxins into the respiratory tract. Often the effect of particles is synergistic with sulphur dioxide, causing irritation of the mucous membrane. The effects of particulate inhalation are especially problematic to people who already suffer respiratory problems.

The effects of sulphur dioxide and nitrogen oxides may be considered together. Exposure to sulphur dioxide can lead to acute effects such as bronchitis and to chronic pulmonary effects, especially in people susceptible to lung dysfunction such as asthma. Exposure to NOx results in similar effects.

4.8.2.2 Environmental Effects

Particles are frequently the cause of complaint, because they are so clearly visible and remain so for a long period after deposition. Thus they tend to accumulate, and quite low deposition rates can lead to soiling of buildings and other prop-

Table 4–1 Emissions Characteristics of Fuels

Fuel	*Sulphur Content*	*Particulate Emissions*	*Other*
Coal	Fairly high	High	
Oil	Fairly high	Low if well-controlled	
Gas	Very low	Very low	
Waste	Source-dependent	Source-dependent	May include metals, volatile organic compounds (VOCs), and toxic materials

erty, and coloration of plant leaves, possibly interfering with photosynthesis. At higher deposition rates, washing is soiled.

Oxides of sulphur and nitrogen cause acidification of rain, soil, and waters, and damage building materials. These acid gases can also lead to leaf collapse in plants. Oxides of nitrogen are implicated in photo-chemical oxidation reactions (see Section 4.8.5).

Carbon dioxide (CO_2) is a product of complete combustion and an important gas in respiration reactions. CO_2 is a greenhouse gas: It has been implicated in climate change (global warming). This phenomenon occurs because the gas allows light from the sun to pass through it and warm the earth. The earth itself then radiates heat, but at a different wavelength. This is absorbed by the greenhouse gases, thus trapping the heat and leading to a surface temperature increase, which in turn may cause significant changes in climatic conditions. CO_2 is an important pollutant that is subject to increasing internationally negotiated controls to reduce emissions and the potential for climate change.

4.8.2.3 Control Techniques

Smoke is usually a result of incomplete combustion, so that attention to the regulation of secondary combustion air is the best preventive technique. Emissions of other particulate matter can be reduced by the use of simple low-cost techniques such as impingement or other momentum-based separators or cyclones. On larger combustion processes, more extensive equipment may be required, and this is described in Section 4.8.3. In the case of smaller processes where it is necessary to control particulates, wet scrubbing techniques may also be worth considering; these may remove acid gases at the same time. The various types of wet scrubber are all designed to make dust particles heavy by wetting them. Table 4–2 summarises the abatement technology options for particulates.

The release of acid gases from combustion must be prevented as much as possible. This is usually achieved by the incorporation of scrubbing equipment in the flue gas cleaning plant. Alternatively, the need for these processes may be obviated by careful choice of fuel. This is essentially an economic decision. A variety of scrubbing techniques are in use, but they can essentially be divided into:

- wet scrubbing, where the acid gases are absorbed into a liquid medium
- dry or semi-dry scrubbing equipment, where lime is injected into the flue gases either as a solid or as a slurry with water; the dry techniques rely on the subsequent removal of the particulate matter

Wet scrubbing has the disadvantage of producing a liquid effluent that may require treatment prior to discharge to a sewer or watercourse, and also has the potential to create a visible steam plume from the chimney. The cost of this equipment is dependent upon the waste gas flow; it can cost from £10,000 to £250,000.

Nitric oxide is produced in any combustion process using air, and is influenced by the temperature of the combustion gases (typically temperatures below 1200° C have an insignificant formation rate). In many cases, it is possible to reduce the emission of nitrogen oxides by conversion to low-NOx burners. Despite the efforts of several development teams, it is still the case that the best way to reduce carbon dioxide emissions is by improvements in energy efficiency over the site as a whole. Very frequently the improvements are self-financing.

4.8.3 Particulate Matter Emissions from Processing

There are many potential sources for the emission of particulate matter from food processing. These include size reduction, raw materials, and product handling. These emissions are generally dry, and the principal effect is deposition outside the factory. Normally, abatement plant will be needed.

Table 4–2 is a comparative list of the different particulate matter arrestment options, detailing approximate arrestment efficiencies and typical particle size collection range. The first four processes listed are dry processes, and the remainder are wet scrubbers. In choosing between these generic types, the disposal issues need to be taken into account. For example, dry processes produce a solid material that needs a disposal route. Its handling needs to be carefully planned if it is not to be released back into the atmosphere. Wet systems produce a slurry, which again needs disposal. Whichever type is chosen, proper attention to inspection and maintenance is essential if nuisance is to be avoided.

Table 4–2 Comparison of Abatement Technology for Particulates

Equipment	*Efficiency (%)*	*Particle Size Removed (lm)*	*Cost*
Settling chamber	> 50	> 50	Low
Cyclone	> 80	> 10	Low
Fabric filter	> 99	> 1	Medium
Electrostatic precipitator	> 99	> 1	High
Fibrous scrubber	> 90	> 1	Medium
Impingement scrubber	> 90	> 1	Medium
Moving bed scrubber	> 90	> 1	Medium
Plate scrubber	> 90	> 1	Medium
Spray tower	> 90	> 2	Medium
Venturi scrubber	> 98	< 1	Medium

4.8.4 Chemical Emissions from Food Processing

4.8.4.1 Sources and Effects

The sheer diversity of the food industry makes it impossible to cover the whole range of possible emissions to the atmosphere. However, one important group of processes uses acids in the processing operations. As discussed in Section 4.8.2, emission of acid gases can lead to acid rain and other acidification, building material damage, and leaf collapse in plants. Other processes may involve the use of sulphides and similar chemicals, and below the occupational exposure limits, the effects of these are primarily odour generation.

4.8.4.2 Control Equipment

Acidic gases and vapours are normally removed using wet scrubbing of some kind. Scrubbers rely on the absorption of the pollutant into a liquid (usually water). The liquor can then be treated to remove the acid simply by pH adjustment, or, more frequently, the scrubber can be maintained at a set pH value by dosing with alkali. There is a wide range of process variations, the main two categories being spray towers (where the liquid is sprayed into the top of a tower, with the air flowing upwards) and packed towers (where the liquid is sprayed onto a packing material that greatly magnifies the surface area of contact between the gas and the liquid). In general, packed towers are the more efficient technique for this application.

4.8.5 Emission of Volatile Organic Compounds

There are many potential sources of emissions of volatile organic compounds (VOCs). They include heating and cooking operations in which organic material present in the food may be volatilised (although in this case odour is usually the main concern). A significant number of food processing operations use organic solvents in the process operations (often for extraction processes) or may emit VOCs as a by-product of reaction (for example, distilling and bread baking).

4.8.5.1 Effects

VOCs have been the subject of increased attention over the past few years and are now recognised as an important group of pollutants. They are known in some instances to have caused pollution at a considerable distance from the sources, possibly in a different country. An important way in which VOCs cause pollution is photo-chemical oxidation by solar radiation in the presence of nitrogen oxides, which produces ground-level ozone.

Whereas there is concern at the loss of ozone from the stratosphere, ozone at ground level (known as tropospheric ozone) is a pollutant that causes eye, nose, and throat irritation; chest discomfort; coughing; and headaches. People who exercise or who suffer from asthma or bronchitis are most at risk. The photo-chemical oxidation process takes several hours, so it is unlikely that there will be adverse effects on local air quality. The local effect of VOC emissions is odour.

4.8.5.2 Control Equipment

Processes to control VOC emissions are very similar to those available for odour control and are addressed in detail in Section 4.12.2.

4.9 MONITORING OF AIR POLLUTANTS

4.9.1 Introduction

Three possible approaches to monitoring and assessment of emissions to the air are:

1. monitoring of emissions at the point of discharge from the process
2. monitoring of ambient air quality
3. monitoring of process equipment or abatement equipment efficiency

Of these, the first two may be said to give quantitative data for assessing the effects of emissions. Monitoring plant performance is nonetheless an important management technique, and may be easier and cheaper to carry out. It may be used to flag-up the need for a more extensive monitoring exercise.

4.9.2 Emission Monitoring

The measurement of the concentration of pollutants in waste gases is central to the enforcement of regulatory standards. The establishment of emission concentration limits provides a performance standard against which the operation and effectiveness of control measures can be assessed.

There are two options for emission monitoring:

1. continuous monitoring involving the installation of a monitoring device in the chimney or discharge point
2. extractive testing, where a measurement of emission concentration for a given pollutant is performed over a discrete time period

In both cases, a wide range of measurement equipment and methodologies are available, and in many cases the technique used for continuous and extractive measurement is the same. Often the methods are specified or recommended by regulatory agencies in support of the air pollution control standards. Twenty years ago, techniques usually involved wet chemical analysis, but since then many instruments have been developed that provide direct read-out of emission concentrations with greater accuracy. The techniques used for particulate matter are usually light-based measurement systems for continuous analysis and extractive sampling, with gravimetric analysis for non-continuous methods.

In planning emissions monitoring, there are several important considerations, including:

1. The method of sampling must ensure that representative samples of the whole flow have been collected or measured.
2. The analytical technique chosen must be sensitive to the pollutant being measured and must have a resolution capable of giving accurate results at the concentrations likely to be encountered.
3. All sampling and analysis systems must be provided with means of calibration, particularly in the case of continuous monitoring devices.
4. Any monitoring must take account of normal process variations and should involve testing over the range of possible processing options.
5. The technique chosen should minimise the possible interference from cross-sensitivity to other species in the gas flow.

4.9.3 Ambient Air Monitoring

Many countries have combined regulatory systems that specify emission concentration limits from process equipment and also establish air quality guidelines or standards for determining the effects of emissions on local air quality. This approach is used in the United States, where in some states the severity of the emission limits depends upon the local air quality and on the foreseen effects of emissions from the process on that quality.

Consideration of the effects of emissions on air quality raises the important issue of dispersion of emissions from a process by use of a chimney or stack. The aim of a chimney is to harness the natural dispersion that occurs in the atmosphere to reduce maximum ground-level concentrations to below specified levels. If the emission rates from a process are known, it is possible to model the expected effect of these on ground-level ambient air quality.

The methods of measurement for ambient air are often based upon similar principles to the methods used for emission monitoring. The main variation is that the concentration of pollutants will be considerably lower than in process discharges, and therefore the technique used must have greater sensitivity. In the case of pollution, compared with odour, short-term variation is of less concern in ambient air, so that monitoring concentrations are frequently expressed as daily averages rather than 30-minute means. In the absence of national standards, it is usual to compare measured values to the guideline values specified by the World Health Organisation. Most countries have a network of ambient air quality monitoring, although currently only the United States uses the results of this monitoring to determine routinely permitted discharge standards.

4.10 ODOUR EMISSIONS

4.10.1 Introduction

Several factors complicate the assessment and regulation of odour compared with pollution:

1. Odour is generally caused not by a single chemical species but by a mixture of many components.
2. The effect being regulated is a human response, involving detection by the nose, which is extremely sensitive to some compounds, especially those derived from the breakdown of food-derived material.
3. Odour is detected rapidly, so that it is instantaneous concentrations rather than averages over time that are of interest.
4. Different people react to odour in different ways.
5. Dispersion of odour, and therefore whether it will reach receptors, is very dependent on meteorological conditions.
6. The generation of odour from a process can vary greatly with raw material quality and with what is momentarily happening in the process.

For good genetic reasons, humans are not only very sensitive to odours derived from food and its decomposition but also find them particularly unpleasant. This exacerbates the odour problems faced by the food industry. The sources of odour are many, but they are often associated with the cooking, heating, or drying of food, and with the storage, handling, and treatment of process wastes. The odorous chemical species are likely to be VOCs such as natural oils, fatty acids, and aldehydes; in the case of wastes, amines, thiols (mercaptans), sulphides, and ammonia may also be important. It is clear that the composition of an odour is likely to be complex and variable.

4.10.2 Odour Sources

The introductory discussion above identified cooking, heating, drying, and waste materials as common odour sources. Another way of categorising sources distinguishes between point sources—such as chimneys and vents, which are essentially known—when specifying the plant, and fugitive emissions, which include building or process leaks. Fugitive emissions are particularly difficult to quantify.

4.10.3 Effects of Odour Emissions

In principle, the effects of odour emissions are very simple: People get annoyed. Indeed, odour is defined solely in terms of its perception. The legislation intended to deal with odours, which is described in Section 4.11, is based on this fact, and on the concept of nuisance.

Because of the strong psychological component in the effect of odour emissions, if a factory begins to cause odour nuisance, or its lesser form: odour annoyance, that fact tends to sensitise the environment to odour. Put another way: If you begin badly, the neighbours will continue to give you a hard time. It follows from this that, if in doubt, it is far better to apply abatement techniques from the beginning, because if you add them later you will have to meet stricter standards.

A good deal of research effort has been put into the measurement of odour, and quantification techniques are described in Section 4.13, but nobody has yet succeeded in devising a reliable method of quantifying offensiveness. This need not worry us unduly. Coffee roasting and bread baking smells have both, for example, been used extensively by retailers as part of their sales technique, but coffee factories and bakeries have frequently been subject to odour complaints. Annoyance is not ultimately linked to how offensive odours are; anything can be a nuisance if you don't have an "off" switch. It is therefore persistence that is the best measure of odour.

Nonetheless, the speed and violence of public reactions to odour emissions certainly are linked to offensiveness. Many of the chemical species that humans find particularly offensive are those produced by the decomposition of food. Hydrogen sulphide, for example, is synonymous with rotten eggs; amines are formed in decaying fish and meat; thiols (mercaptans) are characteristic of rotting vegetables. For good evolutionary reasons, humans can detect all these compounds at very low levels, comparable in some instances with the limits of analytical equipment. Because these odours are associated with unfit food, some people are physically unable to eat even fresh food when they can smell bad food. Under these circumstances, odour incidents are likely to be remembered and complaints initiated. There may also be longer term psychological effects such as stress. Avoiding these smells requires attention to non-production parts of the business: to effluent handling, and to management of returned product and any reject raw materials. Odours from production itself are more likely to comprise other compounds, including partially oxidised ones, such as aldehydes, that are associated with cooking.

If the persistence of an odour defines its strength, this must be related to how much dispersion, or dilution, will cause it to disappear. Actual persistence therefore is partly a function of meteorological conditions. Increasing wind speed gives increasing dilution, so that the times of highest risk to a factory are times of low wind speed and increasing atmospheric stability (poor mixing). Such conditions are often found in late afternoons during August and September, and indeed many sites find that the number of complaints is at its highest then. It is also important to note that persistence is not a function of odour strength per se, but of strength times flow. This latter is defined as the "emission rate" of odour.

4.11 LEGISLATION RELATING TO ODOUR EMISSIONS

In most countries, the control of odour emissions from food processing is a matter for control by reactive legislation (action taken after the event to guard against a recurrence). The legislation is usually based upon the concept of nuisance. Nuisance is a very old legal concept, defined as material harm, inconvenience, discomfort, damage, or interference with someone's right to enjoy their property and common amenities. In the United Kingdom, competence to bring an action for nuisance was therefore originally restricted to those with a proprietorial interest in property (for example, someone using a public highway could not allege nuisance).

The Public Health Act (1936) established the concept of "public nuisance", effectively removing this restriction and giving local government the statutory duty to regulate nuisance suffered by any group of citizens. This act was replaced by the provisions of the Environmental Protection Act, though much of the legislative framework remains the same. The legislation acknowledges that there are some industries in which the production of odour is inherent to the process; these are called "scheduled processes", and they are separately regulated, using a BATNEEC approach. The food industry is not a scheduled process.

Industries not scheduled are regulated in England by local government under Part III of the EPA. These industries operate without intervention on the part of the regulator until a complaint is received. The complaint is then investigated; if it is substantiated, an approach, usually informal, will be made to the factory. If the problem persists, the regulator may believe that a statutory nuisance exists and may serve an abatement notice, requiring the nuisance to cease, and setting other conditions. The operator of the site may appeal, using the defence that he or she is employing the best practicable means (BPM) to abate the nuisance. If the appeal is allowed, the regulator may take no further formal action unless the nuisance gets worse. If the notice is upheld, the operator must comply with the terms or face an enforcement notice, which may result in financial penalties or ultimately in closure of the factory.

The conditions attached to an abatement notice might include specific action that the operator is required to take in abating the nuisance, such as storing reject material in a building or installing some form of odour treatment plant. They may set a limit on odour emission rate.

The BPM concept is important to the planning of odour control measures, even though no abatement notice may have been served. It is commonly regarded as a less demanding form of BATNEEC, but it is important to note that preventive measures such as stock control systems would certainly be considered part of BPM. Again, the extent of BPM includes consideration of what is a reasonable cost. However, in some industries there is no overall consensus as to what BPM includes.

One of the reasons for this, and for the advice not to rely too heavily on the formal provisions of the legislation, is that the forum for hearings is the magistrates' court. Cases are heard by lay magistrates without access to technical assessment, so that even if both parties submit technical evidence, the outcome can be unpredictable.

In some countries, of which the Netherlands is the best-developed example, there are national emission standards relating to odours, and these specify odour strength and in some cases chimney height for a variety of industries. Increasingly, however, the approach is to set environmental standards.

Because of the limits of the legislation described above, planning authorities are careful to ensure that any new development does not cause odour nuisance. To this end, they are very likely, whatever the country of jurisdiction, to set limits on environmental odours and to make these part of the planning consent. Sometimes this is done on the assumption that there is no odour before the development, but that is virtually never the case. If any limit expressed as an environmental concentration is set, it is very important for the factory to undertake a background odour survey before construction begins, and to obtain agreement that compliance with the standard is to be measured relative to the background value. It has frequently been the case that conditions set by planning authorities have been exceeded by existing land users even without the addition of any emissions from the proposed development. Two important sources of odour frequently overlooked are road traffic, which is a substantial source of hydrogen sulphide, and agriculture.

In some instances, and in some countries, new development is not permitted within a specified distance of residential use. These minimum distances are industry-specific.

4.12 CONTROL OF ODOUR EMISSIONS

4.12.1 Introduction

However effective an odour treatment plant may be, the effort is wasted unless the odour can be effectively captured, that is, prevented from escaping into the environment. Design of the ventilation system is therefore of the greatest importance and should be carried out interactively with the choice of treatment process. However, this should always be preceded by a critical examination of the options for preventive measures, which will almost always be more cost-effective than ventilation and treatment. One of the best approaches to prevention is through process optimisation and control. This might include:

- attention to stock control, good housekeeping, and raw material handling practices
- maintenance of equipment to keep it as leakproof and spillproof as possible
- control and minimisation of cooker emissions
- prevention of fugitive loss, as uncontained emissions of this type are not capable of controlled dispersion

In many cases, however, there will still be a need for ventilation and treatment. Ventilation design should seek to minimise air flows with the overall objectives of effective capture and safe and comfortable working environments. This can be best achieved by:

1. Containment of odours within process equipment by good design and extract ventilation where necessary. Ventilate the plant in preference to the room. This has the advantage of making sure that air flows from the working space towards the odour source. It makes the working conditions more pleasant.
2. Separation of contained strong odour sources, such as cooking emissions, from weaker sources. They may even need separate treatment.

4.12.2 Odour Treatment

The common techniques for odour treatment are summarised in Table 4–3. There is, nonetheless, some overlap with air pollution control plant; for example, odours from frying are associated with entrained fat, which can be removed as particulate matter. The odour treatment processes are described below. Caution is needed in interpreting cost data, because of its strong dependence on size.

4.12.2.1 Adsorption

This is a process in which the gas molecules are retained on a solid surface (physi-sorption) or oxidised there (chemi-sorption). Activated carbon is the principal example of the former. The plant takes one of several forms:

1. Disposable cartridge filters for intermittent flows and low odour concentrations. The cost is dependent upon flow rates but typically in the area of a few thousand pounds per plant installed.

Table 4–3 Odour Treatment Processes

Process	*Treats*	*Approximate Efficiency (%)*
Adsorption	VOCs* and odours	> 95
Thermal incineration	VOCs and odours	> 95
Catalytic incineration	VOCs and odours	> 95
Fibrous scrubber	Acid gases, VOCs, and odours	> 90
Moving bed scrubber	Acid gases, VOCs, and odours	> 90
Packed bed scrubber	Acid gases, VOCs, and odours	> 90
Plate scrubber	Acid gases, VOCs, and odours	> 90
Spray tower scrubber	Acid gases, VOCs, and odours	> 90
Sorbent injection	Acid gases	40–80
Biofiltration	VOCs and odours	> 95
Spray drying	Acid gases	> 90

*VOC = volatile organic compound

2. Fixed plant, in which the adsorbent is replaced from time to time, perhaps by the supplier. Sometimes the adsorbent is supplied in the form of pre-filled units. If loose-filled units are used, attention to proper packing down is vital to subsequent performance.
3. Systems involving regeneration. Normally these are only applicable to VOC removal, and the regenerated stream itself requires further treatment for disposal. These are rarely encountered in the food industry.

Generally, moisture is the enemy of adsorption processes. Any plant chosen must as part of its design have means of pre-treating the airstream to remove droplets and ideally to keep humidity below 85%. Elimination of fat is also important.

The cost of fixed-bed and fluidised-bed systems is high, between £50,000 and £500,000.

4.12.2.2 Thermal Oxidation

Heat is used to burn up, partially or totally, the odorous compounds. The advantage of this technique is that it can offer very high destruction efficiencies (often >99%), but because operating cost is proportional to volume, it is used for high-odour-strength emissions. Almost all components can be treated. The destruction efficiency is determined by oxidation temperature, time, turbulence, and oxygen availability in the combustion zone. Typically, a temperature of 800° C is required for efficient control of odours. Several different oxidation systems are available:

1. Thermal oxidisers or after-burners, in which waste gases are heated to more than 800° C by the use of support fuel. The amount of support fuel can be reduced by incorporating heat recovery to pre-heat incoming waste gases.
2. Catalytic oxidisers, in which a catalyst is used to reduce the oxidation temperature, typically to between 350° C and 400° C. The main problem with this type of equipment is the risk of catalyst coating or poisoning, which can lead to catalyst replacement costs of up to 20% of the original capital outlay.
3. Regenerative thermal oxidisers such as the ceramic bed oxidiser. These units incorporate a simple system of passing the air into a pre-heated ceramic bed, where oxidation is completed. The bed acts as a very efficient heat exchanger, and the cleaned air returns energy to the bed before discharge. The airflow is then reversed to use this stored heat. These units can offer 98% thermal efficiency and have an energy equivalent use of oxidising the odours at 30° C to 50° C above the inlet temperature, offering savings of up to 95% running costs compared with the traditional thermal oxidiser.
4. Another option is to pass odorous gases to existing boilers as combustion air. This option is viable only where the boiler is on load continuously or where its use is coincident with the generation of the odour.

Thermal oxidation technology offers extremely high odour destruction for a wide range of odorous emissions, and the low energy use of regenerative oxidisers offers greatly reduced running costs. This technology is amongst the most suitable for very high-intensity odours. The typical capital cost of this technology is between £100,000 and £750,000.

4.12.2.3 Chemical Scrubbing

These processes exploit any solubility of the odorous gases in water. The airstream is blown up through (countercurrent) or into the side of (cross-flow) a container, down through which a liquid stream is flowing. A packing is provided to give a very large surface area over which the gas-liquid stream is flowing. Usually, chemicals such as sodium hydroxide and hypochlorite are dosed into the liquid side to promote solubility and to oxidise the contaminants once they are absorbed. Chemical addition is frequently controlled by the monitoring of pH and rH values. One variant of the process introduces a catalyst into the liquid recycle, in order to speed up the oxidation reactions and ensure that no dissolved odour is returned to the top of the absorber.

Scrubbers have a wide applicability, offer reasonable efficiencies for low to medium odour intensities, and are priced between £40,000 and £350,000. They require regular maintenance and dosing of chemical additives and have the disadvantage that, for some odours, it is necessary to have a number (two or three) of different stages dosed with different chemicals to treat the contaminants concerned. This, of course, has a substantial impact on the price.

4.12.2.4 Biological Treatment

Many odours, especially in the food industry, are the result of natural degradation or are derived from natural materials. It is therefore usually the case that they are amenable to bio-oxidation. There are two basic types of plant: biofilters and bioscrubbers, of which the former are by far the more common. Bioscrubbers are mostly used for reducing high concentrations of hydrogen sulphide, typically in wastewater treatment. Biofilters have a much more general applicability and have even been used with success to treat non-biogenic odours such as toluene.

Abatement efficiencies are good if the plant is well designed and operated, but difficulties may arise with sudden increases in load. However, when production is continuous, they are usually a cost-effective option. Recent developments include the use of media where the presence of carbonates regulates the acids formed during oxidation and of media that can be used at much greater depths than was possible in the past. This latter development is important, because the greatest disadvantage of biofilters is the relatively large size of the plant, which, when it was limited to 1 m in depth at a retention time of about 50 s, precluded its use on some sites. The new media have approximately halved the area needed. Costs range from £40,000 to £200,000, but operating costs are low.

4.12.3 Dispersion of Odours

Even with the application of high-efficiency odour treatment plant, the ability of the atmosphere to dilute and disperse odour is a useful tool to ensure that neighbours do not experience perceptible concentrations. The design of chimneys and the selection of heights to ensure dispersion is a specialist task and is based on mathematical dispersion models.

The same models can also be used to aid the prediction of odour concentrations as part of planning applications, when objections have been or might be raised to a new development. The results are increasingly found in environmental impact statements. Modelling emissions requires that the emissions after treatment and the fugitive emissions from the site can be quantified. This is another area where much development effort has been expended over the past few years.

4.13 MONITORING OF ODOURS

The human nose is a very effective detector of odour and can detect many substances, especially those derived from animal and vegetable protein, at concentrations that rival the sensitivity of analytical techniques. Odour is also a complex and variable mixture of many compounds. As a result of these and other factors, the applicability of chemical analysis for odour is limited, except where one compound, such as hydrogen sulphide, dominates the effect. Odour monitoring is therefore undertaken by olfactometry, in which the human nose is used as the detector.

An olfactometer is an accurate diluting machine. Samples of odorous air are taken into inert plastic bags and transported to the laboratory. The olfactometer has a supply of purified, odour-free air, and this is used to dilute the sample by a known factor, before presentation to panellists. The olfactometer has two or more presentation ports; one is fed with the diluted sample, and the other with purified air. Panellists must indicate which they think is the sample. A panel of six or eight people can accurately establish the strength of an odour, which is defined as being equal to the number of volumes of air needed to make one volume of odour become undetectable to the median panellist. This value is expressed in "odour units per cubic metre" or as "dilutions threshold", which are numerically equal.

Results from this method can be used for regulatory purposes or as inputs to a dispersion model, though the technique may be limited by the accuracy of sampling, especially when ambient concentrations are being measured. It is more accurate to sample at the point of discharge and to calculate ambient concentrations by dispersion modelling.

CHAPTER 5

Food and Wastewater

Andrew D. Wheatley

5.1 INTRODUCTION

Water pollution control is advanced compared with other areas of environmental protection because of its links to disease. Water quality and trade effluent discharges are already rigorously monitored and audited against worldwide standards. Water pollution is also often very visible and persistent; it can generate a high level of public and media attention. Food hygiene means the industry is a large water user, which, in the environmentally sensitive regions of the world, can make companies vulnerable to adverse customer reactions. These are often triggered by accidents or a perception that pollution control policies are not of the highest international standards. Pressure groups have also become adept at attracting media attention, with campaigns such as "name and shame", environmental league tables, and customer boycotts. The food industry, therefore, needs to be well aware of the current environmental issues. Managers must be able to prepare environmental audits, understand the issues of sustainability, and promote environmentally friendly products. Waste management and avoiding contamination will be key issues in the future, both to protect existing business and to develop new products. These are the issues covered in this chapter with the aim of providing training and an introduction to water and wastewater issues in the food industry.

5.2 CONTROLS

Some highly publicised pollution issues, such as global warming and damage to the ozone layer, have put pollution control on the political agenda. Two of the major international trading areas—North America and Europe—already have uniform water pollution standards, although member states still have the freedom to impose more stringent standards, such as those in California and Scandinavia. There are also broader agreements affecting larger regions or sensitive areas, such as the North Sea, the Mediterranean, and so on.

5.2.1 Europe

Within the European Community (EC) new legislation is initiated by the administrative Commission of the European Communities in Brussels. These laws or directives are, after suitable discussion and debate, approved by the Council of Ministers (representatives of national governments) and made law. Prior to approval in the final form, draft directives are circulated and published for discussion and amendment. Outlines of the directives are published in the *Official Journal of the European Community* (series L), and full details can be obtained from government departments, government book shops, the CEC, and local European Union information offices.

Once the directive is approved by the Council of Ministers, it becomes binding on member states; but it is left to the national governments to introduce legislation conforming to the intention of the directive, and the legislation does not come into force until this is done. Directives usually set time limits for governments to comply.

The environmental protection legislation is guided by a basic European Commission policy to promote recovery and recycling of waste materials. It aims to do this by applying three basic principles to waste disposal:

1. *The polluter pays principle:* The waste disposer should bear the cost of environmental restoration, for example, conveyance treatment and monitoring.
2. *The proximity principle:* Waste, hazardous waste particularly, is best treated as close as possible to its source to avoid the generation of diffuse sources of pollutants.
3. *The precautionary principle:* Where there is a lack of toxicity or biodegradability data, maximum permitted concentrations of discharges should be set at the detection limits.

Another aim is to ensure uniform standards among member states. There are objections to this approach, because different discharge standards may be scientifically justifiable in different countries without causing a deterioration in environmental qualities. In the United Kingdom, for example, rivers are short and fast flowing compared with many in Europe and therefore arguably capable of a greater degree of self-purification.

Similar scientific arguments have been presented by the United Kingdom concerning sewage sludge disposal in the sea. In the North Sea it was shown that there was good dispersal and self-treatment with no adverse environmental affects, whereas this practice in either the Baltic or Mediterranean, where there would be poor mixing, would cause environmental damage.

Criticisms have also been levelled at the precautionary principle, which may not provide good value for money in terms of environmental improvement or reduction in risk. These arguments, however, are not understood by public opinion.

The European Commission had hoped to encourage clean technology by increased taxes on non-renewable resources in line with United Nations (UN) recommendations (see Section 5.2.3.3). This has been resisted by some member states, notably Britain, which, for example, has found the introduction of a carbon tax difficult to implement politically. An alternative idea under discussion was to generate a market for tradable pollution credits. The total amount of allowable pollution was to decrease over time and so increase the value of the pollution credits.

The EC directives can be divided into key groups affecting drinking water, natural waters, toxicity and persistence, environmental impact, recycling, and competition (Table 5–1).

5.2.1.1 Chemical Releases

One of the most important characteristics of these directives compared with most previous national legislation was the introduction of quantitative standards for a wide range of parameters. For example, the Dangerous Substances Directive (76/464/CEC) lists 129 substances that are particularly dangerous and therefore priority cases for control; this is known as List 1, the black list. The amounts of these chemicals that can be discharged are limited in two ways. The first is the uniform emission standard which specified limits by maximum concentration and total amount; the second defined an acceptable environmental quality standard that has an upper environmental limit not to be exceeded.

The Dangerous Substances Directive also includes a list of potentially hazardous materials: List 2, the grey list. Discharges of List 2 substances to soil and potentially to groundwater therefore must be monitored and authorised by a national responsible authority.

Fourteen chemical parameters are listed in the National Waters Directive 78/659/CEC, together with minimum sampling frequencies and analytical methods. The establishment of a coarse fishery will be the minimum for all surface waters. Sources of surface water for drinking are classified into three categories—A1, A2, and A3—corresponding to different types of standard treatments (directive number 75/440/CEC). Forty-six substances and parameters are listed in the Surface Water Directive, with maximum allowable concentrations (MACs) and guide levels; 95% of samples must meet these values (Directive number 79/869/CEC). Surface water less than A3 in quality should not be used for drinking purposes. Drinking Water Directive 80/778/CEC lists 60 substances with MAC and guide level

Table 5–1 Major EC Directives Affecting Water Quality

(1) Drinking water	
80/777/EEC	On the approximation of the laws of the member states relating to the exploitation and marketing of natural mineral waters
80/778/EEC	Relating to the quality of water intended for human consumption
(2) Wastewater	
91/271/EEC	Concerning urban waste water treatment
(3) Surface and bathing waters	
75/440/EEC	Concerning the quality required of surface water intended for the abstraction of drinking water in member states
76/160/EEC	Concerning the quality of bathing water
78/659/EEC	On the quality of freshwaters needing protection or improvement in order to support fish life
79/869/EEC	Concerning the methods of measurement and frequencies of samples and analysis of surface water intended for the abstraction for drinking in the member states (reference methods for 75/440/EEC)
79/923/EEC	On the quality required of shellfish waters
91/676/EEC	Concerning the protection of fresh, coastal, and marine waters against pollution caused by nitrates from diffuse sources
(4) Dangerous substances	
76/464/EEC	On pollution caused by certain dangerous substances discharged into the aquatic environment of the Community
This Directive has given rise to several "daughter directives"	
82/176/EEC	On limit values and quality objectives for mercury discharges by the chlor-alkali electrolysis industry
83/513/EEC	On limit values and quality objectives for cadmium discharges
84/156/EEC	On limit values and quality objectives for mercury discharges by sectors other than by chlor-alkali electrolysis industry
84/491/EEC	On limit values and quality objectives for discharges of hexachlorocyclohexane, in particular lindane
86/280/EEC	On limit values and quality objectives for discharges of certain dangerous substances included in List I in the Annex to Directive 76/464/EEC (covers DDT, PCP, and CTC)
88/347/EEC	Amending Annex II to Directive 86/280/EEC relating to aldrin, dieldrin, endrin, isodrin (the 'Drins'), hexachlorobenzene (HCB), hexachlorobutadiene (HCBD), and chloroform
90/415/EEC	Amending Annex II to Directive 86/280/EEC on limit values and quality objectives for discharges of certain dangerous substances included in List I of the Annex to Directive 76/464/EEC
80/68/EEC	On protection of groundwater against pollution caused by certain dangerous substances
(5) Biodegradability	
73/404/EEC	On the approximation of the laws of the member states relating to detergents
73/405/EEC	On the approximation of the laws of the member states relating to methods of testing the biodegradability of anionic surfactants

continues

Table 5–1 continued

82/242/EEC	Amending Directive 73/404/EEC
82/243/EEC	Amending Directive 73/405/EEC
86/94/EEC	Amending Directive 73/404/EEC
(6) Sewage sludge disposal	
86/278/EEC	On the protection of the environment, and in particular of the soil, when sewage is used in agriculture
91/375/EEC	Concerning the protection of waters against pollution caused by nitrates from agricultural sources
(7) Management and policy	
77/795/EEC[a]	Establishing a common procedure for the exchange of information on the quality of surface freshwater in the Community
83/189/EEC[a]	On laying down a procedure for the provision of information in the field of technical standards and regulations
86/85/EEC[a]	Establishing a Community information system for the control and reduction of pollution caused by the spillage of hydrocarbons and other harmful substances at sea
88/346/EEC[a]	Amending 86/85/EEC to include major inland waters
91/692/EEC[a]	On standardising and rationalising reports on the implementation of certain directives relating to the environment
92/446/EEC[a]	Concerning questionnaires relating to directives in the water sector
89/106/EEC	On the approximation of laws, regulations and administrative provisions of the member states relating to construction products
90/531/EEC	On the procurement procedures of entities operating in the water, energy, transport and telecommunications sectors
90/683/EEC	Concerning the modules for the various phases of the conformity assessment procedures which are intended to be used in the technical harmonisation directives
92/13/EEC	Coordinating the laws and administrative provisions relating to the application of Community rules on the procurement procedures of entities operating in the water, energy, transport and telecommunications sector
(8) Other	
82/501/EEC	On the major accident hazards of certain industrial activities
85/337/EEC	On the assessment of effects of certain public and private projects on the environment
85/374/EEC	On the approximation of the laws, regulation and administrative provisions of the member states concerning liability for defective products
(9) Proposed EC Directives	
COM(85)373	On dumping waste at sea
COM(85)733	On water quality objectives for chromium
COM(89)303	On the establishment of the European Environment Agency and the European Environment Monitoring and Information Network
COM(91)347	Amending 90/531/EEC
	Landfill and leachate
COM(97)105	On the backfilling of waste
94/263/EEC	On packaging and packaging wastes

[a]A "Decision" rather than a "Directive".

values. There is a directive on groundwater, 80/68/CEC, that bans the discharge of all List 1 substances, although exceptions are allowed after investigation.

The biodegradability of chemical substances is covered by two groups of directives. The problems of persistent detergents were recognised some time ago and a group of directives (73/404/CEC, 73/405/CEC and 82/242/CEC, 82/243/CEC) specify a biodegradability of 90%. The sub-directives specify methods of analysis. The new substances directives 79/831/671/548/CEC include a series of ecotoxicity and biodegradability standards for all substances produced in quantities greater than 100 tonnes per year. The tests assess breakdown in combination with sewage, in river water, and in anaerobic environments and refuse. The tests are based on bottle tests, respirometry, and simulation trials, for example, soil columns and model wastewater treatment plants.

5.2.1.2 Wastewaters

The Urban Wastewater Directive has had a significant effect on the extent and degree of sewage treatment. In 1990, prior to the directive, only about 45% of the sewage generated in Europe received any treatment. The United Kingdom, Germany, and Holland treat more than 80% of their sewage while Greece, for example, had facilities for only 25% of its sewage (Figure 5–1). This reflected the degree of urbanisation. There are two main provisions in the directive. The first is that sewage collection systems and sewers should be provided for all towns by 1998. A second requirement is to link discharge standards to the pollution sensitivity of the receiving water. Three broad categories of natural waters were envisaged. The least sensitive had large volumes of dilution to enable self-treatment and purification. Typically, these might be coastal waters or estuaries. The most sensitive would be those intended for drinking water, of special scientific or ecological interest, or vulnerable to particular types of pollutant, for example, eutrophication. The directive, therefore, says that a two-stage treatment, biological or the equivalent, be provided for population equivalents exceeding 2,000 when discharging to freshwater or 10,000 if the discharge is to sea. The standards for the most sensitive areas would include 5 mg/L total nitrogen and 1 mg/L total phosphorous (Table 5–2).

Member states have to identify areas that are at low risk of pollution and that would not be adversely affected by stronger discharges of wastewaters. The directive calls for experimental evidence that environmental quality will not be affected by these types of low-quality wastewaters. Requirements for discharge from sewage treatment plants are subject to Articles 4 and 5 of the directive. Phosphorus and nitrogen levels apply to discharges to sensitive areas that are subject to eutrophication.

This directive has stimulated a significant market for sewage treatment in Spain, Italy, France, and Greece. In the United Kingdom there has been significant

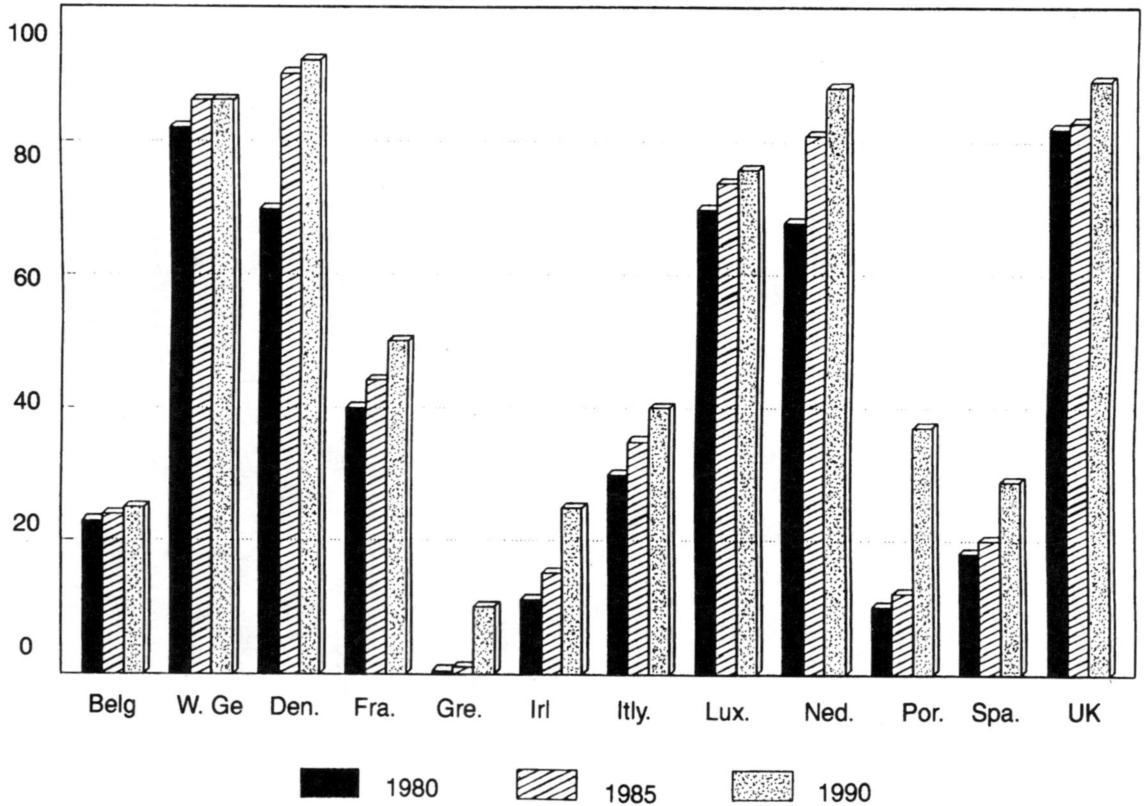

Figure 5–1 Rate of growth in the application of sewage treatment in Europe. Percentage of population with treatment.

Table 5–2 Requirements for Discharge from Sewage Water Treatment Plants Subject to Articles 4 and 5 of the Directive; Phosphorus and Nitrogen Levels Apply to Discharges to Sensitive Areas That Are Subject to Eutrophication

Parameter	*Concentration*	*Minimum Percentage Reduction*
Biochemical oxygen demand (BOD_5 at 20° C without nitrification)	25 mg/L O_2	70–90
Chemical oxygen demand	125 mg/L O_2	75
Total suspended solids	35 mg/L	90
Total phosphorus	2 mg/L	80
Total nitrogen	15 mg/L N	70–80

Note: Standards are expressed as annual mean values for a population equivalent of 10–100,000; greater than 100,000, the standards are more stringent.

Source: Official Journal of the European Communities L135/40 dated 30.5.1991. Council Directive 91/271/EEC covering Urban Waste Water Directive.

expenditure on replacing short outfalls by simple treatments, such as screening and settlement coupled with long sea outfalls. It can be shown by modelling and survey that these long sea outfalls do not cause significant pollution, but this has not always been acceptable to the public, and there have also been a significant number of complete sewage treatment works installed for coastal towns.

The Sludge Directive (86/181/CEC) encourages the recycling of sewage sludge to land, whilst at the same time trying to avoid any harmful affects on soil vegetation or animals and the food chain. Treatment of the sludge (see Section 5.10) is necessary to avoid odour and reduce the potential for contamination by pathogens. In nitrogen-sensitive areas, it will also be restricted by the risk of contamination of surface and groundwaters by runoff. The times of application are therefore regulated. Anaerobic digestion is the most common sludge treatment in the United Kingdom. There are discussions at present on the risks of faults, accidents, and errors with the current sludge recycling procedures. Further pasteurisation at higher temperatures (70° C) may be required in the future to provide additional protection. The directive also requires monitoring, and that records be kept of potentially hazardous material such as metals and persistent organics present in the sludge.

5.2.1.3 Finance and Planning

The Environmental Impact Directive covers planning and new developments (85/337/CEC). It calls for a mass balance of all raw materials, products, and wastes at processing sites. Details of the disposal routes for any wastes, their biodegradability, and the environmental impact are required.

The Procurement and Utilities Directive applies to all water and effluent projects with a value of more than 400,000 ECU (about £265,000). Companies are required to publish basic information about these types of supply contracts, according to specified procedures. For projects valued above 5 million ECU (about £3.3 million), invitations to tender must be published in the *Official Journal of the European Communities*. The aims of the directive are to encourage free and fair competition and to include specifications to European standards.

5.2.2 United States

The United States has a well-established, integrated pollution control policy covering discharges to water and the atmosphere and solids disposal. This policy is managed by the Environmental Protection Agency (EPA). The information provided by the EPA can be very comprehensive and would include advice on the best available technologies necessary to meet the standards required, and their likely costs. The information provided can provide design guidance for other parts of the world. The EPA is responsible for developing new regulations by assessing available technical and scientific information, and by carrying out its own research and experiments. For potentially hazardous or unknown substances, the EPA assumes responsibility for setting, justifying, checking, and enforcing the standards. The EPA has a list of about 130 hazardous or priority pollutants equivalent to the combined European black and grey lists. For routine water quality parameters, state authorities are responsible for maintaining standards, with occasional checks by the EPA.

The comprehensive guidance given by the EPA and the need to justify the standards does lead to extensive arguments and litigation as more knowledge and data become available. This information can, however, be used by any interested parties and can be accessed via databases. More details can be obtained from the EPA Internet site or texts such as La Grega, Buckingham, and Evans (1994).

Water regulations cover drinking water, protection of surface water (Exhibit 5–1), and sludge disposal.

5.2.3 International Agreements

There are also a number of broad agreements or protocols affecting larger regions or sensitive world-wide issues. An example is the North Sea Conference. The first meeting in 1984 agreed to reduce discharges of toxic and persistent compounds; the second, in 1988, agreed to control excessive algal growth in the Baltic. The most recent conference again agreed to reduce sewage discharges and stop

Exhibit 5–1 Water Regulations in the United States

- Safe Drinking Water Act (SDWA): First introduced in 1974 with the federal standards and the formation of the Environmental Protection Agency in 1970. Regularly updated with information on new substances.
- Clean Water Act (CWA): Superseded federal Water Pollution Control Act. Includes standards and treatment procedures. Updated to include toxins and diffuse pollutants; 65,000 chemicals listed. Recent additions cover preservation of wetlands and watersheds.
- National Pollution Discharge Elimination System (NPDES): Regulates industrial municipal and storm water discharges.
- Standards for the Use and Disposal of Sewage Sludge: Standards, analysis, and safety of sewage sludge disposal.

sewage sludge disposal to the North Sea. Similar meetings take place in Barcelona to avoid pollution of the Mediterranean. The Mediterranean Protocol restricts the discharge of toxic and persistent substances, oxygen-depleting organics, and nutrients. It also includes emergency procedures and restrictions on land-based pollution likely to flow into the sea.

The most recent world-wide pollution meetings have attracted media interest and have become significant events. The Montreal Conference met to eliminate ozone-depleting substances, principally the chloroflurocarbons (CFCs). The Rio Summit considered sustainable development and biodiversity, and the Kyoto Summit produced an agreement on greenhouse gas emissions.

Many of the international quantitative standards are set by non-political organisations such as the Organisation for European Co-operation and Development (OECD), the World Health Organisation (WHO), and the UN.

5.2.3.1 Organisation for Economic Co-operation and Development

The OECD is a non-legislative system for co-operation on standards between North America and Europe. It has been a major influence on EC policy. It has the task of setting, justifying, and updating quantitative emission standards.

The OECD has developed a number of standard biodegradability tests and produces a regular state of the environment report. The last report in 1985 noted considerable progress on oxygen balance and the microbiological quality in water. It suggested that new problems were likely to be caused by nutrients and eutrophication. It also called for more research to prevent the transfer of pollutants to the sea, a theme taken up by other regional international groups.

5.2.3.2 World Health Organisation

Much of WHO's work is associated with water supply and sanitation in rural countries. Eighty percent of all disease in the world is linked to poor water quality. The European office is also concerned with long-term hazards and water quality. There are two currently important issues: one is the proliferation of new chemicals (about 1,000 each year), which means that it is impossible to have precise data on their effect on the food chain or human toxicity; the other is the wide range of viruses that it is possible to identify in water as new methods of isolation have been developed. The evaluation of risk, extrapolation of data, and acceptance of risk are used to develop guidelines on drinking water quality. Many of these guidelines have subsequently been included in the EC directives and national legislation.

5.2.3.3 United Nations

In 1987, the UN presented a report on methods of funding additional environmental quality improvements. The report was known as the Brundtland report after its chairperson, the then prime minister of Norway. The report suggested that the present market prices for goods and services ignored the full costs of environmental protection and resource depletion, as much of the cost associated with any subsequent environmental restoration was transferred to the whole community. An example given was that water customers were now paying the costs of removing nitrate pollution from groundwater caused by farming. The report recommended that the economic incentives for clean technology and recycling could be improved by changes in the tax structure, with higher rates charged for pollution or depletion of valuable resources. These suggestions were widely accepted and taken up by the OECD and the EC, who have proposed changes to the value added tax. The first example was supposed to be additions to the value added tax for a carbon tax. Tax increases are very unpopular in Europe, and the plans have been postponed.

5.3 ENVIRONMENTAL AUDITING

Planning and comparing disposal options requires a thorough analysis of the waste-generating processes. A waste audit is a systematic study of the mass balances of raw materials, products, and waste.

Once the sources, nature, and quantities of waste have been established, it is easier to find methods of re-organisation and re-design to reduce losses. Environment auditing can also make contributions to some areas of business activity, such as cleaner and safer products, improved marketing of existing products, and the generation of new, greener products. The sources, scale, and type of waste can be determined by standard analytical techniques published by the Department of the Environment, British Standards Institute, Environmental Agency for England and Wales, or EC. An example of this is the standard methods of water and wastewater analysis (Eaton, Clesceri, & Greenberg, 1995). An environmental management report can then be compiled to answer some key environmental objectives. These are the basic requirements of ISO 14001:

1. identifying existing compliance and consent issues
2. identifying and eliminating any discharges of priority pollutants
3. introducing management systems (ideally software flowchart-based) for the environmental assessment of new processes and practices and to provide emergency response procedures
4. devising and setting appropriate environmental targets
5. determining losses in terms of product and raw materials
6. determining the cost of disposing of this waste and an allocation to a product generating them
7. selecting and pricing treatment alternatives
8. identifying possible control procedures
9. providing for training on waste for key staff

This information will also be of value in fulfilling the terms of the EC Environmental Impact Directive, which calls for a mass balance and declaration of disposal routes for planning consents for large projects. These principles have been included in the UK land and registration procedures for the protection of groundwater and to provide a public register of historical land use. Environmental auditing is a useful system for raising awareness, ensuring legal compliance, limiting liability, assessing costs, and identifying cleaner technology.

One of the aims of integrated pollution control is to promote a rethink of production processes to minimise waste. The onus will be on companies to justify publicly that they are using the best environmental option for waste disposal. At present, only the fuel and power industries are required to apply for registration, but is anticipated that other industries including the food processing industry will eventually need to undertake eco-auditing.

5.4 PREDICTING THE COSTS

Water supply, treatment, administration, and costs are different in the countries making up the United Kingdom. England and Wales are taken together, and water services are provided by two different types of utility companies in the private sector. In Scotland, water services are controlled by three public authorities. In Northern Ireland, water services are controlled by the central government via the Department of the Environment for Northern Ireland. Four regional (East, West, North, and South) divisions are set up to deal with the details of supply and treatment.

In England and Wales, there are 10 regional multifunctional companies, which are privatised public authorities. They supply about 80% of the drinking water and 95% of sewage treatment. There are also 19 separate water supply companies that have been in the private sector since their inception; they do not provide waste treatment facilities. The English and Welsh water companies are regulated by three agencies set up at the time of privatisation. These are:

1. a drinking water inspectorate (regulator of drinking water quality)
2. the Environment Agency (to monitor and regulate natural water)
3. the Office of Water Services (financial control)

The regulators are independent auditors of water company performance, avoiding conflicts of interest between quality and cost.

5.4.1 Sewer Discharge

Until the recent introduction of the Urban Wastewater Treatment Directive only four countries in Europe had more than 80% of their populations served by sewage treatment works. By 2000, sewers and sewage treatment works will be available for the disposal of food industry effluent in most regions of Europe. Usually, the treatability of food industry waters is improved by combination with domestic sewage. Dilution and buffering reduce the impact of cleaning solutions, and the supplementary nutrients available in sewage increase microbial growth rates. Some types of food wastes may lead to a deterioration in treatability because of their poor solubility. The United States and Europe therefore insist that discharges of effluent to sewer or the environment be licensed. This enables a consent to be issued listing MACs to prevent damage to drains and inhibition or overloading the treatment. It also enables the recovery of treatment and monitoring costs and allows public scrutiny.

Corrosive materials—for example, strong acids or alkalis—are prohibited, as are the common organic solvents, which could form explosive mixtures. There are limits on a second category of materials that could form corrosive products within the sewers by cross-reaction, for example, sulphates and high-temperature effluents. There are also restrictions on solids that would accumulate and block drains, for example, silt and fat. A further group of prohibited substances are those that would cause inhibition of the treatment processes or prevent the water companies from disposing of the residues. These include fat, detergents, colour, and very strong wastes, as well as the more familiar list of metals, pesticides, and organochlorines included in the red lists and black lists (see Section 5.2). Since privatisation and the establishment of water companies, licenses to discharge to sewer are also contracts, and water companies may wish to restrict any substance to suit local circumstances. Table 5–3 gives a typical list of commonly controlled substances and their permitted ranges.

Implementation of the EC directives discussed in Section 5.2 was expected to cost £30 billion between 1989 and 2000; this will double the cost of water. The government regulator of water services, OFWAT, reached agreement with the companies on the cost of providing quality improvements by the water companies. Between 1990 and 2000, this factor was 5% to 10% above retail price index. Efficiency savings have enabled these quality and cost targets to be met, and the regulator is now seeking price reductions for the next five year period.

These recent increases in the costs of water have given scope for advanced and complete waste treatment by industrial users. All waste treatment processes are complex, and their apparent cost-effectiveness needs to be assessed to include factors for sidestreams, reliability, maintenance, and manning. Most of the UK food processing sites are too small to justify extensive treatment; for example, of the 5,000 factories in the United Kingdom, only 150 to 200 generate effluent loads greater than 500 tonnes of chemical oxygen demand (COD) per annum, the point at which treatment becomes economically attractive. There are a number of simple pre-treatment processes that even the smallest sites can utilise with advan-

Table 5–3 Typical UK Consent Conditions for Discharge to Sewer

Parameter	*Value*[b]
Temperature	< 40° C
pH	5–9
Metals, total[a]	< 30 mg/L
Cyanide	< 1 mg/L
Phenol	< 500 mg/L
Sulphide	< 10 mg/L
Sulphur dioxide	< 1 mg/L
Chlorine	< 10 mg/L
Oil/grease	< 100 mg/L
Insoluble tar	< 60 mg/L

[a]Totals of individual metals usually specified at 1–5 mg/L except Zn and Fe. Red list and heavy metals (e.g., Pb, Cd, Hg) set at detection limits.

[b]Maximum daily volume and maximum flow are also specified.

Source: Micklewright (1986).

tage to enable favourable terms for discharge to sewer. Flow balancing can, for example, avoid peaks of load, and allow cooling and precipitation of solids at a cost of less than £0.06/m^3 of flow.

5.4.2 Waste Minimisation

Once the sources and quantities of waste have been established, it is usual to find that there are methods of reducing water consumption and waste. Process redesign and re-organisation can also result in improvements in energy consumption, lower product losses, fewer accidents and spills, better public relations, and reduced liability. Exhibit 5–2 contains a basic checklist of procedures to follow.

5.4.3 Calculating the Costs

The charge for treatment in Europe and the United States is based on the ratio of the organic strength of the food effluent to the local sewage, and there are significant regional variations. The cost of discharge to sewer is based on the strength and volume of waste discharged. A formula known as the NWC/CBI or Mogden equation is used to link the costs of the sewers and the various types of treatment to an overall discharge cost. The equation includes elements of charge for the sewerage system (reception and conveyance), solids separation (Volumetric charge), a charge for biological treatment (B), and a charge for solids treatment (S). The costs of treatment are averaged across a region and may not precisely reflect the costs incurred at the local works. The water companies can make allowances if the trader agrees to contribute to capital extensions of the treatment works. Industry can then compare the options on site treatment with discharge to sewer or a capital contribution to the regional water company. There have been a number of changes to the basic formula (Table 5–4) to take into account additional costs of treatment, such as nitrification and stormwater treatment, or savings, for example, from long sea outfalls with no biological treatment.

The 1996/97 OFWAT annual report, for example, called for greater simplicity in the system of trade effluent charges, particularly highlighting what is included, such as costs for surface water and nitrification (Office of Water Services, 1996). In Germany, for example, there is also an assessment of treatability that is used to increase charges for poor biodegradability. In France, there is also a standard charge for oxidising and removal of nitrogen. In the United Kingdom, this is obscured within the Vb or Bv element (see Table 5–4) but is not always based on an analysis of ammonia. Additional solids costs may be incurred in the future from phosphate removal. Solids or sludge disposal charges have risen more steeply recently than oxidation charges because of sludge disposal problems. Urban Thames Water, for example, has the highest charge for solids.

Exhibit 5–2 Waste Minimisation Procedures

1. *Material handling:* Material handling procedures should be designed to avoid damage contamination spoilage and to improve the quality of storage. If hazardous materials are used they should be kept separate and contained for the possibility of accidents.
2. *Control:* The aim of better control is to deliver precise amounts of raw materials, reactants, catalyst, physical change, or water at exactly the right time. This avoids wasted energy, reactants and water. Improvements in monitoring and analytical equipment will be effective if they reduce waste.
3. *Product formulation:* Changes to an alternative cleaner raw material can reduce the amounts of waste produced and improve the quality of the product.
4. *Process modification:* Different processing operations may be used to avoid waste. For example, dry processing will reduce water consumption without affecting efficiency and quality. Very strong streams (fat, solids, etc.) can usually be easily separated and disposed of differently.
5. *Equipment design:* Changes in the design of existing machinery may also reduce pollution without interfering with quality and efficiency; for example, lower operating temperature, less water, and less mechanical damage.
6. *Plant layout:* Better integration of processes can reduce the amount of equipment and pipe runs; this usually means less cleaning. For example, heating and pumping operations should be in one location to avoid unnecessary pipework.
7. *Recycling:* Damaged raw materials or products can often be recycled without adverse effects. A common possibility is to use a high-grade wastewater from cooling or final product washing for a lower grade use such as primary washing.
8. *By products:* Some by-products can be recovered, such as:

   a. animal feeds
   b. waste paper or card
   c. fats and oils
   d. fertilisers and land irrigation

   The simplest by-products are:

   a. animal feeds
   b. biogas and combustion of waste-derived fuel
   c. fats and oils
   d. fertilisers

Table 5–4 Trade Effluent Charges (1992–1993)

	Average Regional Strength		*Charge Factors*						
	O_s (mg/L)	S_s (mg/L)	*R* (p/m³)	*V* (p/m³)	*VB* (p/m³)	*VM* (p/m³)	*M* (p/m³)	*B* (p/kg)	*S* (p/kg)
Anglian	426	371	6.68	10.46	2.02	5.19	0.42	11.39	5.61
Northumbrian	386	187	13.96	6.84	0.00	0.00	0.00	10.84	5.50
North West	389	234	7.30	5.90	0.80	0.00	5.50	6.40	4.30
Severn Trent	767	323	9.88	9.07	0.00	0.00	0.00	10.67	5.84
Southern	452	512	15.06	11.01	1.78	7.63	1.95	14.51	9.94
South West	767	487	22.42	22.69	0.00	0.00	3.94	36.58	25.38
Thames	445	336	5.54	6.81	0.00	0.00	0.00	10.74	13.68
Dwr Cymru	500	350	6.75	5.45	2.17	5.77	8.29	12.58	8.86
Wessex	802	313	5.56	8.95	0.85	5.13	9.65	19.81	11.45
Yorkshire	971	321	14.13		0.00	0.00	0.00	14.78	8.64
Northern Ireland	480	340	10.0	8.00	0.00	0.00	0.00	10.00	8.00

Note: Total average charge is for average strength effluent. Trade effluents are charged using the following formulae:

$$C = R + (V + VB) + B\,\frac{O_t}{O_s} + S\left(\frac{S_t}{S_s}\right)$$

$$C = R + (V + VM) + B\,\frac{O_t}{O_s} + S\,\frac{S_t}{S_s}$$

$$C = R + (V + M) + B\,\frac{O_t}{O_s} + S\,\frac{S_t}{S_s}$$

C = total charge for trade effluent treatment
R = reception and conveyance charge
V = volumetric and primary treatment cost/m³
VB = additional charge/m³ where there is a biological treatment
VM = treatment and disposal charge for sea outfalls
M = treatment and disposal charge for designated long sea outfalls
O_t = the chemical oxygen demand (COD) of effluent after 1 h quiescent settlement at pH 7
O_s = COD of crude sewage after 1 h quiescent settlement
B = biological oxidation cost/m³ of settlement sewage
S_t = total suspended solids (mg/L) of the trade effluent at pH 7
S_s = total suspended solids (mg/L) of crude sewage
S = treatment and disposal costs of primary sludges/m³ of sewage

5.5 CASE STUDIES

5.5.1 Yeast-Processing Effluent

A yeast-processing company, which in 1988 was paying £132,000 per annum to discharge aerobically pre-treated effluent to sewer, would have been faced with a half million pound charge in 1993 if the existing treatment were not replaced (Table 5–5). These figures do not take into account inflation. Calculations (shown in Table 5–6) indicate that significant savings could be made by the installation of either aerobic or anaerobic processes on site: The running costs of anaerobic digestion were much lower than those of the aerobic process. Added to this was the potential benefit of fuel savings through use of the gas; these were likely to be greater than predicted because of the continuing rises in energy prices.

5.5.2 Starch Waste Example

Tunnel Refiners process maize and wheat flour to produce gluten and starch. A separate process is used to hydrolyse starch to produce various glucose syrups. The waste from these processes contains starch, sugars, proteinacious materials, and minerals from the corn and flour. An anaerobic digestion treatment system was installed, as detailed in Table 5–7. The waste contains significant amounts of sulphur (SO_3, 100 mg/L) and ammonia (640 to 1700 mg/L total N).

The simple payback on the capital of the plant was 1 year, but there have been two unforeseen running costs. There were some problems resulting from hydro-

---

Table 5–5 Example of Effluent Disposal Charges for a Yeast Processing Company (Severn Trent Area)

Volume	1,000 m^3/day
COD	5,500 mg/L
Suspended solids	400 mg/L
Cost of disposal (real cost)	
1988	£311,400
Projected	
1989	£327,170
1990	£350,100
1991	£435,250
1992	£470,100
1993	£507,700

Source: Adapted from Smith et al. (1988).

Table 5–6 Example of the Costs of Anaerobic Compared to Aerobic Treatment (Discharge to Sewer Without Treatment: £311,400)

Treatment Type	*New Biotower, New Foundations (Option 1)* £	*Refurbishment of Existing Biotower (Option 2)* £	*New Anaerobic Filter (Option 3)* £[a]
Capital	655,200	523,500	630,000
Plant running costs and effluent disposal	224,612	226,000	143,026
Annual saving	86,780	85,400	168,297
Simple payback in years	7.55	6.03	3.74
Payback in years by tax computation for cash flow method	7.67	6.25	4.08

[a]The value of surplus gas (option 3), estimated at £12,000 p.a. in terms of coal equivalent, has been ignored.

Source: Smith et al. (1988).

gen sulphide (H_2S), with complaints being received about odour, so the clarifier was roofed. The gas contained 1% H_2S and the effluent 1mg/L. Ferric chloride was added to the treated effluent as it was discharged to sewer; this precipitated sulphide as iron sulphide, which helped to control the odour. It has also been noticed that when the protein content was high, this interfered with biomass settlement and density. Lime was used to improve settlement and to neutralise the waste. The resultant average total alkalinity was 1,000 mg/L. Nickel and cobalt are also added as micronutrients.

5.5.3 Brewery Waste

An expansion of the Mortlake Brewery (Brooking, Buckingham, & Fuggle, 1990) meant that changes would be needed to the on-site effluent treatment plant. This was a 10-year-old high-rate biotower. Seven options were shortlisted, comparing discharging the whole flow untreated to sewer (option 1, Table 5–8) with several different treatment plants to remove 80% to 90% of the applied COD. The high-rate biotower was designed to remove 70% of the applied COD at a flow rate of 2,500 m^3/day.

Conventional activated sludge was not considered as suitable because it had a poor reputation with strong industrial wastes. Other breweries reported difficulties with settlement, particularly controlling dissolved oxygen and sludge return rates.

Table 5–7 Example of Anaerobic Plant Treating Starch Waste

Reactor type	Anaerobic mixed flow contact process
Size	5,710 m^3
Biomass concentration	0.5% at optimum conditions
Clarifier volume	2,500 m^3 volume
Effluent flow	4,678 m^3/day
COD in	7,000–8,000 mg/L, max 20,000
Total solids in	12,000 mg/L
Total solids out	500 mg/L
Suspended solids	660 mg/L
Loading rate	5 kg COD/m^3-day
	1 T COD/T Biomass-day
Daily load	4.8–28 T COD/day
COD removal	93% on total COD
pH_{in}	8.0
pH_{out}	7.0
Temperature	
Waste	37° C
Tank	30° C
HRT	29 h
R (recycle ratio)	0.34 (60 m^3/h) 1:3 recycle to feed
Biogas production	4,000–10,000 m^3/day
	avg. 5,000 m^3/day, 0.38 m^3/kg COD removed
Biogas use	Hot water generation
Capital cost	£1.75M, payback period 1.5 years
Location	Tunnel Refiners, South East London
Contractor	Biomechanics, built in 1986

Duplicating the existing biotower system would have been attractive because there was 10 years of successful operating experience with this system, including overcoming an odour problem (option 3). This was, however, shown to be impractical because of the available space on site. Option 6, a new low-rate biofilter, was ruled out on similar grounds. There was little experience with anaerobic treatment systems in the United Kingdom in 1986, and this was, therefore, judged to be unproven technology. A review of treatment methods at other breweries showed that pure oxygen-activated sludge was in use at two other breweries in the United Kingdom, where it had overcome the difficulties normally associated with activated sludge and was working well. Oxygen systems were more readily controlled because of the "turn-up/turn-down" facilities offered by the stored oxygen. This was especially suited to the variable nature of the effluent. The oxygen-activated

Table 5–8 Possible Treatment Options for Brewery Waste

Option	*Description*	*Estimated Capital Cost (£000s)*	*Operation Cost (£000s)*	*Cost (p/m³)*
1	Discharge total flow to sewer		954	52.3
2	Discharge excess flow to sewer		774	42.5
3	Extensions to existing biotowers	895	465	25.5
4	Anaerobic filter and existing biotower	1,113	388	21.3
5	Pure oxygen-activated sludge	1,100	507	27.8
6	Completely new biofilters for existing and new flow	1,480	467	25.6
7	Completely new pure oxygen-activated sludge	1,855	593	32.5

Source: Brooking et al. (1990)

sludge had the additional advantage of being totally sealed and was therefore unobtrusive. Thus, option 5—combining the existing high-rate biotower with a new oxygen-activated sludge plant—was chosen as the best combination of established technology, capital, and running costs.

The flow rate during 1988–1989 unexpectedly decreased (Table 5–9) because of changes in production. It meant that the biotower could be switched off and the whole load passed to the oxygen-activated sludge plant, and costs could then be compared. The oxygen-activated sludge plant performed better than the biotower combination at the expense of slightly more sludge and the extra cost of the oxy-

Table 5–9 Design Characteristics of New Effluent Treatment Plant

	Design	*Actual*
Flow rate	4,000	2,269
Influent BOD (mg/L)	2,000	—
Effluent BOD (mg/L)	100	95
Influent suspended solids (mg/L)	320	—
Effluent suspended solids	100	179
Power consumption (M/Wh week)	15.9	19.8
Sludge production dry solids (tonne/day)	3	1.2

Source: Brooking et al. (1990).

gen—about £100,000 per annum. These costs were more than offset by the improvement in effluent quality and discharge costs to sewer. The biotower was demolished and contingent extra oxygen capacity provided. This also released land for further production developments at the site.

5.6 WASTEWATER CHARACTERISTICS

Wastewater is the largest of the major categories of waste (Table 5–10). In the United Kingdom, there are about 18 x 10^6 m^3 each day for treatment (Figure 5–2). About 5,000 sewage works serve 80% of the population; 1,000 of these are for populations over 10,000, and 17 take the waste of over 300,000 (Table 5–11).

Wastewater is contaminated with organic matter and bacteria, some of which may be pathogens. The traditional aims of wastewater treatment have been to remove organic matter and reduce the number of faecal organisms to allow the wastewater to be safely discharged into the environment.

Municipal wastewaters contain excreta, household wastes, industrial effluents, and storm run-off. They are usually the most complex of the wastewaters, containing different micro-organisms and a wide range of suspended and dissolved materials. The range of contaminants and the pollution load is defined by a number of tests designed to assess these physical, chemical, and biological characteristics. The methods of analysis used are similar. In the United Kingdom, they are published by the Department of the Environment, Transport and the Regions (Department of the Environment, 1979); in Europe, they are appended to EC directives; and in the United States, they are jointly published by the water institutions (Eaton et al, 1995). The analyses are then used to design the most suitable combinations of treatment equipment. Food, domestic, agricultural, and a wide range of other industrial wastewaters contain high concentrations of organics. This organic ma-

Table 5–10 Quantities of Waste in Europe, 1986 (Quantities Were Increasing at the Rate of 3% per Year)

Millions of Tonnes	*Type of Waste*
120	Household waste
950	Agricultural waste
160	Industrial waste
300	Sewage sludge
250	Mining industries
170	Demolition and building waste
120	Consumer durables, cars, electrical goods, etc.
130	Other

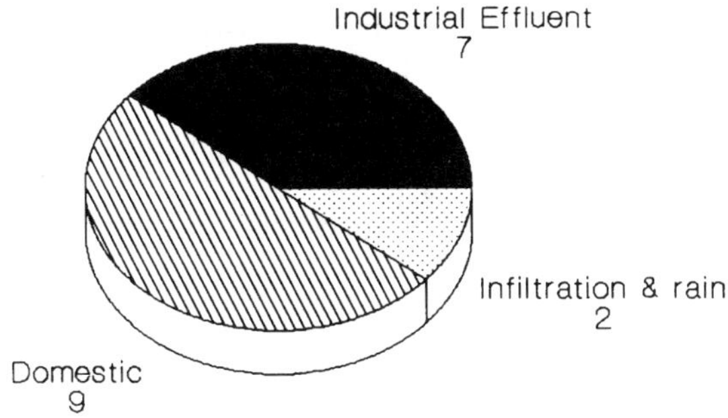

Figure 5–2 The United Kingdom's daily effluent treatment (M tonnes)—the largest controlled industrial process in the United Kingdom.

Table 5–11 Sewage Works in England, Wales, Northern Ireland, and Scotland in 1978 with a Design Dry Weather Flow Greater than 100,000 m^3/day

Name of Sewage Plant	*Water Authority or River Purification Board*	*Dry Weather Flow (000s m^3/day)*	*Population (000s)*
Beckton	Thames	912	2,250
Crossness	Thames	505	1,550
Minworth	Severn Trent	384	1,050
Davyhulme	North West	314	714
Mogden, Isleworth	Thames	439	1,330
Deephams	Thames	171	685
Stoke Bardolph	Severn Trent	146	465
Blackburn Meadows	Yorkshire	128	466
Derby (Raynesway)	Severn Trent	108	210
Knostrop (High Level)	Yorkshire	150	481
Finham	Severn Trent	112	339
Maple Lodge	Thames	113	434
Avonmouth	Wessex	160	500
Rye Meads	Thames	78	322
Dalmuir	Strathclyde	219	440
Dalmarnock	Clyde	136	300
Belfast (Duncure Street)	Northern Ireland	110	530

Source: The Water Data Unit (1979).

terial stimulates the growth of bacteria and fungi naturally present in water, which then consume dissolved oxygen. Normally, the oxygen demand in natural waters is less than 2 mg/L biochemical oxygen demand (BOD); the most common sign of extensive water pollution by sewage, industrial effluents, or agricultural waste is a reduction in dissolved oxygen (oxygen sag).

The organic content of the wastewater is determined by three commonly used tests: COD, total organic carbon, and BOD (Table 5–12). In rivers and streams, the water undergoes continuous mixing and turbulence, and some oxygen (i.e., 2 mg/L) can be replenished (Figure 5–3). The original standards for discharges to rivers in the United Kingdom were set to avoid raising the BOD of the river above

Table 5–12 Tests for Organic Carbon

Test	*Method*	*Use*
Chemical oxygen demand (COD)	Sample is oxidised with boiling chromic acid. The amount of organic carbon is equivalent to the chromic acid reduced. Test takes 3 hours.	Very common test used for all of wastewater to assess total organic carbon. Apparatus and techniques available in most laboratories.
Total organic carbon (TOC)	Sample is ignited and the carbon dioxide produced measured by infrared. Very rapid test, 5–10 min.	Used by the larger laboratories with large numbers of samples. Requires specialised expensive equipment. Is becoming common as a process monitoring instrument.
Biological oxygen demand (BOD)	Sample is incubated with a culture of micro-organisms. Their metabolic rate is measured by oxygen consumption over 5 days and is proportional to the biodegradable organic carbon.	Very common test, used for all types of waste to assess biodegradability. Equipment simple, but requires some understanding of microbiology. Some automated equipment available.
Permanganate (PV)	Sample is oxidised by permanganate over 4 min or 4 hours. The amount of organic carbon is equivalent to the permanganate reduced.	Many of the organic compounds are not oxidised by the mild conditions and the test has largely been superseded by the COD. Is useful for field work, apparatus simple.

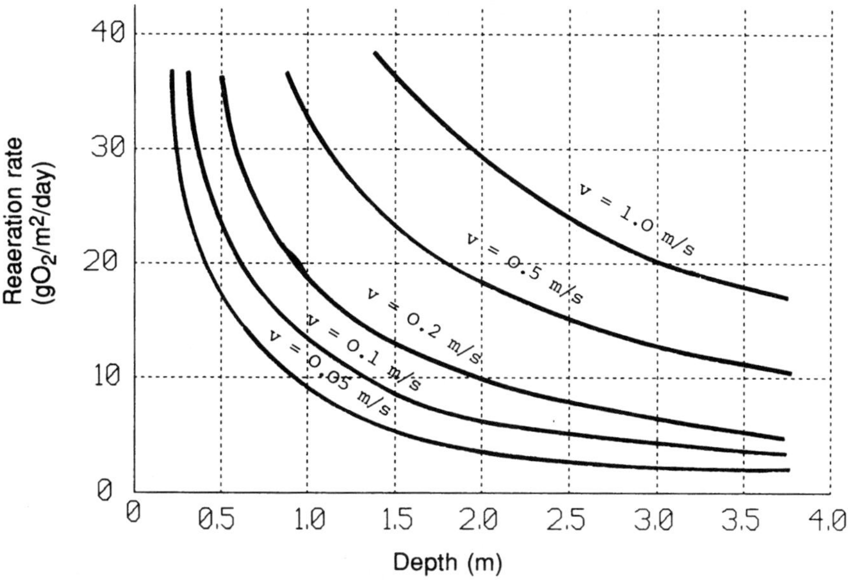

Figure 5–3 Oxygen replenishment in rivers and streams according to depth and velocity.

4 mg/L. In lakes, the water is static and higher standards are required. If the BOD exceeds 4 or 5 mg/L, changes in dissolved oxygen will occur. More recently, it has become recognised that an average BOD value may be insufficient to protect the ecosystem of the receiving water, and standards have been restated in terms of 95 percentile compliance values. The EC Urban Wastewater Directive defines discharge limits in similar terms: a guideline value and a MAC. Standards are also stated as COD values, but the COD test is not very precise at low concentrations (i.e., <100 mg/L), and BOD or total organic carbon is preferred.

Another very common constituent of wastewaters is solids. Suspended solids are a source of organic matter; they interfere with light penetration in receiving streams, and settleable solids precipitate and cover the natural habitat of the benthic or bottom-dwelling invertebrates and algae. Measurement of suspended solids is by filtration or light-scattering (Exhibit 5–3).

For design purposes, a number of average values have been developed from experience to describe the daily pollutant load per person:

- wastewater litres/h/day: 230
- solids in sewage grms/h/day 30% water: 200
- oxygen required for treatment grms/h/day: 60

Exhibit 5–3 Classification of Solids Found in Domestic Wastewater

Total 700 mg/L	Suspended 200 mg/L	Settleable 100 mg/L	Organic 75 mg/L Mineral 25 mg/L
		Non-settleable 100 mg/L	Organic 75 mg/L Mineral 25 mg/L
	Filterable 500 mg/L	Colloidal 50 mg/L	Organic 40 mg/L Mineral 10 mg/L
		Dissolved 450 mg/L	Organic 160 mg/L Mineral 290 mg/L

Note: Dry weights determined by warming at 105°–110° C until constant weight. Inorganic content measured by heating to 650° C and weighing the residue. Settled samples settled for 20 min. Filtered samples filtered by GFC paper (1.5 μm).

Water consumption, and therefore wastewater generated, depends on income. The United Kingdom is at 130 litres per head per day, above average for Europe but less than Germany, where 200 litres per head is normal, and the United States, where consumption may be as high as 500 litres in the warmer states. To design waste treatment plant, it is usual to double these consumption figures to include unmetered discharges, infiltration, storm drainage, and expansions in population. The cost of treatment of industrial wastes is often expressed in terms of population equivalents.

Except for sewage, very few wastewaters have been analysed in detail, but all the major components of food wastes—carbohydrate, fat, and protein—can be fermented to carboxylic acids in the anoxic or anaerobic conditions in drains and sewers. These acids can then cause odour and corrosion.

Other pollution problems have become apparent in recent years, particularly from inorganic pollutants and persistent or recalcitrant materials. Nitrogen is a good example. Many agricultural wastes contain significant amounts of organic nitrogen as protein or urea. These are rapidly converted to ammonia by the anoxic conditions in drains. Ammonia is very toxic: 2 mg/L are sufficient to eliminate most fish, and 1 mg/L has an effect on invertebrates. If effluents are to be recycled for public supply, residual ammonia can generate hazardous by-products during chlorine sterilisation, such as the chloramines. Effluent treatment can be designed to oxidise ammonia to nitrate, but there are also problems with excessive concentrations of nitrates.

Two potential health problems are associated with nitrate. First, it may interfere with the oxygen-carrying capacity of the blood in infants (known as infantile cyanosis, methaemoglobinaemia, or blue baby syndrome). Second, nitrate can, under certain circumstances, be reduced to potentially mutagenic nitrosamines in the stomach. WHO has recommended a limit of 50 mg/L nitrogen (as nitrate) in drinking water, and this recommendation has been included in the EC Drinking Water Directive (80/778/CEC).

Nitrate accumulation in the environment has generated a further difficulty: In combination with phosphate nitrate, it can lead to nutrient enrichment and excessive algal growth (eutrophication). These algae are unsightly and may also cause taste, odour, oxygen depletion, filter blockage and occasionally microtoxin problems if the water is to be reused for domestic supply.

Phosphorus is actually often the limiting plant nutrient for the growth of phytoplankton (algae and aquatic plants). For example, 1mg of phosphorus can generate 100 mg of algae (as dry solids) and eventually, on degradation, 140mg of oxygen demand.

Phosphate, like nitrate, is a residue from the biological oxidation of organic matter, but about half is also derived from domestic and other detergents. Restrictions will be imposed on the release of nitrate and phosphorus to surface waters in the future (EC Directive on Wastewaters in Sensitive Areas). There is no really effective and efficient method of removing phosphate from effluents at present that does not generate large quantities of sludge. The most cost-effective method of reducing phosphates in water might be a voluntary code to eliminate phosphate in detergents. There has been a special ban on phosphate detergents in Switzerland since 1986 because most sewage discharges are to lakes. Results show (von Gunten & Lienert, 1993) that this is successful, and there have been other benefits as well, including significant reductions in organic matter and therefore metals normally solubilised by the reducing conditions in lake and stream muds. The alternative to the phosphates—water softeners—may themselves be refractory and toxic.

The accumulation of other inorganic salts—for example, sodium chloride and calcium and magnesium—can also cause problems in arid and semi-arid climates where water is recycled many times. Satisfactory methods of recycling these inorganic materials that do not generate other residues are also necessary.

Another common problem with wastes is that they may contain significant quantities of metals. There have been dramatic improvements in the amounts of metals released into the environment, partly brought about by the recession of old heavy industries but also because of the imposition of more rigorous control by authorities. Some problems remain, for example, metals in landfill leachate and zinc from cosmetics in domestic sewage. These will require further restraint by voluntary recycling or code of practice.

Zinc, copper, nickel, and boron are, for example, phytotoxic, their effects are additive, and there are risks of accumulation in the food chain. Cadmium lead and chromium are more mobile, and there is a direct risk from the consumption of contaminated food. Cadmium causes the most concern because it is readily absorbed by plants and hence into the food chain. There are other potential risks from mercury, molybdenum, arsenic, fluorine, and selenium. Most of these elements can also be derived from aerial deposition and from applications of surplus sludges to soils and leachate contamination of water. Routine monitoring of soil is necessary to ensure that total inorganics are kept within safe limits.

Modern and more sensitive chemical analysis has also revealed some accumulation of persistent organic micropollutants in the environment. The major areas of concern at the moment are the pesticides, organic solvents, and residues from plastics, such as the polychlorinated biphenyls, and dioxins. These chemicals are in common use by industry and agriculture, and there are standards to prevent them from entering general mixed wastes (red list and black list substances).

Future practice covered in present European and US legislation will call for better identification of persistent organic micropollutants by test work prior to licensing and use of new chemicals.

Many dyes and colours and some modern detergents are also only poorly degraded by conventional waste treatment processes. These two types of compounds have, however, been biodegraded under carefully controlled conditions, and these less toxic but more concentrated organic pollutants are likely to be the most amenable to short-term biotechnological solutions.

5.6.1 Microbiological Characteristics

Domestic wastewaters contain a large number of enteric bacteria (10^3 to 10^6/mL). There are also a wide variety of organisms present, including the fungi, protozoans, nematodes, and viruses. There are about 40 potential pathogens that are waterborne or can contaminate water (Horan, 1990), and wastewater treatment processes are effective at reducing numbers of bacteria by 50% to 90% (Table 5–13). It would be usual to disinfect treated sewage effluent if it were to be used immediately for irrigation (World Health Organisation, 1973). Treated sewage effluent has routinely been chlorinated in the United States prior to discharge to surface waters, but in Europe there is concern about the persistence and hazards associated with residual chlorinated organics. Other disinfectants are less persistent and not as cost-effective. The excreta of infected individuals, discharges from abattoirs, and run-off from grazing land provide a constant source of these potential pathogens. Faecal contamination of water and potential risk are indicated by

Table 5–13 Summary of Pathogens Removal by Various Sewage Treatment Processes

Organisms	*Stage in the Treatment Process*[a]	*Primary Sedimentation*	*Trickling Filter with Primary and Secondary Sedimentation*	*Activated Sludge with Primary and Secondary Sedimentation*	*Oxidation Ditch with Sedimentation*	*Septic Tanks*	*Land Application or Slow Sand Filtration as Tertiary Treatment*	*Anaerobic Digestion (30°–40° C)*
Enteric viruses	A	10^3–10^5	10^3–10^5	10^3–10^5	10^3–10^5	0–10^9	10–10^4	May survive over 3 months
	B	10^3–10^5	10^2–10^4	10–10^4	10–10^4	10–10^3	0–10^2	
	C	0–30	90–95	90–99	90–99	50	99–100	
Salmonellae	A	10^3–10^4	10^3–10^4	10^3–10^4	10^3–10^4	0–10^9	10–10^3	May survive for several weeks
	B	10^2–10^3	10^2–10^3	10–10^3	10–10^3	0–10^8	0	
	C	50–90	90–95	90–99	90–99	50–90	100	
Shigellae	A	10^3–10^4	10^3–10^4	10^3–10^4	10^3–10^4	0–10^9	10–10^3	Unlikely to survive for more than a few days
	B	10^2–10^3	10^2–10^3	10–10^3	10–10^3	0–10^8	0	
	C	50–90	90–95	90–99	90–99	50–90	100	
Escherichia coli	A	10^6–10^8	10^6–10^8	10^6–10^8	10^6–10^8	10^7–10^9	10^4–10^7	May survive for several weeks
	B	10^5–10^7	10^5–10^7	10^4–10^7	10^4–10^7	10^6–10^8	0–10^3	
	C	50–90	90–95	90–99	90–99	50–90	99.99–100	
Cholera vibrio	A	10–10^3	10–10^3	10–10^3	10–10^3	0–10^8	0.1–10^2	May survive for 1 or 2 weeks
	B	1–10^2	1–10^2	0.1–10^2	0.1–10^2	0–10^8	0	
	C	50–90	90–95	90–99	90–99	50–90	100	
Entamoeba histolyca cysts	A	10–10^4	10–10^4	10–10^4	10–10^4	0–10^5	10–10^3	May survive for 3 weeks
	B	5–10^4	5–10^3	5–10^3	5–10^3	0–10^5	0	
	C	10–50	50?	50?	50?	0?	100	
Hookworm ova	A	10–10^3	10–10^3	10–10^3	10–10^3	0–10^4	10–10^2	Ova will survive
	B	10–10^2	10–10^2	10–10^2	10–10^2	0–10^3	0	
	C	50?	50–90	50–90	50–90	50–90	100	

continues

Table 5–13 continued

Organisms	*Stage in the Treatment Process*[a]	*Primary Sedimentation*	*Trickling Filter with Primary and Secondary Sedimentation*	*Activated Sludge with Primary and Secondary Sedimentation*	*Oxidation Ditch with Sedimentation*	*Septic Tanks*	*Land Application or Slow Sand Filtration as Tertiary Treatment*	*Anaerobic Digestion (30°–40° C)*
Ascaris ova	A	$10–10^3$	$10–10^3$	$10–10^3$	$10–10^3$	$0–10^4$	$0–10^2$	Ova will survive
	B	1–10	$0–10^2$	$0–10^2$	$0–10^2$	$0–10^3$	0	for many
	C	30–80	70–100	70–100	70–100	50–90	100	months
Schistosome	A	1–100	1–100	1–100	1–100	1–100	1–10	Ova may survive
ova	B	1–10	1–10	1–10	1–10	1–10	0	up to 1 month
	C	80?	50–99	50–99	50–99	50–90	100	
Taenia ova	A	1–100	1–100	1–100	1–100	0–103	0.1–50	Ova will survive
	B	0.1–50	0.1–50	0.1–50	0–50	0–50	0	for a few
	C	50–90	50–95	50–95	50?	50–90	100	months

[a]Number of organisms: A, in raw sewage; B, in treated effluent; C, removal % numbers of organisms per litre.

the isolation of harmless bacteria that are reliably present in excreta. Three different groups are commonly used (Table 5–14) (Department of the Environment, 1982).

There is a continued health risk from pathogens in recycled waste. There are few reliable data on the infective doses of the traditional bacterial and protozoan pathogens, nor is there an easy method of determining viability, as classical culture techniques are not suitable indicators of the potential to cause infection. Even less is known about the survival and persistence of viruses. There is, therefore, a continued need to develop test procedures for the rapid identification of all types of potential pathogens, including newly evolved strains.

5.6.2 Biological Indicators

Changes to the natural environment, such as pollution, cause alterations in the ecology of rivers and streams. All the major biological groups are affected and have been used as biological indicators of pollution. The benthic or bottom-dwelling macroinvertebrates are the most popular because their habitat and low mobility mean they are rapidly affected by pollution. They are abundant and easy to sample and identify, which also makes them suitable for qualitative analysis. Several biotic indices of water quality have been developed for different habitats based on species diversity and abundance. Standard methods of analysis have

Table 5–14 Indicator Organisms Commonly Used to Detect Faecal Contamination

Test	*Method*	*Usage*
Faecal coliforms	Generates gas and acid from the fermentation of lactose peptone media. Test performed at 37° C and confirmed at 44° C.	Most commonly used test, occasionally produces false positives, but very sensitive and most useful indicator.
Faecal streptococci	Growth in a sodium azide, glucose, and bromocresol purple media with acid production. Test performed at 45° C.	Some streptococci are exclusively human and can be used as confirmatory after the coliform test.
Clostridia	The reduction of sulphites to sulphide in litmus milk media. Incubated anaerobically at 37° C.	It is used to indicate intermittent pollution in the absence of other indicator organisms.

Source: The Bacteriological Examination of Drinking Water (HMSO, 1982).

been published (Hellawell, 1978). The principal influence on ecology is an alteration of the dissolved oxygen, and three broad divisions of organic pollution and consequent dissolved oxygen have been recognised (Table 5–15 and Figure 5–4).

5.6.3 Measuring Treatability

To avoid damage to the environment, sewers and sewage works discharges of industrial effluents are by licensed agreement in all the European countries and the United States (see Section 5.2). The hazardous nature of some substances is now quite well known.

The Dangerous Substances Directive (76/464/CEC) lists 129 substances that are particularly dangerous and therefore a priority for control: the black list. Attempts have been made in the United Kingdom to set priorities for restricting the release of some of these compounds; the procedure used is based on the following four criteria:

1. acute toxicity tests
2. chronic toxicity tests
3. bio-accumulation
4. quantities produced in the United Kingdom

In the United States, the EPA is responsible for controlling a similar group of substances (see Corbitt, 1989).

Using these procedures, 23 substances were selected for priority action in the UK red list. They were all on the original EC List 1, except chloroform and carbon tetrachloride, which did not show either high toxicity or accumulation by the Department of the Environment procedure. An additional 23 substances were identified as possible red list candidates after further test work. Many of the substances on the red list are pesticides originating from very diffuse sources.

In practice, most synthetic compounds (xenobiotics) have been found to be biodegradable, but there are four common problems in effluent:

1. chlorinated solvents
2. biocides and pesticides

Table 5–15 Saprobic Zone System of Biological Indicators of Pollution

Zone	*Dissolved Oxygen (mg/L)*	*Species*
Oligosaprobic	8–9	Salmonid fish, stone flies and may flies
Mesosaprobic	5–7	Cyprinid (coarse) fish, caddis fly and freshwater shrimps
Polysaprobic	1–2	Tubificial worms, chironomids, leaches

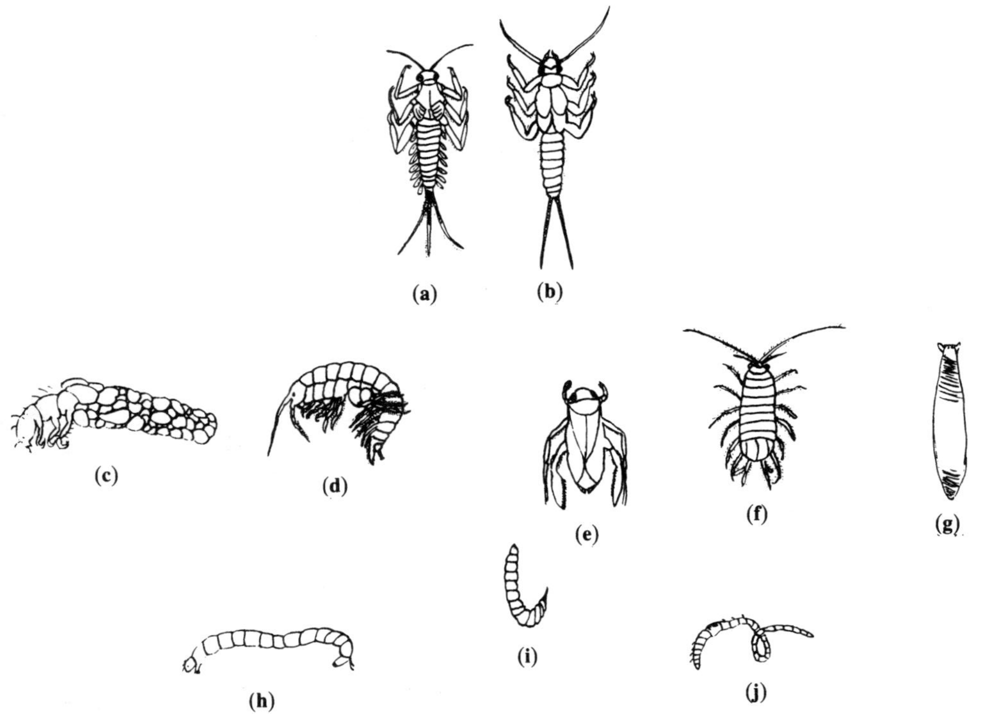

Figure 5–4 Ecological indicators of pollution. (a through b) Typically oligosaprobic invertebrates; (c through g) mesosaprobic invertebrates; (h through j) invertebrates associated with significant pollution, i.e., polysaprobic. (a) *Chloropera torrentium* (stone fly); (b) *Baetis rhodari* (may fly); (c) *Stenophylax* (caddis fly); (d) *Gammarus pulex* (fresh water shrimp); (e) *Corixa dorsalis* (water boatman); (f) *Asellus aquaticus* (water louse); (g) *Polycelis felina* (flat worm); (h) Chironomid larvae (filter fly also known as blood worm); (i) *Psychoda alternata* (filter midge); (j) *Tubifex tubifex* (sludge worm).

3. detergents
4. ammonia and sulphur

These compounds can be tolerated at low concentrations, but trials and test work are appropriate if these materials are known to be present.

For a number of wastes, treatability data are available for the design of treatment plant without intermediate laboratory and pilot studies. This is true of domestic wastes and many food waste applications, where suppliers have already installed such plant at several similar factories and can use common design principles. There are still many wastes for which treatability data are not available or where the processing operations are significantly different.

The most frequently encountered problems with food wastes compared to domestic wastes are a lack of nutrients, variability, and persistent toxins. Many of these difficulties can be predicted and overcome by laboratory treatability trials before design. Such tests are essential if there are any doubts about the characteristics of the waste.

Treatability trials are conducted at three scales:

1. bottle tests (250 mL reactors)
2. laboratory scale (5 to 10 litre reactor size)
3. pilot scale (10 to 20 m^3 reactor size)

5.6.3.1 Bottle Tests

The bottle tests (OECD die-away tests) measure microbial activity or substrate uptake under standardised and ideal conditions. They can be used to measure biodegradability and identify toxins when compared with well characterised substrates such as glucose and yeast extract by using standard micro-organisms taken from a sewage works. The tests in common use are BOD and biochemical methane potential (BMP). The ratios of BOD and BMP to COD will give an indication of biodegradability because the COD test measures all the organic carbon. The COD to BOD of starch wastes, for example, is about 1:2:1; of dairy waste, 1:5:1; and for domestic sewage, around 2:1.

5.6.3.2 Continuous Simulation Trials

Continuous culture trials in larger reactors (10 to 15 litres) can use the real waste as a substrate and can be run continuously for several months to identify nutrient deficiency, toxicity, and reliability problems. These difficulties are not usually apparent from the initial analysis. The laboratory study can also investigate competing reactor types and likely process loading rates.

There are some standard test units for continuous culture work, but these are not really suitable for food effluents. The standard activated sludge models, for example, are the Husmann or Porous pot, which recommend synthetic totally soluble substrates and do not effectively measure settlement. Model activated sludge units with sludge return can be workshop made and are more suitable for continuous trials. Similarly, the standard percolating filter model is a rolling table (Gloyna, Comstock, & Renn, 1952), whereas a model rotating biological contractor (RBC) is easier to construct and a better model of a percolating filter.

A laboratory study can never totally simulate the conditions at full scale owing to the highly variable nature of industrial effluents. On-site pilot trials are recommended in addition to laboratory studies whenever there are doubts concerning treatability. Advantages include more representative data and better information on operating problems with industrial scale ancillary equipment. Factory staff can also be familiarised with the process. It is important that pilot plant is of a realistic size to allow reasonable extrapolation to full scale. This normally means a reactor 10 m^3 to 20 m^3 in size. Most contractors will offer potential customers treatability trials at cost or with a discount against eventual equipment purchase. This type of arrangement reduces the independence of the advice given.

5.6.4 Sector Review

The food industry is the largest water user in Europe and the United States. Average water consumption in the food industry expressed as water to product is 10:1 (Table 5–16). For the chemical industry, it is about 5:1; and it is about 2:1 in paper and textiles. The hygiene and quality controls required in the food industry lead to a high water usage and few opportunities for recycling.

Most of the UK food processing sites are too small for economic on-site treatment. There are, for example, about 5,000 food processing sites, but only 150 to 200 have loads greater than 500 tonnes per annum, the point at which treatment becomes attractive. For cost-effective anaerobic treatment, wastes also need to be warm and strong (e.g., >20° C and COD >3000 mg/L). Treatment plants are common in the four largest sectors (Table 5–17):

1. dairies
2. breweries
3. sugar and starch processing
4. distilling and fermentation

5.6.4.1 The Dairy Industry

The dairy industry is currently the largest single source of industrial effluent in Europe. Sites are often rural and discharge to municipal sewer is impossible. The milk industry spent about £60M a year on effluent treatment in 1990, about 50% of the costs for sludge disposal. Some of these costs could be avoided by different treatment. A typical European dairy processes 500 m^3 of milk per day and gener-

Table 5–16 Strength and Wastewater Generated in the Food Industry

Industry	*Strength (typical BOD) (mg/L)*	*Water to Product Ratio (L/unit of Product)*
Dairy	500–750	12:1
Butter and cheese	1,500–2,500	3:1
Distillery	1,500–2,000	20:1
Brewery	500–1,500	8:1
Maltings	2,000–3,000	15:1
Food canning	100–1,300	10:1
Frozen peas	1,000–2,000	12:1
Chips and other vegetables	1,000–1,500	20:1
Chickens	600	30:1
Slaughterhouse	1,000–2,000	20:1
Sewage	250	130 per head

ates a similar volume of effluent. The waste is from the washing of tanks, evaporators, and transport. Average strengths of dairy effluents can vary from 1,000 to 5,000 mg/L COD. The effluents from dairy products—for example, cheese, evaporated milk, and so on—are warm and strong, and thus ideal for anaerobic treatment. Traditionally and prior to 1950, most effluent plant was modelled on sewage treatment and based on percolating filters. Percolating filters eventually became blocked with growth, debris, and age. The successor to filters for domestic wastes—conventional activated sludge—proved very difficult to operate, because of the warm temperatures, ease of fermentation, and surplus sludge production. Plastic media filters then become the favoured form of treatment, but these tower structures often gave problems with odour. More recently, low-rate extended aeration with standby oxygen injection has been the treatment of choice.

Typical problems with the treatment of dairy waste are the high concentrations of fat, protein, shock loads, and cleaning agents. Highly alkaline cleaning aids based on caustic soda are used, which can raise the pH to toxic levels. A two-stage balancing/neutralisation process is essential, and stream separation of the caustic clean-downs will also be beneficial. An ideal treatment process would include fat separation, flow balancing, dump tank, and two-stage biological treatment. The first biological stage should be a fixed film process, followed by extended aeration or percolating filters. Dilute wastes COD ~ 1,000 to 2,000 are suitable for this type of treatment; stronger wastes would benefit from a third anaerobic pre-treatment step. There are 10 full-scale anaerobic plants in Europe: 3 in France, 2 in Holland, 2 in Ireland, and 2 each in the United Kingdom, Germany, and Spain.

Table 5–17 Survey of Effluent Treatment in the Food Industry

	Sector Comparison by Size				*Effluent Characteristics*					
Industry	*Production (000 Tonnes)*	*Sales (£M)*	*Added Value (£M)*	*No. of Employees (000s)*	*COD (mg/L)*	*SS (mg/L)*	*Volume (M^3/Annum, 000s)*	*Cost (p/m^3)*	*Water Usage to Produce Ratio*	*No. of Sites Surveyed*
Meat and meat products					2,468	712	135	26	10	10
including poultry	495	3,046	174	103	500–8,632	75–4,500	26–482	17–60		
Oils and fats	523	568	955							
Milk and milk products	1,600	4,045	855	42	4,500	820	280	54	1.5/1.0	100
					80–9,500	24–4,800				
Fruit and vegetables	1,657	1,035	223	36	3,500	500	175	25	15/20	6
					1,617–11,147		14–633	1.3–79.0		
Potato products					2,360	656	414	6.0	20/30	4
					2,300–2,500	329–900	373–600	4.7–58.0		
Bread, biscuits,	2,000	2,669	1,142	139	5,076	3,144	40	4.2		
confectionery					275–9,500	215–7,960	14–88	1.3–12		2
Sugar and syrups	2,260	1,042	219	7.9	2,200	250	365	24		1
					1,400–3,000	200–300				
Cocoa, chocolate,	838	1,718	683	62	9,500	500	150	45	4	12
confectionery					19,000–30,000	75–1,500	3–1,000	6–113		
Starch cereals		1,202			1,855	394	223	29		1
					1,300–2,575	300–670				
Brewing	280		1,072	55	2,105	837	441	4.3	4	9
					1,500–3,500	598–1,294	77.7–1909	1.1–7.2	1.5–10	
Malting					2,195	189	81	7.5		3
					1,500–3,910	184–220	68–93	4.5–14.2		

Source: Gorsuch (1986).

5.6.4.2 Starch Wastes

Starch effluents are strong (CODs >10,000) and warm, 20° C to 25° C, too strong for aerobic treatment but ideal for anaerobic digestion. A high proportion of the COD is colloidal solids. These solids reduce biodegradability and favour solids-tolerant processes, that is, the contact process. There are seven anaerobic plants—three in the United Kingdom, two in the Netherlands, and one each in Belgium, Eire, France, and Germany. Four are based on upflow anaerobic sludge blanket reactors, three on contact digesters, and two on filters. Operating data on one of the UK contact digesters treating starch waste has been published regularly (Butcher, 1988). The most common problems with starch wastes are the slow breakdown of solids and low concentrations of additional nutrients. Two-stage treatment with separate prehydrolysis and balancing (10 to 20 hours retention) is a common method of overcoming treatability problems. More dilute wastes (i.e., COD <2,000) favour aerobic processes directly, but solids-tolerant fixed-film processes would still be preferable as the first stage. A typical treatment process would be screens, flow-balancing anaerobic treatment or high-rate fixed-film followed by extended aeration activated sludge.

5.6.4.3 The Sugar Processing Industry

The sugar industry is a major water consumer, with usage rates between 20 and 30 times the product produced. The industry has had to take steps to conserve water, and most processing factories now have extensive recirculation. This generates a concentrated effluent, well-suited to anaerobic pre-treatment followed by aerobic polishing. The sugar industry in Europe is based on beets. It is rural and seasonal, and requires on-site treatment. The largest volume of wastewater is from washing and rinsing the beets. The wastewater contains 10 to 20 g/L COD and deteriorates rapidly. Lime is used to counteract the acidity and prevent malodour. Lime is used extensively in the sugar processing and is generated on-site at the larger factories (Huber & Metzner, 1986). The lime also precipitates solids, reduces acid corrosion, and allows some of the water to be recycled. Surplus waste wash water is combined with other clean-down solutions and condensates for dilution before treatment. The waste is still strong (COD 3 to 10 g/L), with total carboxylic acid concentrations up to 6 g/L. The pH, however, is usually between 5 and 6 because of the lime treatment. The alkalinity and calcium in the lime aid anaerobic treatment. The biomass in the anaerobic digester flocculates around the lime and promotes settlement in the contact and upflow anaerobic sludge blanket processes. A disadvantage of lime is that it accumulates during the season, thus reducing the space in the digester for active biomass. It also causes pipe scale. Solids-tolerant digesters, the contact process, and upflow anaerobic sludge blanket reactors are essential. The anaerobic treatment of sugar wastes has been very successful: There are 20 plants in Europe, most with aerobic extended aeration final treatment before discharge to a water course. This combination is now established as the best treatment for this type of waste.

5.6.4.4 Confectionery/Soft Drinks

Much of the waste from the confectionery industry is sugar-based and CODs are large: 5,000 mg/L to 10,000 mg/L. The wastes are from cleaning; they are periodic and usually hot. Both aerobic and anaerobic treatability are low because of acidity and lack of additional nutrients. The chemical costs required to correct these problems mean that dilution and mixing domestic sewage is frequently the best method of treatment. There are occasions, however, when—because of the strengths of the waste—some form of pre-treatment is required by the controlling authorities. Three anaerobic treatment plants have been built recently, one of which is in the United Kingdom (Housley & Zoutberg, 1994). Running costs for alkalinity and nutrient additions have been high. Long retention times (more than 5 days for contact digesters and the same for extended aeration) are preferred if treatment is inevitable.

5.6.4.5 Breweries

Brewing is the second largest source of industrial waste after dairy effluents. Breweries are also large and can show economic benefit from on-site effluent treatment. Unlike the dairy industry, most breweries are urban and do have the option to discharge to mains drainage; many do so to avoid the complications of running effluent treatment plant. Wastewaters are weak—COD 500 mg/L to 1,500 mg/L—with a neutral pH and cold; they are often best treated aerobically (Figure 5–5).

Aerobic treatments have been successful and include two oxygen-activated sludge plants. Increased water and effluent charges will encourage further applications. More efficient water usage may also make wastewaters more concentrated and improve the energy balance for anaerobic digestion. Small anaerobic digesters could then be utilised on stronger, warmer process streams, ignoring most of the cool wash waters for direct discharge to sewer or combined aerobic treatment (Oliva, Jacquart, & Prevot, 1990; Wheatley, Fisher, & Grobicki, 1997).

5.6.4.6 The Distillery and Fermentation Industry

Distillery wastes are frequently very strong (COD values 10,000 mg/L to 60,000 mg/L, solids 10,000 mg/L). Water consumption is on average 10:1 (water to product), but volumes are smaller than either the dairy or brewery industries. Many distilleries—particularly in the United Kingdom, where they also are small and rural—have been able to dispose of effluents by land irrigation or discharge to sea. A small Scottish highland distillery, for example, will produce only 500 m^3 of effluent a week. Large industrialised distilleries, however, have major waste dis-

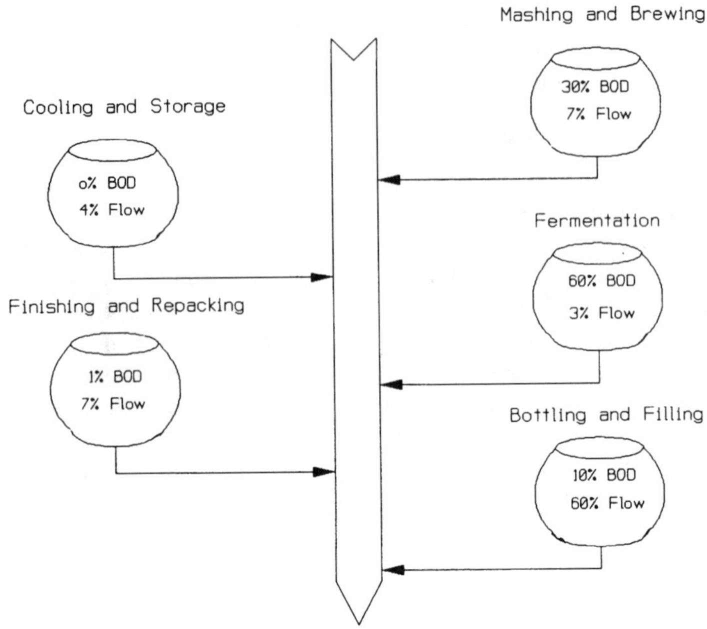

Figure 5–5 Typical sources of waste volume and load from a brewery.

posal problems. One French cognac producer, reported in the literature, generated 10,000 m^3 of effluent a week (see Section 5.2), equivalent to the waste from a population of 500,000 people (Racault, 1990). In these circumstances, on-site treatment is inevitable, and there is extensive experience of waste treatment and by-product recovery within the distilling industry.

Wastewaters have previously been treated aerobically but only after dilution with other wash waters and recycle. Aerobic effluent treatment plants have always been difficult to operate because of the acidity of the waste, high temperatures, and high oxygen demands. Robust biological filters have always been used as the first stage of treatment and have usually worked very well (Smith, 1986). Distillery wastes are ideal for anaerobic treatment; there are few alternatives. The sector has the largest number of installed anaerobic plants (25 to 30), most of which are in Italy and France.

Distillery and fermentation wastes from molasses contain high concentrations of sulphate. Typically, sulphate concentration is 3,000 mg/L to 8,000 mg/L. Sulphate is reduced to sulphide under anaerobic conditions, and the sulphide is toxic and malodorous. These types of effluent are unsuitable for anaerobic treatment.

5.6.4.7 Vegetable Processing

The vegetable and fruit processing industry, like the dairy and sugar industries, are rural. They may, like the sugar industry, also be seasonal, although many process different crops at different times of the year. Effluents are, however, dilute and cold, which has favoured aerobic treatment. Typical treatments are high-rate biofiltration followed by extended aeration-activated sludge (Figure 5–6).

There are two exceptions—potatoes and peas—and both have easily soluble starches and produce a higher strength waste. Anaerobic digesters have been built at 11 potato and potato snack factories. They are mostly the sludge blanket type or contact stirred tank reactor (CSTR) to cope with the high concentrations of colloidal solids from potatoes.

Pea processing wastes are also strong (COD >4,000 mg/L) and could be treated anaerobically. However, the season is so short, 10 to 12 weeks on average, that the capital costs of anaerobic treatment compared with aerobic treatment are uncompetitive.

Vegetable wastes often contain significant amounts of large solids, and primary treatment is used. A typical vegetable effluent plant, therefore, would consist of

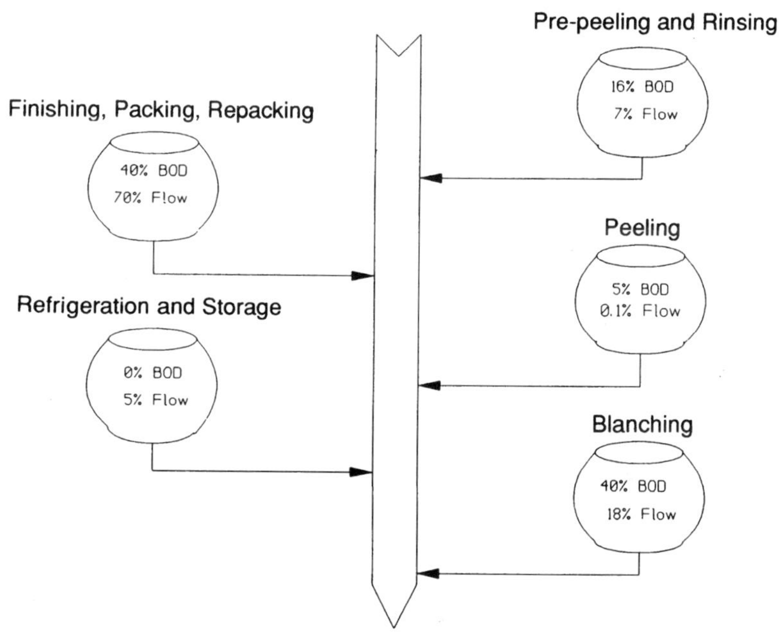

Figure 5–6 Sources of waste volume and load from fruit and vegetable processing.

screening, mud separation, balancing, settlement, and two-stage biological treatment.

5.6.4.8 Meat Processing

Meat processing wastes are significantly different from other food processing wastes. They are very strong, containing grease, blood, and faeces. Significant amounts of recalcitrant organic matter, straw, and hair are also present (Figure 5–7).

The wastes frequently include high concentrations of biocides and disinfectants such as hypochlorite. There are also problems arising from malodours and low temperatures. These are, therefore, very difficult wastes for any form of treatment, and good separation of the various process streams is the best form of waste management. Only solids-tolerant, long-retention-time reactors are likely to succeed, and the best treatment option, where possible, is dilution and mixing with domestic sewage. Two stages of screens may be necessary, followed by flow-balancing flotation and settlement. Two or three stages of biological treatment may then be

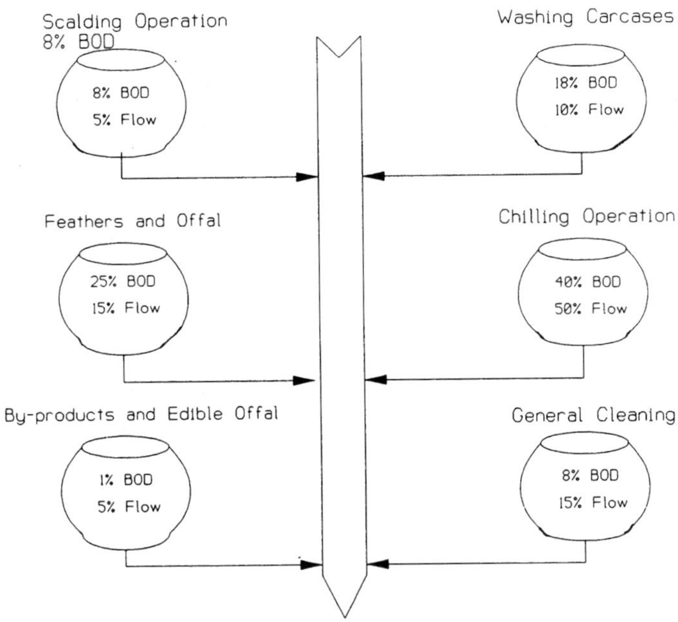

Figure 5–7 Sources and strength of wastes in chicken processing, typical of meat processing operations.

necessary, starting with anaerobic digestion and finishing with extended aeration-activated sludge.

5.6.5 Conclusion

In principle, food industry wastes should be treated easily. In practice, there are a number of difficulties, despite the fundamentally biodegradable constituents. The main problems are the high strengths and temperature compared with domestic sewage. Cleaning acids and sanitizers cause further problems. The high strengths of the waste also normally make them deficient in additional nutrients, particularly nitrogen and alkalinity. Analysis and characterisation of the effluent are essential prior to any decisions on a process design for a waste treatment plant.

5.7 TREATMENT PROCESSES

5.7.1 Preliminary Treatment

Food processing effluents are a complex mixture of floating, settleable, suspended, and dissolved materials. Complete treatment, therefore, requires a combination of physical, chemical, and biological processes (Figure 5–8). Biological processes can deal with both suspended and dissolved or colloidal materials, but normally it is cheaper to remove easily settleable or floatable material physically or chemically. Biological treatment usually costs 2 or 3 times as much as physical separation processes, but there are occasions when the cost of the disposal of the separated solids is more than capital and running costs associated with oxidising them biologically (in, for example, biological filters and extended aeration as discussed in Section 5.8). Sludge and solids disposal is getting more expensive because of increasing restrictions on the traditional land disposal. Some pre-treatment prior to biological or secondary reactors is essential, otherwise there is always a risk of packaging and other debris damaging or blocking mechanical equipment. Pre-treatment is also necessary if recalcitrant or toxic materials are present; for example, high pH due to caustic cleaning. These may be made more treatable by chemical neutralisation.

5.7.1.1 Flow Balancing

It is usual for food industry wastes to vary in strength and flow. Drains have to be designed to be large enough for periodic storms and changes of flow. Designs include those for a 4:1 peak to average for up to 10 m^3/hr and those for about 2:1 for flows greater than 20 m^3/hr. Biological treatment can cope with a wide range of loads, including some shocks, but most equipment works more efficiently when the flow and load are controlled. It is cheaper to balance surges than to design the

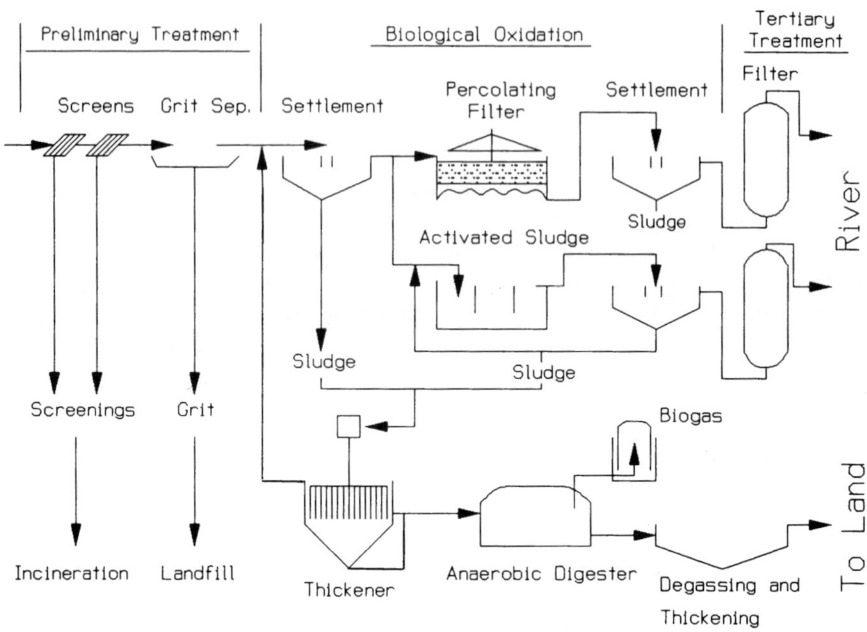

Figure 5–8 Schematic of sewage treatment.

biological reactor for maximum load. Water companies frequently specify a maximum flow rate as well as the total amount of effluent to be discharged to sewer, and flow balancing may be required to conform to these consent conditions. Flow balancing also evens out the chemical composition of the waste, and it is quite common for water companies to specify a range of MACs (see Section 5.2). The two most common difficulties associated with food wastes are pH and fat . Water companies usually insist that the fat should not exceed 150 mg/L and the pH should be between 6 and 9. Excessive fat can interfere with oxygen transfer. A wide range of food industries require some form of balancing and fat removal to achieve these standards consistently. Sudden discharges of very strong streams such as lost product, sludge returns, or hazardous discharges (e.g., caustic cleaning) should be collected in a separate dump tank. They may usually be slowly reintroduced back into the main stream for combined diluted treatment.

Balancing tanks are designed with a constant forward flow and an emergency overflow to treatment. They usually have a holding capacity of up to a maximum of one day's average flow. The design of the right-sized balance tank is difficult because too long a storage has an adverse effect on aerobic treatment. The usual problem is septicity: As the effluent is stored, natural microbial activity occurs and the dissolved oxygen is depleted. This is accompanied by a drop in pH (down to a minimum of 3.5) and bad smells associated with the sulphides and amines. Balance tank design depends on the effluent analysis and flow measurements described in Section 5.6. If there are high sulphate concentrations in the effluent from hard water, or sulphuric acid from boiler water treatment or from flocculents, balancing tanks will need to be smaller and will require roofing and extra air.

Some of these problems can be overcome by mixing, particularly with air, which then also counteracts the septicity. It is essential to be able to control the amount of air introduced because there is virtually no way of calculating how much will be required to avoid anaerobic conditions. Some effluents—particularly those that contain an active inoculum such as the yeast in brewery effluents and lactic acid bacteria in milk wastes—have high oxygen demands (>20 mg/O_2 gm dry solid/hour). Major seasonal fluctuations in load or flow would require standby plant.

5.7.1.2 Anaerobic Treatment

There are fewer problems in designing balance tanks for anaerobic biological treatment. Pre-acidification in an anoxic or anaerobic balancing tank can be beneficial, leading to accelerated hydrolysis and solubilisation. The balancing tank then becomes the first stage of a two-stage design. One reactor is for acid fermentation and solids breakdown, the other for methanogenesis from the solubilised solid waste. For some industrial wastes that are predominantly particulate (e.g., starch and paper), pre-hydrolysis would be very important to accelerate treatment.

Up to 3,000 mg/L to 4,000 mg/L total volatile acids can be obtained at retention times of up to 2 days. The pH after this first stage of the anaerobic process will be about 4 to 5.5, but if the wastes generate surplus alkalinity in the second stage anaerobic reactor, the system can tolerate acid feeds. In practice, it is useful to have some form of pH control between the stages. Anaerobic pre-acidification tanks will always need to be roofed.

Sludge scum and foam formation are common difficulties with both balancing tanks and acidification reactors. Methods for their removal need to be incorporated into the design; these can be simple draw-off valves.

5.7.1.3 Screens

Screens are designed to strain out large pieces of debris such as feathers, rags, string, plastic film, cardboard, and other packaging material as well as any stray raw materials or product. Normally the minimum aperture size is 1 mm to 5 mm; smaller sizes block too quickly even for continuous cleaning. Sedimentation or flotation can be used for smaller particles.

It is prudent to install a simple form of screen for even apparently solids-free wastes such as dairy effluents to prevent packaging materials from accidentally getting into the biological reactor and ancillary equipment. Some wastes—for example, meat processing waste and vegetable waste—may need several coarse and fine screens in series because of the quantities of large debris. Coarse screens have spacings of 25 mm to 50 mm.

Screening of food processing effluents, notably in the meat and vegetable industries, can remove significant amounts of the polluting load. In the past, the recovered screenings may have been used for animal feeds, but in most cases they now need to be landfilled or incinerated. A variety of proprietary self-cleaning screens are available based on bars, mesh, and wedge wire.

Bar Screens. These are the simplest type of screen, with parallel bars at between 30° and 60° to the flow. If the effluent is comparatively clean, they can be "home-made" and manually raked. If there is a significant solids capture, purpose-built units are necessary with automatic brushes or rakes. The rakes protrude through the grid bars and are scraped or wiped clean at the top of their travel (Figure 5–9). The smallest bar screen is 10 mm, and small particles, paper film, and so on pass easily through it.

Figure 5–9 Typical bar screen arrangement.

Mesh Screens. For finer screening (5 mm and less), mesh sieves or wedgewire are used. There are three common types: brushed sieve, disposable bags, and drum screens. In the brushed sieve (Parkwood type, Figure 5–10) the effluent flows over a partly cylindrical screen and passes through slots or holes into a collection sump or channel beneath. The solids collecting on the perforated plate or mesh are swept away by revolving brushes into a separate collection channel and are periodically flushed into a skip.

For low solids loads, disposable filter bags have become common. Various mesh sizes are available, and the bags are effective at removing plastic, cardboard, string, and fat balls. Typically the bags will last from 4 to 6 weeks.

Different types of rotating drum screens are also available. The effluent enters the centre of the drum and passes out though the slowly rotating mesh-covered

Figure 5–10 Brushed screen type arrangement.

drum. Water jets are used to keep the mesh clean. Traditionally, drum screens have been used for final straining of biologically treated effluents to avoid the worst problems of screen blinding, which would occur with raw wastes. They are efficient at removing solids but are more complex than the fixed screen types.

Wedgewire Screens. Wedgewire screens are popular for food industry effluents because they offer fine screening and are largely self-cleaning. The name is derived from the shape of the parallel bars used: The wider space is on the screened side to avoid blockage. There are three basic types: parabolic, cylindrical, and inclined to the flow.

The parabolic wedgewire is 2 m high with a top feed and wire box to achieve good distribution across the width of the screen (Figure 5–11). The wedgewire bars are transverse to the flow, and the effluent cascades down the parabolic curved surface, driving the solids off at the bottom by impact and gravity. The effluent passes through the wedgewire openings. A disadvantage of the parabolic screen is that there is a significant head loss (3 m to 5 m) through the screen. This usually requires pumping.

The alternative types are either semi-cylinders (rotamat and rotascreen) or moving inclined screens; both are designed to fit into an open channel. The cylindrical screens lie at a shallow angle in the channel, and screenings are removed and drained as they are lifted by a rotating internal archmedian scroll conveyor. The

Figure 5–11 Parabolic wedge wire screen.

inclined moving wedgewire screens are cleaned by water jets at the top of their travel (Figure 5–12). In both systems, the screenings are discharged by a chute into a skip.

Cost. It is usual for screening and preliminary treatment to represent about 5% to 10% of the capital costs of total treatment but for running costs to be high (20% of capital per year). Typically, unit costs are between £10,000 and £40,000.

5.7.1.4 Grit Removal

Vegetable processing wastes require mud and soil removal to reduce wear on pumps and mechanical equipment. Large quantities of grit and soil can be removed from root crops such as sugar beet and potatoes. The separation process relies on the denser mineral particles settling at a much faster rate than organic

Figure 5–12 Inclined wedge wire screen.

solids. Effluent velocity is controlled to about 0.3 to 0.5 m/sec in special channels, chambers, or hydrocyclones. The deposited grit is sucked or scraped from the separator and landfilled. Grit separation is also a standard feature of domestic sewage treatment, to remove road grit. The grit collected may need washing with final effluent to prevent odour.

5.7.2 Primary Treatment

5.7.2.1 Settlement

Settlement is most useful for the removal of solids larger than 10 mm. Sedimentation of vegetable and meat processing wastes can remove 35% of the polluting load by precipitation of 60% of suspended solids. Particles less than 1 mm will not settle by gravity, and soluble and colloidal materials in soft drinks or dairy industry effluent are unlikely to benefit from settlement without the addition of chemicals.

Traditionally, settlement was modelled on the density difference between the particles and water, the velocity of flow, and temperature, that is, Stokes Law. It has since been shown that the concentration of the suspension has a marked effect on settling rate. Various modifications to Stokes Law have been introduced to take account of the interference by particle, particle interaction, and water retention. Most of the changes were based on laboratory tests. Six effects were observed.

1. Large particles settle relative to a suspension of smaller ones, so that the effective density and viscosity of the fluid are increased.
2. The upward velocity of the fluid displaced during the settlement of a concentrated suspension is significant, and the apparent settling velocity is therefore less than the actual velocity relative to the fluid.
3. There may be significant cooling from the surface of the tank, which can set up eddy currents and circulation.
4. The velocity gradients in the fluid close to the particles are increased as a result of the change in area and shape of the void spaces.
5. The smaller particles tend to be dragged down by the motion of the large particles, and settlement is accelerated.
6. Flocculation and aggregation occurs and the effective size of the particles is increased.

These results make prediction of the rate of sedimentation of mixed particle sizes difficult. Agitation and stirring also increase the rate of settling. The stirring releases bound water held in place by the thick sediment at the bottom of a tank. During these final stages of settlement, liquid is squeezed out from a packed bed of particles.

Four types of settlement behaviour are listed below. The first two are the predominant mechanisms in primary settlement, but two other conditions exist where there is significant hindrance of settlement in concentrated biological suspensions such as activated sludge.

1. Settlement according to classical Stokes theory, dependent on density difference.
2. Settlement of flocculent particles with increased velocity compared with theory.
3. Hindered settlement (or "zone settlement") occurs when a stable suspension settles as a unit and there is a reduced velocity compared with theory as a result of inhibited water displacement (interparticle forces hold the suspension together).
4. Compressive settlement is similar and caused by the particles' being mechanically supported by each other; compression and consolidation of the particles then takes place. This occurs at very high solids concentrations, that is, in sludge consolidation tanks and sludge thickening in the bottom of settling tanks.

Design. Laboratory studies could be used to measure likely settling velocities, but scale-up factors such as wall effects, turbulence, and short circuiting limit their usefulness. Settlement tanks are designed to produce quiescent conditions. This is achieved by generating low velocities and a uniform flow, and by preventing resuspension of settled particles.

Most tanks are circular because this makes scrapers cheaper and more reliable, but rectangular tanks are common at domestic sewage plants to reduce land area. Deep hopper bottom tanks have been used in the past to eliminate the scraper mechanism completely and reduce sludge production by anaerobic resolubilisation. These tanks do not work because of poor sludge consolidation and anaerobic fermentation causing rising sludge. They are also expensive because of the cost of excavation. The best tanks are usually 2 m to 4 m sidewall depth with a 1 in 5 to 1 in 10 slope to the base.

Loading Rate. Typical loading rates are shown in Table 5–18. Reduced loading or increased retention time may have adverse effects owing to anaerobic fermentation, that is, a reduction in pH, gassing, and malodour. Additional features are usually incorporated into the tank to reduce the possibility of turbulence and short circuiting. One of the main features is a diffuser drum for dissipating the feed over the plan area of the tank.

The diffuser drum occupies about 30% of the depth of the tank (Figure 5–13) and slots are often cut into the base of the drum to reduce bottom scour. The diffuser drum is galvanised or plastic to prevent corrosion. A bottom scraper ensures

Table 5–18 Settlement Tank Design

Type of Waste	*Retention Time (h)*	*Surface Load (m^3/m^2 per Day)*
Meat processing	2–3	45
Vegetable	2.5	25
Cannery	4	40
Typical	4	40

that the sludge is collected into a hopper at the centre of the tank. The gentle stirring from the rotating scraper prevents solids from accumulating on the tank walls and assists with flocculation. A corrosion-resistant material, usually galvanised steel, is used for the scraper, which is fitted with plastic or rubber replaceable blades. The bottoms of the tanks are reinforced concrete with a 5° to 10° slope to the central sludge hopper. Sludge accumulating in the hopper of the tank is removed periodically by the hydrostatic head above the sludge through a screw-threaded bell valve. The bell mouth allows inspection of the sludge during decanting to avoid drawing off clean water. A scum baffle and upper scraper are also necessary with fatty wastes, to collect floating material. Scum and fat are then

Figure 5–13 New settlement tanks showing diffuser drum and scraper.

removed by a slot valve, to be combined with the sludge. An adjustable peripheral overflow weir is also required to ensure an even flow through the tank and compensate for later tank movement. It should be corrosion-resistant, that is, plastic.

Traditionally, settlement tanks have been built of reinforced concrete. These are too expensive and too large for most industrial applications. Standard size prefabricated tanks are the most common. Smaller tanks are made of plastic, larger tanks from enamelled steel. Buoyancy valves are required if the tank is below ground.

5.7.2.2 Flotation

A wide range of food processing effluents contain fats and oils. Most of the greases and oils are readily biodegradable, provided they are soluble or sufficiently dispersed to give a uniform concentration accessible to the bacteria. Most fats, oils, and greases, however, tend to separate from the main effluent stream and solidify, blocking or blinding biological reactors and process instruments.

High concentrations of fat and oil can also interfere with gas transfer during aeration. The maximum concentration for aerobic treatment is 150 mg/L. Often this standard can be achieved by reducing the flow in a simple chamber and trapping the oil between underflow weirs. The oil floats to the surface by density difference. This means separation can take place in balancing, first stage reactors, or settlement tanks. A scum baffle or boom is linked to the sludge scraping mechanism, and fats, oils, and floating sludge are collected by a scum or slot valve.

Unfortunately, wastes also often contain detergents and surface active materials, which can emulsify the oil and reduce the efficiency of separation. The efficiency of the separators can also be reduced by warm temperatures. There are three methods of overcoming this by flotation: fat traps, dissolved air flotation, and inclined plate separators.

Fat Traps. Fat traps are designed to produce a slow and gentle uniform flow through a tank. This allows density difference to bring the fat to the surface without disturbing any accumulated scum and sludge. An even distribution of flow can usually be achieved by a baffle at the inlet to the tank. Slotted inlet pipes can also be used with baffles to assist the flow pattern. Pumping prior to the fat trap should be avoided or special care taken to use pumps that do not generate turbulence and emulsification.

Solids and sludge accumulate in the tank as a result of settlement, and it is essential to incorporate some form of drain valve. Ideally, this should be complemented by a sloping tank bottom (i.e., 1 in 5). In home-made interceptors, fat removal can be done manually, but it is important to provide good access to allow this to be done safely. Removable and adjustable retaining baffles can also then be hung on the walls of the tank. Purpose-built units normally include a moving scraper on an endless chain with a trigger to activate a scum-removal valve.

Dimensions and Loading Rates. Fat traps are designed according to general settlement principles. Typically, they have a length-to-width ratio of 2:1, a retention time of 10 to 40 minutes, and a loading of 0.4 $m^3/m^2/hr$ to 3 $m^3/m^2/hr$ at maximum flow rate. Table 5–19 shows some typical examples from the literature.

Dissolved Air Flotation. Common problems with fat-containing effluents are precipitation and emulsification due to pH, temperature, pumping, and detergents. In these circumstances, gravity separation with a simple trap rarely gives satisfactory results, and assisted flotation has to be used. Dissolved air flotation relies on the introduction of supersaturated water into the effluent. There is a hydrophobic mutual attraction between the bubbles and the suspended particles, and this leads to the flotation of the solids. The main advantage of dissolved air flotation is its speed. Flotation rates are 2 to 3 times those of gravity, and tank sizes are much smaller. It also has the advantage of avoiding potential anaerobic conditions. The efficiency of the process, like sedimentation, is dependent on particle size, but it is also affected by bubble attachment and the efficiency of gas transfer. It is, as a consequence, more complex and less reliable than simple gravity separation.

Two types of flotation systems, dispersed and dissolved flotation, have evolved more or less separately. In dispersed air flotation, the bubbles are generated physically. This is the oldest type of flotation, developed at the turn of the century. It is extensively used in the metal-refining industries. It is not suitable for effluent treatment because of the shear forces produced by mechanical agitation.

Dissolved air flotation is preferred and is based on the release of air from a supersaturated solution by pressure reduction. Two different methods of pressure release have been developed. The earliest was vacuum flotation, in which air dissolved under atmospheric pressure was liberated by reducing the surface pressure. In the second, more common method, pressure is used to supersaturate the effluent and is then released to reprecipitate the dissolved air.

Treated effluent is recycled and pressurised to provide the supersaturated process water. This eliminates the necessity to aerate the crude effluent.

Table 5–19 Examples of Loading Rates to Fat Traps

Product	*Retention Time (min)*	*Surface Loading (m^3/m^2 per Hour)*
Meat processing	20	2
Margarine processing	20–40	3
Soap	40	1
Milk, butter, and cheese	30	1
Milk processing	20	0.4

Design. Design is based on upward flow velocity and retention time, as in settlement, but additionally there is a need to define a minimum air-to-solids ratio. It is difficult to set the air-to-solids and recycle ratios theoretically, and optimum conditions are often set during commissioning. In most cases, however, because of the variability of bubble attachment and gas transfer, purpose-built units include considerable flexibility. Periodic adjustments are necessary. The following are some typical design characteristics:

- surface load: 30 to 60 m^3/m^2/day; 1 to 3 m^3/m^2/hr
- retention time: 20 to 30 minutes
- recycle rate: 25% to 100%
- pressure vessel: 3 to 6 atmospheres

Surveys of the dissolved air flotation have shown that reliability is poor and operation difficult. Operating costs have been shown to be higher than gravity separation. The relatively complex technology is not well-suited to effluents nor to the frequent changes in effluent character that occur. Typical capital costs are £1 to £2/m^3/year, with running costs between 10% and 30% of capital per year. Performance from a well-operated unit is, however, much better than that from gravity separation (Table 5–20).

Inclined Plate Separators. These are settlement tanks containing a series of inclined parallel plates. The plates are corrugated to give the maximum amount of surface within the tank. Inclined plate separators have been introduced recently for the treatment of food wastes. There is so far little case study experience, but

Table 5–20 Examples of the Performance of Ideal Dissolved Air Flotation

	Influent			*Effluent*			*% Removal*		
Effluent Type	*SS*	*Fat*	*BOD*	*SS*	*Fat*	*BOD*	*SS*	*Fat*	*BOD*
Cooking oil	230	460	2,900	20	25	94	91.3	94.6	96.9
Margarine	5,000	3,900	—	200	40	—	96.0	99.0	—
Cosmetics	15,000	5,405	24,500	1,800	485	5,880	88.0	91.0	76.0
Laundry	3,469	3,014	—	281	475	—	91.9	84.2	—
Wool scouring	8,700	4,650	2,820	81	20	268	99.1	99.6	90.5
Abbatoir	7,428	3,110	—	712	97	—	90.4	96.9	—
Poultry waste	1,690	331	1,075	275	74	86	83.7	77.6	92.0
Gelatine waste	2,680	2,825	—	458	315	—	82.9	88.9	—
Jam/pickles	1,350	—	790	270	—	315	80.0	—	60.1
Metal plating	1,700	—	6,170	127	—	3,000	92.6	—	51.4

inclined plate separators have become more common in recent years, and in some cases there will be operating data on a similar waste. Maintenance and cleaning costs will be high, but capital costs much lower than conventional settlement.

The principle relies on the mutual hydrophobicity of surfaces and particulates, so that the particles adhere to surfaces despite high upward flow velocities. Solids that collect on the surface may then be retained long enough for coalescence and for the larger particles to float to the surface or settle. This aggregation allows the tank size to be reduced. Retention times are 30 minutes, and surface loadings can be increased compared with passive settlement and flotation (2.5 to 3.0 m^3m^2hr).

5.7.3 Chemical Treatment

Chemical adjuncts can improve the performance of most of the physical and biological processes used in effluent treatment. Traditionally, the most extensive use of chemicals has been to correct pH and to improve settlement rates by increasing particle density. The chemical oxidation of waste is not difficult, but even the cheapest oxidising agent, chlorine, is 20 times the cost of biological treatment. Another problem with chemical treatment is that it requires good control to ensure optimum performance. This increases running and maintenance costs. Chemical treatment is useful when problems occur—e.g., because of shock loads and seasonal changes—and it can be made available on standby.

5.7.3.1 pH Neutralisation

The two most common neutralising agents are sodium hydroxide and sulphuric acid. They are normally the cheapest reagents when handling is taken into account (£200/tonne). It is usual to purchase and use up to 50% solutions. Lime is a cheaper alkali but is very difficult to prepare because of its poor solubility. Commercial slurries are available, but there are additional difficulties because of the large volumes of sludge generated and the formation scale. Magnesium oxide is a safer alternative. If anaerobic biological treatment is in use, sulphuric acid may add too much sulphate. This can cause malodour and inhibition of anaerobic digestion. Phosphoric acid or carbonic acid would then be preferred.

5.7.3.2 Coagulation and Flocculation

Chemical treatment can improve settlement and flotation rates, and can be used to upgrade existing solids/liquids separation. In some cases, where particle size is very small (i.e., less than 5 mm), it is the only way to produce reasonable separation.

The process of particle destabilisation is very complex but can be simplified into three basic mechanisms:

1. Charge neutralisation by adsorption of counter ions. Most natural particles develop negative charge in solution and can be aggregated by positive counter ions. This process is frequently called coagulation.
2. Bridging of particles by polymer chains. Each polymer may attach itself to several particles, causing aggregation. This process is frequently called flocculation.
3. Physical enmeshment and straining by large masses of precipitate, a sludge blanket.

The most important mechanisms in effluent treatment are bridging and enmeshment. The most common reagents therefore are polymeric, such as polyacrylamide, aluminium hydroxide, and ferric hydroxide. Sulphate salts should be avoided if there is a possibility of anaerobic conditions and therefore sulphide formation. Sulphides in effluents will lead to corrosion and malodour.

In most cases, it is not sufficient to simply add the aggregating agent, because good contact between the chemical and particles is required. This means a separate rapid-mix tank to completely mix the reagent with the waste (Figure 5–14). The precipitate or flocs when formed are also fragile but still need gentle mixing to ensure good enmeshment. This is carried out in a second, slow-stirred tank (stirrer speed, 1 to 2 rpm) before final settlement or flotation. The synthetic organic polymeric materials cannot be stored, because they deteriorate when diluted. This means an additional rapid-mix tank for dispersion (working solution, 0.05% to 1%) and a slow-mix tank for unravelling and activating the polymer. For aluminium and iron, dilution to a working strength of 1% to 2% is sufficient. Typically, 1 to 5 mg/L organic polymers are required and/or 10 to 100 mg/L as Fe or Al. It is essential to use laboratory or pilot tests to establish the correct dose and combinations of flocculents.

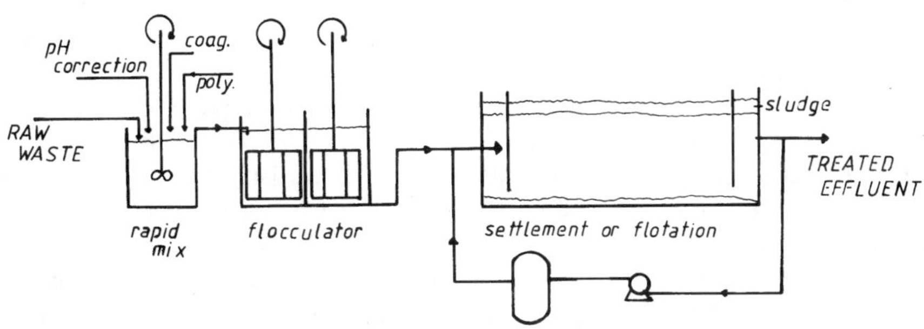

Figure 5–14 Typical flocculation system.

A range of food-grade flocculents are also available if the solids material is to be recycled, for example:

1. cellulose
2. starch
3. lignosulphonic acid
4. calcium polyphosphate

They are not usually as efficient as aluminium, iron, or polyacrylamide. The economics of chemical treatment have to be considered very carefully. Normally, it is a simple cost-benefit analysis for sludge consolidation or concentrated suspensions. For dilute effluents, the running costs (maintenance, chemicals, and sludge disposal) counteract the benefit.

5.7.3.3 Oxidation

Chemical oxidation can be justified when short-term problems arise, such as a very short operating season, toxic wastes, or a biological malfunction. It is a useful adjunct before biological oxidation when insufficient aeration is available. In these cases, it is usual to use pure oxygen, peroxide, or ozone, although chlorine, permanganate, and sulphur dioxide have also been used. In general, the sulphur compounds should be avoided if possible. Powerful oxidising agents—chlorine, hypochlorite, and ozone—are commonly used to sterilise recycled effluent.

5.7.4 Nutrients

Few wastes, even food processing wastes, are balanced nutritionally, and supplementation with nitrogen, phosphorus, trace metals, and vitamins will improve performance. For successful aerobic treatment, it is usual to establish a carbon (as BOD), nitrogen (as inorganic nitrogen), and phosphorus (as inorganic phosphorus) ratio of 100:5:1 (see Section 5.8). In practice, if organic nitrogen is present as protein, it can be used as a nitrogen source by the micro-organisms. If only nitrogen is required, it is usual to add urea. If both nitrogen and phosphorus are required, inorganic fertiliser (Nutraphos) is added.

Once the biomass is established, nutritional requirements are lower because most of the nutrients are recycled. Requirements can be controlled by ensuring that there is a small residue of nutrients in the treated effluent (e.g., 1 to 2 mg/L).

The growth rate of anaerobic bacteria is lower than aerobes, and nutritional requirements are less. Ratios of 300:5:1 are usual. The anaerobes are, however, vulnerable to shortages of transition metals— iron, cobalt, nickel, and manganese—because of the reducing environment and their precipitation as sulphides. These materials are available commercially and are commonly added to industrial digesters. Additional nutrient will also benefit aerobic treatment, providing complex nutrients or as biaugmentation. Iron additions can also be an effective method, counteracting mild sulphur toxicity in treatment by precipitating sulphide from solution.

5.7.5 Conclusion

Preliminary treatment such as screening, balancing, settlement, and flotation will often be beneficial compared with the discharge of untreated effluent. Preliminary treatments are essential prior to biological or chemical treatment to prevent damage and clogging.

The more complex preliminary processes, such as flotation and chemical flocculation, are difficult to operate and maintain. These extra costs need to be considered when calculating the benefits of treatment.

5.8 AEROBIC PROCESSES

The most common types of treatments for removing organic matter from food industry effluents are aerobic (with aeration) and anaerobic (no oxygenation). Bioreactors are used for food industry effluents. Providing aeration for strong wastes (i.e., with a COD in excess of 5,000 mg/L) is expensive. Roughly 1 kWh is required for each kg of BOD; also, the efficiency of aerobic growth means that there are more surplus solids to dispose of than from the equivalent anaerobic process (Figure 5–15). Disadvantages of anaerobic treatment are that it is slow and unable to produce a high-quality treated effluent. Anaerobic treatment can only be used as a pre-treatment process. In some areas of the world, where land is relatively cheap and factories are rural, long retention time in natural process lagoons or lakes is possible. This is not attractive in Europe because of the cost of land, odour, and flies.

Aerobic biological waste treatment reactors are normally open-topped mixed-microbial cultures that also acquire a diverse population of grazing organisms. There are two basic types of treatment. The first type, plug flow fixed-film systems (known as bacteria beds, or percolating, biological, or trickling filters) include the following:

- percolating filters
- biological or trickling filters
- biotowers
- rotating biological contacts
- submerged aerated filter
- biological aerated filter (BAF)
- expanded and fluidised beds

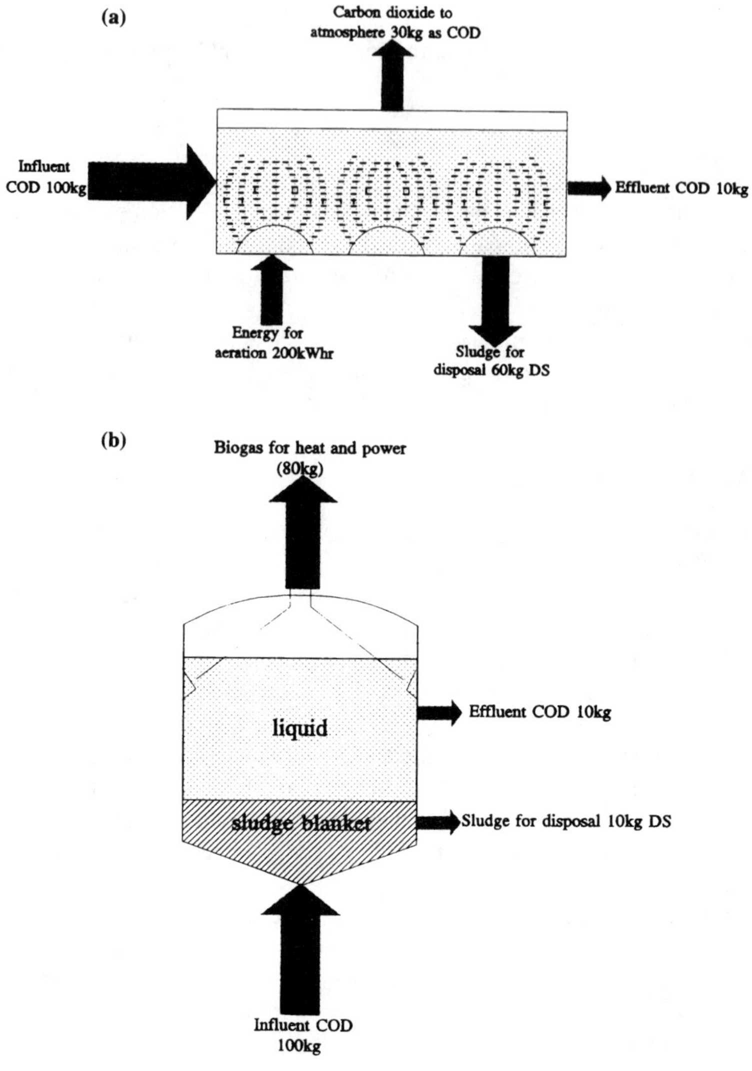

Figure 5–15 Comparison of the mass balance of carbon utilisation in (a) aerobic and (b) anaerobic processes.

The other type of treatment involves completely mixed flocculent processes, or activated sludge. The earliest type of reactor was the trickling filter, which developed directly from experiments carried out on the treatment of sewage by percolation through soil. Trickling filters are a tank containing a graded biological support medium—usually clinker, slag, stone, or gravel, depending on local availability. They are easy to maintain, incur low running costs, and have a plant life of between 30 and 50 years. Percolating filters have proved to be very reliable, and the majority of European and US treatment plants use this type of system. The alternative is activated sludge, which is a mixed process, aerated and stirred by air diffusers or surface paddles. The retention time in the aeration tank is 6 to 15 hours, after which the effluent is separated from the active sludge by settlement (2 hours), and most of the sludge is returned to mix with the incoming effluent. The process is more efficient than percolating filters and is able to treat 10 times as much waste; it is therefore cheaper to build. Activated sludge is, however, more difficult to operate and maintain, and it incurs higher running costs because of the mixing and aeration. The largest sewage works in Europe are activated sludge plants. Table 5–21 compares the advantages and disadvantages of the two processes.

Experience has shown that the plug flow fixed-film process is more robust than activated sludge and is better suited to industrial wastes, particularly those containing only slowly biodegradable components. The plug flow means that the slowly biodegradable components may be retained and held for long periods in the biofilm. This gives more time for breakdown and also means that potential toxins are not evenly spread throughout the reactor. A second advantage is that fixed-film reactors are not dependent on successful settling of the biomass. Surveys have shown that most activated sludge plants suffer from poor settlement at some stage, and this is particularly true when dealing with food processing wastes.

5.8.1 Percolating Filters

The basis of reactor design is to provide an environment in which the waste, the organisms responsible for purification, and air for aerobic treatment are brought into contact. The initial step is adsorption of the substrate onto the biological surface—typically, 50% of the substrate is colloidal. This is followed by breakdown of the adsorbed substrate by extracellular enzymes and absorption of dissolved material into cells and biomass growth. Biological filters immobilise the biomass responsible for treatment on an inert support medium, either stone or plastic. The wastewater is sprayed over the surface of the filter by nozzle or moving distributor and allowed to percolate down through the bed. Good distribution of the waste over the surface of the filter is very important to provide intimate contact with the biofilm covering the support. The most popular system is a circular filter, to give a simple rotating distributor (Figure 5–16). The distributor arms are drilled at intervals, to provide an even spread over the surface of the filter. Usually the force of the water issuing from the holes is sufficient to drive the distributor around, and no power is required.

Table 5–21 Simple Comparison of the Features of Activated Sludge and Percolating Filters

	Activated Sludge	*Percolating Filters*
Capital costs	Low	High
Land area	Low	High
Energy requirement	High for mixing aeration, impossible without power	None, if necessary gravity flow, reaction distributor
Sludge production	Large amounts, high water content difficult to dewater	Small volume, higher solids content, well stabilised
Hydrostatic head	Small for bioreactor but pumped sludge return equivalent in volume to feed	Head loss though media normally 2m to 3m
Maintenance	Cleaning of diffusers, blowers monitoring equipment; not very robust, mechanical failures	Distributor needs cleaning daily; sturdy, few difficulties with breakdowns
Control	Requires skilled operation and monitoring, liable to upset and poor settlement	Simple but little turn up or control possible
Environmental	Few odours but may produce microbial aerosols; can be noisy; no flies	Odour problems common and severe fly problem in local vicinity in summer
Seasonal	Not really affected by season or weather	Affected by temperature, poor nitrification in winter and seasonal, usually spring, discharges of solids
Waste	Sensitive to changes in nature of waste; industrial effluents usually cause a deterioration in settlement; foam is also a problem with fatty or detergent-containing waste	Industrial effluents treated easily, able to withstand changes in loading; no problems with settlement or foam
Treated effluent quality	Good when working well, flocculates and removal of solids to a low concentration	Not as good effluent quality, difficult to achieve very high removals of solids
Nutrients	It is possible to remove nitrogen and phosphate by advanced control	Not possible to remove either nitrogen or phosphate by changes in operation

Figure 5–16 A high rate plastic media biofilter treating dairy waste.

Cleaning the distribution holes is one of the major maintenance problems with biological filters. They may require cleaning 2 or 3 times a day with a poorly screened effluent.

Two other important features of biological filter design are the surface area of the media, which governs the amount of biomass, and the space between the support media. The voidage of the media controls the ventilation, and therefore oxygen available for microbial growth, drainage, and discharge of solids. Ventilation through the media is assisted by the temperature difference between ambient and the wastewater. There is a continuous cycle of biofilm growth and detachment taking place, partly due to autolysis and partly due to the action of the grazing organisms. The debris or humus flushed from the biological filter is collected in a final settling or humus tank.

In the spring, with the first increase in ambient temperature, very large amounts of biofilm and debris, which have accumulated during the winter, can be dislodged by the grazing organisms. This can cause overloading of the humus tank (Figure 5–17).

5.8.1.1 Process Design

Early in the development of the percolating filter, it became apparent that the loading rates were limited by excessive growths on the media. The voidage be-

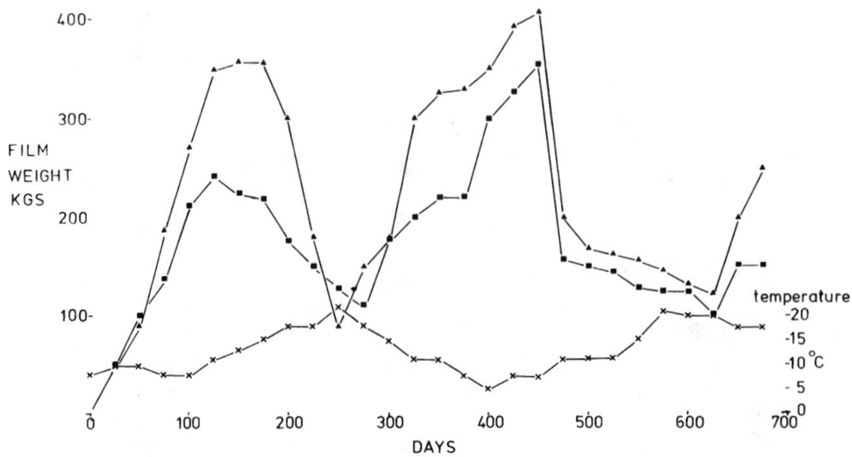

Figure 5–17 Seasonal changes in film weight on Flocor E, 150 m²/m³ (■) and Flocor M, 200 m²/m³ (▲).

came filled with biomass, and the filter was blocked or ponded. Treatment efficiency is proportional to the amount of biomass and increases in process intensitivity or loading will increase the amounts of active biomass.

The amounts of waste that could be treated and the aeration or ventilation required to ensure reliable treatment were at first established empirically in 1908 (Royal Commission on Sewage Disposal), but there have been continuous experiments to improve the process. Tests in 1931, for example, (O'Shaughnessy), noted that BOD removal improved with dilution by recycle, which led to better mixing; with the recycle there was a more even spread of biomass through a reactor. Measurements on the distribution of biomass showed that there were two distinct zones. The system was plug flow, and the surface of the filter received and removed the greatest amount of the applied substrate. The most active organisms were in the top 0.5 m of the bed, known as the heterotrophic layer (Figure 5–18). Below this was the autotrophic zone, where the bios is mostly debris and humus from the more active upper layers. The humus occurs in flocculated deposits rather than as prostrate slime adhering closely to the support media as in the case of surface growth. There is little BOD removal in these layers but considerable autotrophic nitrification. Higher hydraulic loads, such as from recirculation, increase efficiency by dissipating the load deeper into the bed. Dilution of the waste by recirculation also reduces the strength of the waste, reaerates it, and recycles valuable nutrients. Wishart and Wilkinson (1941) described other experiments in which they successfully controlled biomass by washing a winter-blocked filter

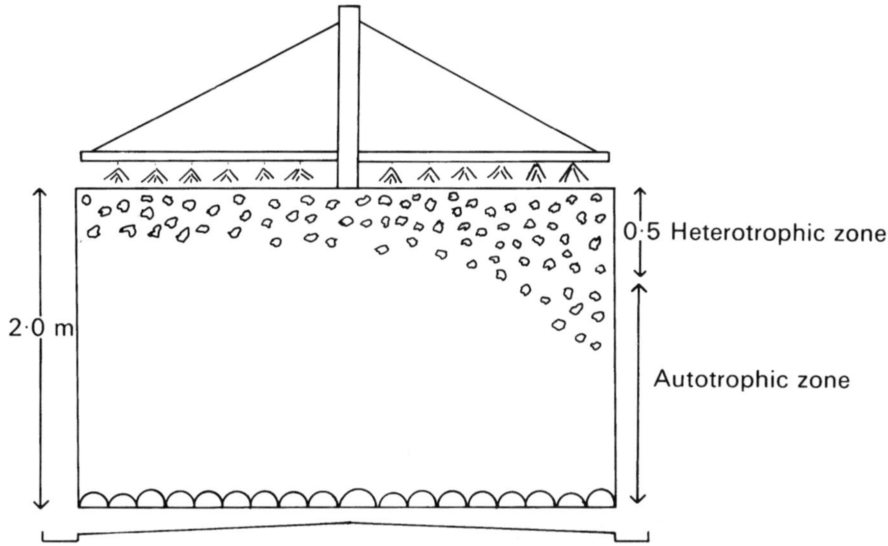

Figure 5–18 Ecological zones in biological filtration. *Source:* Reprinted with permission from A.D. Wheatley, Biotechnology and Effluent Treatment, in *Biotechnology and Genetic Engineering Reviews*, Vol. 1, G.E. Russell, ed., p. 271, © 1984, Intercept.

with treated effluent. They suggested a scheme of alternating double filtration, in which the waste load normally applied to two filters in parallel was instead fed to two filters in series. Biomass accumulated in the first filter as it received and consumed virtually all of the BOD load. The order of the filters was then reversed before the first filter ponded. The original primary filter then received treated effluent that had passed through the new primary filter. This procedure resulted in rapid autolysis and consumption of the starved biomass in the secondary filter. Alternating double filtration enabled filters to be operated at more than twice the usual loads with fewer problems of blocked media. The best frequency of alternation was found to be weekly, and alternating double filtration became the common method of operating large biological filters.

Other work used similar principles. The following are some methods of operating percolating filters:

- single filtration
- recirculation
- alternating double filtration
- low-frequency dosing with electric drive

- two-stage filtration with high-rate primary filter
- plastic media filters

Hawkes and Shephard (1972) and others showed that slowing down the distributor on a biological filter also improved performance by encouraging an even spread of biomass through the filter depth. Double filtration with different-sized media is an example. The first filter in the series of two contains a large medium (150 to 300 mm) to accommodate more biological growth, but at a lower BOD removal efficiency (about 70%). The second filter contains conventional media (50 to 75 mm) and completes the removal of organic waste to 95% (Table 5–22). Overall, the loading is higher. Traditionally, single-pass percolating filters were 1.8 m (6 feet) deep, to achieve 95% BOD removal from domestic sewage. Recirculation and higher loadings required deeper filters: 2 to 3.5 m deep filters for 95% BOD removal. The advantages of double filtration led to the introduction of plastic media. Plastic materials can be moulded into shapes with optimum surface area and voidage. To be effective, however, the much greater surface area needs to be uniformly wetted. Two types of plastic media are used: modular self-supporting blocks and random rashig rings (Figure 5–19). Much taller biological filters were possible with the lightweight plastic filter media, and simple prefabricated cladding on a concrete base could be used for the self-supporting modules. The modular synthetic media have specific surface areas of 100 to 150 m^2/m^3, about the same as 50 mm blast furnace slag.

Filters were built up to 6 m high (biotowers), and high rates of recirculation (usually between 6:1 and 10:1) were used to generate complete mixing and ensure adequate utilisation of the available surface. The high rates of treatment (Table 5–22) with plastic media enabled major savings, and the process became widely used

Table 5–22 Typical Biological Filter Loadings

Type of Filter	*BOD Load (kg/m³ per Day)*	*Hydraulic Load (m³/m² per Day)*	*Retention Time (min)*	*BOD Removal (%)*	*Conversion of BOD Applied to Sludge (%)*
Septic tank unattended system	0.05	1–1.5	100–200	95	5
Low rate traditional	0.05–0.1	1–2	60–90	95	15–20
Mid-range double filtration ADF or recirculation	0.01–0.05	2–5	30–60	85–90	20–50
High rate filtration	1–2	5–10	10–20	50–70	50–90

(a)

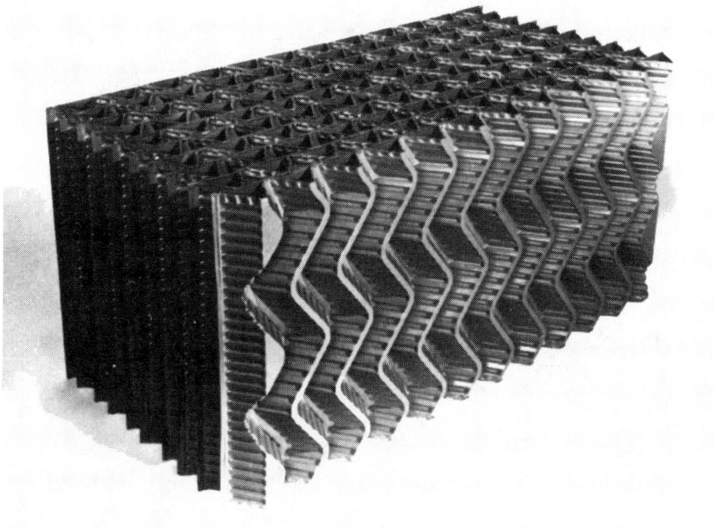

(b)

Figure 5–19 The main types of plastic media. (a) High-rate modular medium Flocor E, module size 1200 × 600 × 600 mm. (b) The new random medium Flocor RC prior to loading into the experimental filters. *Source:* A.D. Wheatley, The Treatment of Industrial Effluent, *Critical Reports in Applied Chemistry*, Vol. 31, pp. 172–220, © 1990, Society of Chemical Industry. Reproduced with permission.

to treat strong industrial wastes. The tall towers were, however, vulnerable to odour. The voidage and height meant they acted like chimneys, distributing the odour from the surface irrigation over a wide downwind area.

The random plastic media are shaped with corrugations or vanes to promote an even flow pattern over the higher specific surface; up to 250 m^2/m^3 are available. Lower recycle rates (e.g., 2:1 or 3:1) could then be used and some plug flow retained. Like the modular media, random plastic filters can then be taller (i.e., 2 to 4m) and utilise cheaper methods of construction—mainly prefabricated coated steel tanks. Unfortunately, the cost of plastics has risen sharply recently, and there are now fewer advantages to using plastic media.

5.8.1.2 Rotating Biological Contractors

A common problem with biological filters is the frequent blockage of the distributor holes. The RBC overcomes this difficulty. A series of plastic sheets are rotated through a tank of the wastewater (Figure 5–20). About 40% of the surface area is submerged. RBCs, or biodiscs, have similar ecology to biological filters but make more efficient use of the support, which is alternately completely wetted and then aerated. The slow rotation of the discs (1 to 2 rpm) ensures uniform concentrations of substrate and dissolved oxygen. The tank beneath the discs can be baffled to produce a plug flow of waste, and then nitrification can be achieved. To consistently remove 95% of the applied BOD, the maximum organic loading rate is 5 g BOD/m^2 of specific surface per day (Pike, Carlton-Smith, Evans, & Harrington, 1982). This is about 5 times the load on a biological filter to achieve the same standard. RBC plants are compact and have become very popular as a package plant for small flows. Typical prefabricated units are 1.5 m in diameter by 2 m long; larger units have been built for food industry wastes up to 5 m in diameter by 8 m long. Applications to food effluents have not been very successful. The treatment of strong wastes is limited by oxygen transfer, and experiments have been conducted with faster speeds of rotation and forced aeration of the effluent tank (Wilson, Hamoda, Islam, & Buranasin, 1988). Maintenance problems have also been encountered: The rotating shaft and bearings are subject to a very corrosive environment. The shaft and discs of the RBC must be kept turning at all times; otherwise, biomass accumulates in one position and imbalances occur. This can distort and damage the shaft and media.

5.8.1.3 Submerged Filters

Totally submerging the biological support media significantly improves the efficiency of use. Fluidised beds, for example, combine the advantages of the percolating filter process with those of activated sludge. Biological fluidised beds are able to retain about 5 times the concentration of micro-organisms compared with conventional plant, and significant savings in capital cost are likely. The support

(a)

(b)

Figure 5–20 (a) Rotating biological contactor (RBC) with (b) close-up photograph of the support media. *Source:* Reprinted with permission from A.D. Wheatley, Waste Water Treatment and By-Product Recovery, in *Topics in Waste Water Treatment*, J.M. Sidwick, ed., pp. 68–105, © 1985, Blackwell Science, Ltd.

particles (sand or plastic) can be much smaller than in percolating filters because there is no risk of clogging. The supports are expanded and kept well mixed by aeration or oxygen injection combined with effluent recycle, often 4:1 or 5:1. Bed expansion is 20% to 25%. Most of the work reported on fluidised beds is from laboratory or pilot scale tests; the majority are for the anaerobic treatment of waste (Heijnen et al., 1985). Difficulties have been encountered with keeping the bed at the correct expansion; there are no aerobic examples in the food industry, and widespread application may depend on better control systems.

A different and more robust high-rate, fixed-film process is submerged filtration, or BAF. There is no intention to completely expand the carrier bed in submerged filtration; the system works rather like sand filtration and includes periodic backwashing. The forced aeration and fine carrier gives a high rate of performance and solids retention, so that most BAF units work without a clarifier.

BAF units work well as simple filters, and as a consequence process air requirements are less than for mainly biological oxidation processes such as activated sludge (about 0.5 kg O_2 per kg BOD). However, because there is less metabolism, sludge production is high: 0.8 to 1 kg DS per kg BOD removed (Table 5–23).

The solids accumulate in the filter until eventually they reach a critical concentration. This is monitored by head loss through the system. The filter is then cleaned by backwashing and air scouring. Repeated backwashings are used until the system is clear and free of solids. The wash water has to be collected and balanced before recycling to avoid hydraulic surges. Volumes of wash water used are between 2% and 8% of flow rates. The wash water is returned to the front of the treatment plant for primary settlement.

Both upflow and downflow units are available. Downflow systems are thought to work with less dispersion and mixing. They are therefore claimed to have advantages for nitrification and for avoiding clogging of the distribution systems.

Table 5–23 Biological Aerated Filter (BAF) Loading Rates Compared to Conventional Percolating Filters and Activated Sludge

	To Achieve 90% BOD Removal (kg/m³ per Day)	*To Achieve 70–80% Nitrification (kg/m³ per Day)*
Trickling filters	0.1–0.4	
Activated sludge	0.3–0.6	
BAF	0.7–3.0	
Nitrifying filters		0.6
BAF nitrification unit		0.6–1

Upflow systems with greater mixing can work at higher loads; they are also less odorous because only treated effluent is exposed. In downflow systems, there is the potential to strip volatile compounds from the raw waste above the media (see Stephenson, Mann, & Upton, 1993, for a table of proprietary processes).

A number of European BAFs have been built indoors to avoid odour and aerosol problems in urban areas. This is possible because of the small plan area that BAFs occupy. The number in the United Kingdom is growing rapidly (Table 5–24). Ammonia can consistently be reduced from 20 to 30 mg/L to 1 to 5 mg/L (Lilly et al., 1991). The process has also been shown in the United States to be a cost-effective method for the production of high-quality effluents at smaller works where activated sludge is difficult to operate (Bishop & Kinner, 1986).

Sand shale and gravels are thc most common support media but plastic media, both modular and random, have also been utilised. In the case of the modular media, air is injected through diffuser plates beneath the pack to provide mixing and aeration; separate clarification is necessary. Hydraulic retention times in BAF units are between 30 and 100 minutes.

5.8.2 Activated Sludge

The large areas of land required for percolating filters and the capital costs mean that activated sludge is the preferred type of treatment for loads greater than 8 tonnes of COD per day. Activated sludge is a completely mixed process with biomass recycle after settlement. The development of colonies of flocculated bacteria are crucial to allow the successful separation of the biomass or sludge by settlement. Activated sludge has become the standard wastewater treatment system for population equivalents greater than 10,000 and has been successfully adapted for the removal of nutrients and recalcitrants (Cooper & Downing, 1998). There are two basic methods of aeration: diffused air and mechanical surface aeration.

Table 5–24 United Kingdom and Some French Operating Biological Aerated Filter Units

Location	*Capacity (m^3/Day)*	*Standard SS/BOD (mg/L)*	*Start Up*
North Brierley (Bradford)	10,000	10/10	1990
Silchester, Gloucester	14,000	10/7	1991
St. Austell, Cornwall	25,200	35/25	1993
Nimes	70,000	30/30	1991
Monaco	31,000	15/20	1990

The most common type of diffuser is a ceramic air stone fixed on the base of the tank. The rising bubbles also serve to mix the tank. Oxygen transfer efficiency from bottom diffusers has been measured to be better than surface aeration because bubble size can be controlled (Figure 5–21). There may, however, be problems keeping the diffusers clean: They can easily become clogged with fat and scale, and may need frequent cleaning. Another difficulty with diffusers is that achieving good overall tank mixing requires more aeration than would be required for effective waste treatment alone. Coarse diffusers are also used for industrial wastes or in individual tanks to enhance mixing. Diffusers are now available made from light synthetic plastics, which enables them to be withdrawn from the tank and cleaned without taking the tank out of service (Table 5–25).

There are a number of proprietary surface aerators, many with draft tubes to ensure good mixing. The oxygen transfer characteristics of mechanical aerators can be varied by adjusting the rotor height. Surface aerators are simpler and easier to maintain than diffused air but are mostly used by the smaller works because of their lower oxygen transfer efficiency and increased costs.

5.8.2.1 Process Designs

A number of activated sludge process designs have been developed (Table 5–26), most aimed at reducing energy demands and improving settlement. A com-

Figure 5–21 Diffused, air-activated sludge plant treating dairy waste.

Table 5–25 Characteristics of Some Available Aerator Systems

Aeration System	*Typical Transfer Efficiency (%)*	*Transfer Rate (kg O2/kWh)*	
		Standard[a]	*Field*[b]
Diffused-air systems			
Fine bubble	10–30	1.2–2.5	0.7–1.5
Medium bubble	6–15	1.0–1.6	0.6–1.0
Coarse bubble	4–8	0.6–1.2	0.3–0.9
Turbine sparger system	10–20	1.2–1.4	0.7–1.0
Static-tube system	7–10	1.2–1.6	0.7–0.9
Jet	10–25	1.2–1.4	0.7–1.4
Pure oxygen systems			
Mechanical surface aeration	79	—	1.4–1.8
Venturi systems	90	3.0	1.5–2.5
Turbine sparger	80	1.2–2.4	1.2–1.5
Impeller systems			
Low-speed surface	10	1.2–2.0	0.7–1.3
Low-speed surface with draft-tube	10–20	1.2–2.0	0.7–1.3
High-speed floating aerator	10–20	1.2–2.0	0.7–1.4
Rotor-brush aerator	10–20	3.0	0.7–1.3
Bayer turmbiologie	90	5.0	—
ICI deep shaft	90	—	—

[a] Standard shop test under controlled conditions (ASCE, 1984).
[b] Average of field data available.

mon method of design is tapered aeration, which is intended to match microbial oxygen demand with aeration. Most activated sludge plants are plug flow, and the oxygen demand at the outlet is much less than the inlet, so aeration can be progressively reduced along the length of the tank. This is accomplished by using fewer diffuser domes in the base of the tank towards the outlet. A similar modification for use at surface aerated plants is step aeration, which introduces some waste at intervals throughout the length of the tank. Aeration of the returned sludge without the addition of more waste is also used to encourage the bacteria to utilise any stored nutrient. The activated sludge is then able to assimilate greater amounts of substrate or waste when reintroduced into the main treatment tank. This process is known as contact stabilisation. Oxygen availability in the activated sludge process can be increased by using pure oxygen instead of air. There are two proprietary systems. One system uses completely sealed tanks and is very similar to conven-

Table 5–26 Design and Performance of Activated Sludge

Type of Activated Sludge	*Substrate Loading, F/M (kg BOD/kg MLSS per Day)*[a]	*Biomass Concentration, MLSS (g/L)*	*Hydraulic Retention Time (HRT) (h)*	*Biomass Growth Rate, Doubling Time or Solids Retention Time (SRT) (Days)*	*Removal of Organic Matter, BOD Removal (T)*	*Organic Loading Rate (kg BOD/m^3)*	*Biomass Growth Rate Surplus for Disposal Conversion to Sludge (%)*[d]
Extended aeration	0.02–0.06[b]	4–10	100+	10+	95+	0.1–0.2	5–20
Conventional design practice	0.2–0.5[b]	3–4	10–20	5–10[c]	95	0.6–1.5	25–50
High rate treatment	1.0–2.0	4–5	5–10	2–5	70–75	4–10	50–75

[a] Assumes 80% volatile or organic suspended solids.
[b] These are averages; reactors need to be plug flow so initial loading rates are much higher, i.e., up to 100.
[c] To achieve microbial nitrification minimum solids retention is 10 days.
[d] Often expressed as kg dry solids/kg BOD removed.

tional activated sludge. Surface mixers are used and the head space above the liquid is enriched with oxygen, either from a storage tank or from on-site production. The second system uses an open tank with external recirculation pumps to keep the tank mixed. High-pressure oxygen is injected into a venturi throat on the outlet side of the pump. This dissolves the oxygen at high pressure. Oxygen utilisation efficiency is further increased by releasing the oxygenated mixture through outlet nozzles in the base of the tank. Pure oxygen plants are able to work with larger biomass concentrations and therefore higher loads than conventional plants. The capital costs of the plant are lower, but the running costs higher. The process is ideal for strong seasonal wastes.

A totally different method of increasing oxygen solubility is to increase the hydrostatic pressure in the reactor by building tall reactor tanks. The best known of these is a tube sunk 50 m to 100 m into the ground. The process is called the "deep shaft". The shaft is 0.5 m to 2 m in diameter and divided into two concentric tubes. The waste is introduced into the inner, downflow section, together with the recycled activated sludge and process air. In the outer, upflow section, bubbles of dissolved gases are released as the pressure decreases, and this reduces the density of the circulating water. It is this difference in density between the water in the downflow and upflow tubes that produces the circulating force for the shaft. The velocity generated is much higher than the rise rate of the air bubbles in the downflow section, and the air is carried down. Dissolved oxygen and turbulence in the shaft are high, which permits a high concentration of very active biomass to be obtained. To avoid flotation, the treated effluent has to be vacuum-degassed before settlement. The system has low running costs but can be very expensive to build because of the structural costs.

A slightly different approach to reduce these construction costs has been to utilise tall, above-ground tanks. The tower tanks are 20 m to 30 m high in protected welded steel. Air is injected through an array of draft tubes arranged around the base of the tank or at high pressure directly into a multiport waste inlet manifold. The technology has been used successfully on chemical wastes, and there is one unit in the United Kingdom (Chartered Institution of Water and Environmental Management, 1997).

An unwanted consequence of intensifying aerobic waste treatment using the activated sludge process has been an increase in the amount of surplus biomass (sludge) to be disposed of. Extended residence times (e.g., >10 days) and low loads can be used to aerobically digest or autolyse most of the biomass produced. This process is known as extended aeration, and the oxidation ditch is the most common example. The oxidation ditch was originally designed to be a maintenance-free, cheap-to-run system for small communities. They are oval-shaped shallow ditches (Figure 5–22), run at a very low load and without primary settlement. Mixing and aeration are provided by a horizontal brush paddle at one or

(a)

(b)

Figure 5–22 Typical oxidation ditch (a) with close-up of the turbulence caused by the rotor (b).

both ends. They incorporate separate settlement of the treated wastes and recycle the sludge as usual.

5.8.2.2 Nutrient Removal

Most large domestic activated sludge plants now recover some energy by anoxic denitrification. The heterotrophic bacteria in activated sludge are able to utilise nitrate for respiration in the absence of oxygen. In a conventional activated sludge plant, at least 50% of the flow is recycled; this is enriched with nitrate from the oxidation of ammonia and organic nitrogen. Therefore, if the first portion of a plug flow activated sludge plant is made anoxic or has no free dissolved oxygen, the high substrate concentration will encourage the micro-organisms to use nitrate instead of oxygen for their respiration. Waste treatment will continue, without the cost of aeration. The transient exposure to anoxic conditions also has beneficial effects on sludge settlement, and there is the added benefit of removing nitrate as nitrogen gas. In practice, up to the first 30% of the activated sludge tank (i.e., 4 to 5 hrs) can be anoxic.

In a typical activated sludge plant with a 1:1 recycle, only about 50% denitrification should be achievable. Cooper, Drew, Bailey, and Thomas (1977) observed 57% nitrate removal in a standard plant. Increased recycle rates are necessary to improve denitrification, but carbon-to-nitrate ratios of 3:1 (BOD:NO_3) are necessary and the BOD must be readily degradable. Increased recycling will improve the degree of mixing and therefore encourage poor sludge settlement.

Some nutrient removal will be required in sensitive locations in the United Kingdom to meet the new EC Urban Wastewater Directive (see Section 5.2). Phosphate removal is common in warmer climates; for example, South Africa, Australia, and the United States. There are two approaches: biological and chemical phosphate removal. Biological phosphate removal uses a three-stage activated sludge plant, starting with the usual anoxic zone but then including an anaerobic section before the final reaeration zone. It is the totally anaerobic zone that is necessary to encourage phosphate uptake. The theory is that normally aerobic bacteria can derive some energy from polyphosphate storage compounds (volutin granules), while their aerobic metabolism is blocked. The organisms release phosphate and become phosphate deficient, so that on reaeration they absorb and store much more phosphate than originally released in "luxury uptake". Good denitrification is important to ensure anaerobic rather than anoxic conditions in the phosphate-removal stage. The totally anaerobic environment of the phosphate-removal step will also lead to settlement problems. Nutrient-removal plants are therefore notoriously difficult to operate, which has led some to suggest sidestream processes with chemical back-up (Upton & Churchley, 1993). Adapting the activated sludge plant for nutrient removal will present major design and operational challenges for the future. The alternative and more common approach, certainly in the

United Kingdom, will be chemical treatment with iron or aluminium flocculents to precipitate the phosphate.

5.8.2.3 Sludge Settlement

The degree of mixing and organic loading have been found to exert a major influence on activated sludge settlement. The better the mixing, the greater the exposed microbial surface to substrate and oxygen. The well-mixed activated sludge process is therefore able to support a larger population of actively growing bacteria per unit volume of bioreactor than traditional percolating filters. It has also become evident, however, that good mixing, which leads to low residual substrate concentrations, can lead to poor settlement.

The micro-organisms that grow in wastewater treatment processes are selected by their ability to degrade various substrate components in the waste, their ability to grow in oligotrophic (nutrient-deficient) environments, and their retention within the system by bioflocculation. The ecology is also influenced by reactor design. The activated sludge process is totally aquatic, while percolating filters utilise a thin film of effluent flowing over the surface of the bios.

The reduced ecological diversity of the activated sludge process compared with percolating filters contributes to process instability, increases sludge production, reduces the resilience to shock or variable organic loads, and makes the process more difficult to manage (Hawkes, 1963). The bacteria are the basis of both processes and the zoogloea (gram-negative, non-sporing, non-motile, capsulated rods) are assumed to be the most important group and responsible for much of the treatment. There are still some doubts as to whether the zoogloea are a defined group, a polymorphic variety, or a strain of pseudomonas that develop a slime layer because of the particular conditions in the process (Pike, 1975). Classical microbiology by media isolation does not give any information on which are the important bacteria in a wastewater treatment plant. A very wide range of bacteria can be isolated from wastewater, but few of these are actively growing. Many of the bacteria in fresh sewage, for example, are enteric, including some potential pathogens. These bacteria have no role in the treatment process, and their numbers fall rapidly (Pike & Carrington, 1979). The organisms that colonise treatment processes are derived from aerial inoculation and infiltration water. Zoogloea, pseudomonas, chromobacter, achromobacter, flavobacter, and arthrobacter are thought to be the principal groups (Pike, 1975).

Experiments have shown that only a small proportion of the sludge floc or slime layer is actively growing, but most of the biomass can still carry out biochemical reactions via extracellular enzymes and biophysical adsorption (Pike & Carrington, 1972; Bishop & Kinner, 1986). The production of extra-cellular bacterial polymer is a most important feature of the activated sludge process. The detail chemistry of the polymer is complex, but it is predominantly polysaccharide. A number of soil and aquatic bacteria produce extra-cellular polymer as a response to oligotrophic (nutrient-limiting) conditions. The extra-cellular polymer has evolved as a binding agent, primarily to collect or fix to solid substrates but also to adsorb dissolved materials at the surfaces.

In a well-settling sludge, at least 25% to 30% of the total biomass is thought to be extra-cellular polymer. Bacterial polymers are not as strong as polyelectrolytes or resistant to shear as polyacrylamide polymers, for example, but the overwhelming concentration makes activated sludge an effective binding agent for most of the solids in the wastewater. Other bacteria, stalked protozoans, detritus, and debris can all be observed adsorbed onto the polymeric floc. Soluble and hydrated inorganics are also bound to the floc by the slightly negative surface charge from the carboxyl and amine groups present in the polysaccharide and proteins of the polymer (Brown & Lester, 1979).

Significant growth of bacterial filaments beyond the floc structure can cause settlement problems. The filaments prevent coagulation of the individual bacterial flocs by stopping close contact between them. Water is then held between the flocs rather than being squeezed out, sludge density is reduced, and there is no consolidation. This condition is known as bulking. The deterioration in settlement is linked to the filament length protruding beyond the floc structure. Twenty-two morphologically different filamentous bacteria have been observed (Eikelboom, 1975), and a simple key system has been developed to identify them (Chambers & Tomlinson, 1982). The most commonly occurring ones are shown in Table 5–27. The organisms are not well characterised, and many of the bacteria are identified only by number. It seems likely that some are the same polymorphic organism (Postgate, 1989). Traditionally, most filamentous organisms were identified as

Table 5–27 Common Filamentous Organisms in Activated Sludge

Organism	*Frequency of Occurrence*
Microthrix parvicella	Very common
Type 021N	Very common
Type 0041	Common
Sphaerotilus natans	Occasionally
Beggiatoa	Occasionally
Leptothrix	Occasionally
Nocardia (actinomycete)	Occasionally
Fungi	Rarely

Source: Chambers and Tomlinson (1982). There are differences between countries, particularly warmer climates.

Sphaerotilus or the sulphur bacteria, and older texts refer only to these. All the causes of bulking are not known, but there are five common triggers: low dissolved oxygen, low pH, septic wastewater, nutrient deficiency (e.g., N and P), and low loading rates. Each results in problems from a different group of organisms. Low dissolved oxygen is associated with Sphaerotilus; septic effluent with Thiothrix and the other sulphur bacteria; low nitrogen and phosphorus with Sphaerotilus; low pH with the fungi; and Microthrix 021N and 0041 with low loading rates. The floc-forming bacteria have the ability to rapidly absorb and store nutrient compared with the filamentous bacteria. To avoid filamentous bacteria, it is necessary to ensure that the activated sludge is rapidly mixed with the raw waste. This is designed for by a plug flow reactor incorporating denitrification or a pre-mix tank prior to aeration (known as a selector). The subsequent aeration at low substrate concentrations allows the substrate to be biologically oxidised and to regenerate the absorption capacity of the floc. Filamentous bacteria are much more efficient at taking up nutrients at lower concentrations, possibly because of their larger surface area. Therefore, completely mixed reactors, which have a lower and uniform concentration of substrate, promote filamentous growth. Measurements on the kinetics of the organisms have confirmed these observations. Zoogloea flocs have a higher maximum growth rate but a low substrate affinity, whereas filamentous organisms have a lower maximum growth rate but a higher affinity for the substrate.

Most activated sludge plants suffer from bulking or settlement problems at some time, particularly if treating industrial effluents. Much research has been conducted in this area, but there are still occasions when the cause of filamentous growth is not understood.

A floc loading term has been introduced to the design of activated sludge to take account of the degree of mixing. Like the F:M ratio, floc loading is a measure of the ratio of substrate to sludge solids; but unlike the sludge loading rate (or F:M ratio) it is the instantaneous loading occurring at the point where the substrate and returned sludge meet. Floc loading is the ratio of substrate to returned activated sludge. Typically, floc loading needs to be 100 to 200 kg BOD/kg mixed liquor suspended solids (MLSS) for a minimum of 10 minutes. It can be designed for by either the degree of mixing or selectors to induce a very high substrate concentration at the inlet. The technique of floc loading was successfully used to eliminate bulking problems at Halifax Sewage Treatment Works (Jones & Franklin, 1985).

Stable foam formation is also a common problem with activated sludge plants. Present problems are not linked to synthetic detergents but the presence of natural surfactants such as the long-chain fatty acids and proteins. These cause the accumulation of solids, air, and liquid films at the air-sludge interface. Activated sludge plants with foam also contain filaments. Fatty wastes (e.g., dairy), high temperatures, poor sludge recycle rates, and high dissolved oxygen are contributory factors (Foot, 1992). The foam, if unchecked, quickly dries to form dense mats, which are further stabilised by the opportunistic growths of a filamentous actinomycete within the foam and at the interface. This type of dense foam is dark in colour and known as chocolate mousse. Interestingly, stable foams were not recognised as a problem until 1970 and can be associated with low loading rates and nitrification and also possibly with the much wider use of long-chain vegetable oils, although this needs to be scientifically established. Foaming is always worse at high temperatures and is prevalent in warm climates and occasionally in gas-mixed sludge digestion.

The stabilising actinomycetes are all from the genus Nocardia. The actinomycetes, like the fungi, are truly branched and themselves generate long-chain fatty acids during growth to produce hydrophobic hyphae and aid bubble attachment.

Significantly, Nocardia has not been isolated from any other environment, and it is possible that it is a polymorphic form of a standard activated sludge microbe (Horan, Bu'ali, & Eccles, 1988). Microthrix is also occasionally linked to foaming.

Unlike the filamentous bacteria, Nocardia is not a major difficulty in settlement because it does not extend far beyond the bacterial floc structure. Foaming is an operational difficulty because foam and solids can be carried over into settlement and be discharged with the final effluent. One of the simplest and best ways of controlling foam is to avoid its formation by water sprays on the aeration tanks and clarifiers. This prevents thick foam formation and the ingress of Nocardia. Spraying and antifoams are ineffective against stable foams. Thick foams can be reduced by increasing loading by reductions in sludge age or MLSS concentrations (Foot, 1992). Microthrix and Nocardia have rapid growth rates, and sludge ages may have to be reduced to 1 or 2 days. These growth rates often lead to an unacceptable final effluent quality and loss of nitrification, but reductions in MLSS may improve redox and help oxygen transfer in warm weather, even at much higher sludge ages.

5.9 ANAEROBIC BIOLOGICAL TREATMENT

5.9.1 Types of Anaerobic Digester

There are several commercial designs of anaerobic reactor; the choice among these versions depends on the type of waste to be treated. The organics in domestic sludge and agricultural wastes are mainly as solids, whereas the organic pollutants in food effluents are in solution or colloidal suspension. Effluents are therefore more amenable to treatment. Many of the internal structures within reactors used for the treatment of dilute wastes could not be used for wastes with a high solids content. Large amounts of solids would lead to clogging and blockage. Reactors for the treatment of sewage sludge and agricultural waste are stirred tanks or long-retention lagoons mixed by the methane generation. Retention times are governed by the rates of solids hydrolysis. Rates of solids breakdown are rapid for up to 15 days retention, but then become slower as the easily metabolizable carbohydrates, proteins, and fats are exhausted. There is always an indigestible residue of fibre of 10% to 30% of the volatile feed solids, even at retention times greater than 30 days (see Institute of Water Pollution Control, 1979). Typical retention times for solids and slurry digestion are between 12 and 20 days; this is sufficient for microbial growth to replace organisms passed out with the treated waste. If there is no system for keeping or reusing bacteria, the minimum retention time in the reactor is limited by microbial growth rate. The doubling time of the methanogens is 5 days; this is too long for a reasonable-sized, commercially viable anaerobic reactor for industrial effluents. Different reactor designs have therefore been developed for food wastes that hold back biomass within the reactor or recycle the bacteria after separation. Biomass retention is then no longer dependent on liquid retention time, and high bacterial concentrations can be obtained. Micro-organisms are retained by the standard methods (Figure 5–23). These are:

1. Physical separation of the biomass from the effluent by filtration or sedimentation. The active sludge is then recycled back to the completely mixed reactor. This type of anaerobic system is usually known as the CSTR.
2. Retention and attachment of the bacteria to an internal support medium. The packing increases head loss and provides areas of low upflow velocity where bacteria can accumulate. This type of reactor is called an anaerobic filter. The flow of waste can be upward or downward.
3. Natural bacterial flocculation assisted by low upflow velocities to generate a sludge bed or blanket in the bottom of the reactor. This reactor is known as an upflow anaerobic sludge blanket reactor. This type of biomass retention is unusual to anaerobic treatment and is made possible by the accumulation of inorganics with the reactor.
4. The flow through a flocculated sludge blanket may also be horizontal and the biomass retention assisted by a series of baffles. This reactor is known as an anaerobic baffled reactor.
5. Attachment of the bacteria to small support particles and fluidisation of the particles by gas and effluent recycle. The particles and biomass can be completely mixed when the system is known as a fluidised bed or less vigorously mixed in an expanded bed.

Low-rate large lagoons have occasionally been used for the anaerobic digestion of industrial and domestic wastes, but their size and lack of control makes them unpopular in the United Kingdom.

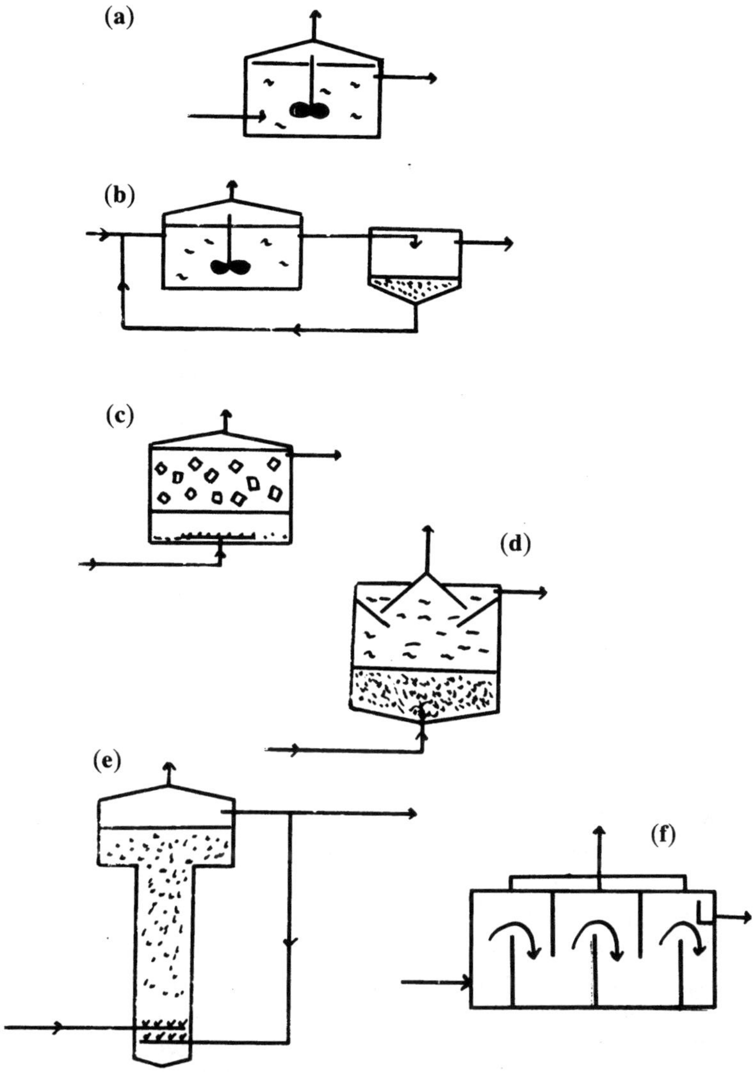

Figure 5–23 Types of anaerobic reactor: (a) see text; (b) contact stirred tank reactor (CSTR); (c) anaerobic filter; (d) upflow anaerobic sludge blanket (UASB); (e) fluidised bed; (f) anaerobic baffled reactor (ABR).

5.9.1.1 Sewage Sludge Digesters

Anaerobic digesters for the treatment of domestic sludges and farm slurry are large, completely mixed, continuously heated tanks. The most important features of design are the efficiency of mixing, the effectiveness of heating, and (because most of the substrate is solid) the solids retention time.

Mixing. Surveys of traditional concrete sewage sludge digesters (Brade & Noone, 1981; Swanwick, Shurben, & Jackson, 1969) have repeatedly found much of the reactor ineffective because of poor mixing. Most problems were traced to blockage, fouling, and corrosion of mechanical mixers. Mixing is required to ensure dispersion of the fresh raw sludge within the digester and to promote good contact between the waste and the bacteria. Mixing is also necessary to avoid dead zones, thermal stratification, and grit and scum accumulation. Gas production within the digester causes some stirring, but not enough for the complete mixing of sewage sludge digesters. Industrial digesters operating at high rates and producing more than 1.5 times the volume of the digester in gas each day can be completely mixed by gas production (Tilche & Vierra, 1991). Reactors that operate at a very low load—typical of rural digesters—can also be self-sufficiently mixed by gas production. There are three methods of mixing sludge digesters: recirculation of sludge, recirculation of gas, and mechanical mixing.

External recirculation of the sludge, although easy to maintain, does not produce thorough mixing. The digester contents are turned over "on block" rather than completely mixed, and short circuiting and dead zones occur. Sludge recirculation is not used as the sole method of mixing; traditionally, this is assisted by internal mechanical mixing. Sludge is sucked up a fixed tube by an impeller and sprayed out over the digester surface. This action avoids scum formation. Mechanical mixers are the most efficient type of mixer when new (Brade & Noone, 1981) but rapidly become choked with rags and eventually damaged by corrosion. Cleaning impellers is expensive and unpleasant. Current preferred practice is to use gas-lift mixing. Gas is collected at the digester surface, cleaned, dewatered, and compressed. The gas is then reintroduced at the base of the digester. The gas may be injected into a vertical tube inside the digester, which acts as a gas-lift pump (confined gas-lift mixing), or simply released at intervals around the base of the tank (unconfined mixing). Experiments have shown that unconfined gas mixing from evenly spaced distribution points around the base of a tall, narrow tank gave the best dispersion and least scum (Baumann & Huibregtse, 1981; Rundle & Whyley, 1981). Unconfined gas-lift mixing was also significantly cheaper than other types of mixing.

Many older digesters were too large for easy mixing: They were over 1,000 m^3 in volume. Reactor shape was dictated by providing the largest volume at the cheapest cost. Reactors were large-diameter, shallow tanks. These were difficult to mix and were based on storage tank design rather than biological reactors (Figure 5–24). Standard prefabricated digesters based on silos have better aspect ratios and are better suited to gas-lift mixing. Larger sizes can be provided for by several digesters. This modular approach has maintenance advantages as well, as each digester can be serviced in turn without totally disrupting treatment.

Loading Rate. Loading rates are defined by the rates of breakdown of the substrate, the solids in the sludge, and the growth rates of the mixed culture of anaerobic organisms. These can be used to fix the retention time of the digester. Design loading rates have been established empirically over time. It was observed from batch experiments that sludge digestion occurred rapidly during the first 10 to 15 days, then slowed down between the next 15 to 30 days. There was little further degradation beyond 30 days. The polysaccharide and cellulose polymers are the most abundant natural solids; they are derived from plants and bacterial cell walls (Figure 5–25).

Most of these polymers are degradable quickly (i.e., within 10 days), but cellulose bound to hemicellulose and lignin, or fibre, is resistant to anaerobic breakdown. On average, 30% to 40% of the volatile solids introduced into a domestic digester are lignocelluloses and undigestible. The rates of hydrolysis of the simple polymers, such as the oligosaccharides and starches, are rapid. Many older digesters were designed with retention times of 20 to 30 days or 0.2 m^3 reactor volume per person per day (Ministry of Housing, 1954). This was to achieve a 50% to 60% reduction in the organic solids in the raw sludges, and included a significant contingency for poor mixing and dilute feeds. Improvements in the design of ancillary

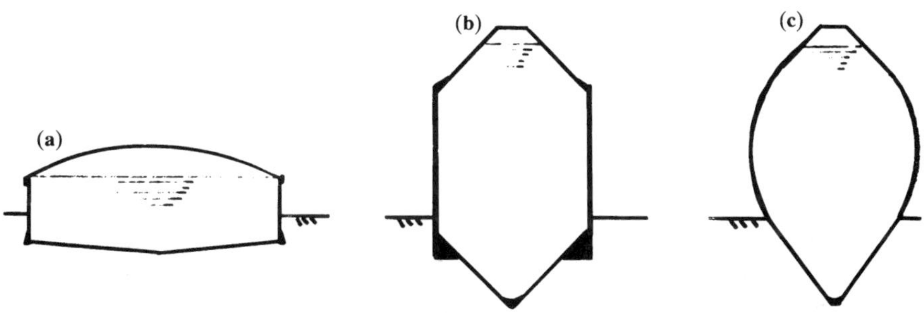

Figure 5–24 The shapes of digesters. (a) Anglo-Saxon, (b) European standard, (c) Teutonic. *Source:* Reprinted from P.N. Hobson and A.D. Wheatley, *Anaerobic Digestion: Modern Theory and Practice*, p. 138, © 1993, with kind permission from Kluwer Academic Publishers.

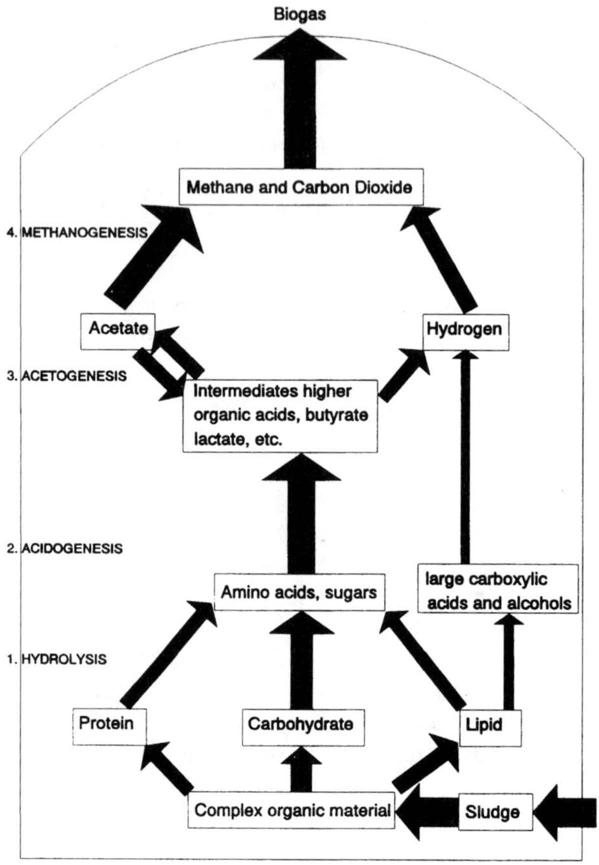

Figure 5–25 Biochemical steps in anaerobic digestion.

equipment and the understanding of anaerobic digestion have enabled retention times to be reduced to 12 to 15 days (Noone & Brade, 1985). Typical loading rates are shown in Table 5–28.

Measurements on microbial growth rates have established that the doubling time for methanogens is about 5 days. This, then, is the critical minimum retention time to avoid washout (Henze & Harremoes, 1983). These figures were based on laboratory experiments with perfect mixing and ideal temperatures, so in practice 12 to 15 days is good design to allow for imperfections in the process.

Table 5–28 Operating Conditions of a Domestic Sludge Digester

Parameter	*Units*	*Typical Value*
Biomass VSS	mg/L	2,000–6,000
Loading rate	kg VS/kg VSS per day	1
Retention time	days	15
Performance removal of solids applied	% removal of VS	50–60

5.9.1.2 The Contact Stirred Tank Reactor

The CSTR was the first type of anaerobic reactor to be tested for the treatment of dilute effluents. The reactor was developed from those used for sludge digestion, and experiments were tried on very strong industrial wastes that were difficult to treat aerobically (Department of Scientific and Industrial Research, 1960). The principle of operation is like that of the activated sludge process. Active anaerobic bacteria are separated by settlement or filtration and recycled back into the reactor. Anaerobic bacteria do not produce much extra cellular polymer, and the anaerobic bacterial consortia do not precipitate as easily as aerobic activated sludge. Settlement is made worse by the warmth of the mesophilic effluent, which causes thermal eddies in the clarifier. Methane gas given off also makes the solids buoyant. In most designs, settlement is assisted by degassing and by filtration or inclined plates within the clarifier.

The Bioreactor. Both mechanical and gas mixing are used for different proprietary CSTR systems. The more efficient the mixing system, the greater can be the biomass concentration in the reactor. Solids contents up to 25 to 30 kg/m^3 can be achieved with gas injection compared to 5 to 10 kg/m^3 with mechanical mixing (Nahle, 1991). The power required for gas injection is 5 to 10 W/m^3 and for mechanical mixers, about 2 to 5 W/m^3. Reactor design is based on the biomass loading rate; the optimal loading rate varies according to the characteristics of the waste (Table 5–29). Typical loadings are shown in Table 5–30.

Degassing. There are four common methods of removing the biogas from solids prior to settlement according to the proprietary system (Figure 5–26). Degassing units have a retention time of about 10 to 15 minutes, and the separated gas is collected. Most systems have a vacuum pump to assist separation.

Settlement. Sludge separation and recycle during settlement are in most cases assisted by plate clarifiers. Settlement rates are significantly lower than aerobic

Table 5–29 Typical Organic Loading Rate for a Mesophilic Contact Stirred Tank Reactor

Load	0.5–2.5 kg COD/m per day
Retention time	1–5 days
COD removal	80–90%
Solids concentration	5,000–10,000 mg/L

biomass, and upflow rates are designed at 0.1 to 0.2 m/hr. The high concentration of inorganic materials in some wastes—from sugar processing, for example—can improve rates of settlement; other wastes—from polysaccharide oils or protein processing—generate a poorly settling sludge. Clarifiers need to be roofed to avoid odour problems.

Applications. The absence of internal fittings makes CSTR digesters suitable for the treatment of wastes with a high solids content. Loadings are usually modest, but high COD removal efficiencies between 85% and 90% can be achieved easily because of the high degree of mixing. At typical retention times (2 to 5 days), the reactor does not require close supervision and is quite resilient. The CSTR system is still cheaper than aerobic treatment because of the low sludge production and mixing energies. The CSTR is a popular design because of its similarity to conventional sludge digestion. Thirty percent of the industrial digest-

Table 5–30 Examples of Working Contact Stirred Tank Reactors Treating Food Wastes (taken from Nahle, 1991)

Type of Effluent	*Organic Loading Rate (kg COD/m³)*	*F:M (kg COD/kg TS)*	*COD Removed (%)*
Sugar processing	0.6–13.0	1.3–3.0	90–95
Distillery	1.5–2.5	0.2–0.25	90–95
Citric acid	1.3–4.0	0.2–0.3	75–80
Yeast production	3.0–4.0	0.2–0.4	77–80
Sauerkraut	1.5	0.4	96
Vegetable cannery	2.0–4.2	0.1–0.3	90–95
Pectin factory	1.7–5.3	0.03–0.2	88–93
Starch factory	3.6	1.4	65
Meat processing	0.8–4.8	0.5–1.1	90–95
Cellulose condensate	1.3–1.8	0.1–0.2	95–98

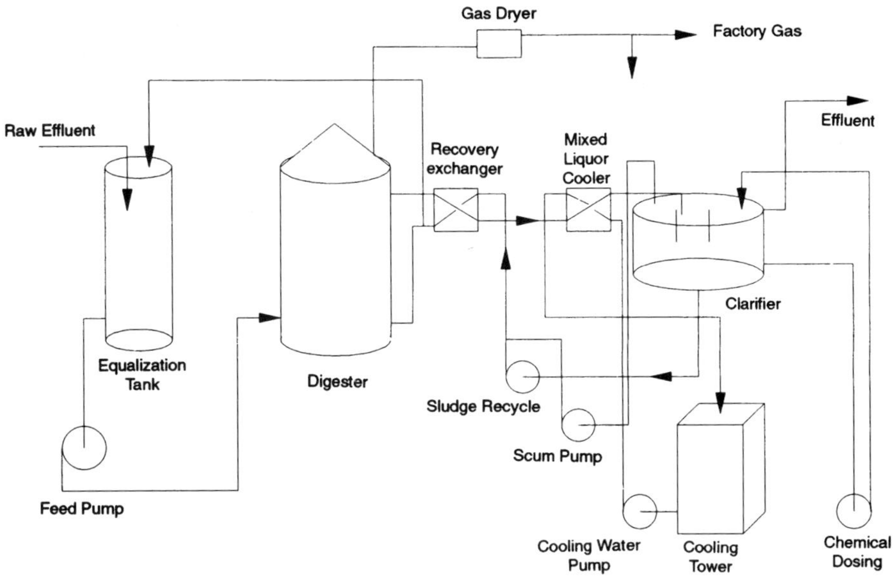

Figure 5–26 Typical process flowsheet for an anaerobic digester treating industrial effluent.

---

ers in Europe are of this type, about half of them do not bother to recycle biomass, and retention times are greater than 7 days. The most common application is for the treatment of distillery effluent (30 full-scale plants, 15 in Italy) (Pauss, Depaepe, & Nyns, 1990). Treatment rates with this type of waste are low because of the lack of alkalinity, and retention times are reported to be over 10 days. There are also a number of contact digesters that treat sugar processing wastewater. Data have been published on UK examples (Shore, Broughton, & Bumstead, 1984; Smith, 1984). Organic loads are 2.5 kg COD/m^3/day at retention times of 2.5 days to achieve 86% COD removal. The system included a flocculator and degasser.

Other successful examples are for the treatment of starch processing wastewater (Butcher, 1988; Butler, 1984; Morgan, 1980). The starch waste had a COD strength of 2,000 mg/L, and the reactor was loaded at 2.5 kg COD/m^3/day with a retention time of 24 hours. COD removal was 70% to 80%. Biomass separation was assisted by a membrane filtration unit that filtered the sludge from the five 200 m^3 digesters on rotation. This type of separation was used to improve performance compared with simple settlement. UK operating data have also been published on the treatment of soft drinks, wastewaters, and brewery effluent (Housley & Zoutberg, 1994; Wheatley et al., 1997).

Many older digesters therefore have been significantly over-designed and include spare capacity that could be used to treat very strong food industry waste streams (i.e., >10,000 mg/COD). Centralised anaerobic digesters for combinations of wastes have become popular in Italy and Scandinavia recently to exploit grants available for non-fossil fuels such as biogas. Centralised anaerobic digester plants typically treat a mixture of sewage sludge, strong food processing streams, agricultural waste, and source-sorted putrescible municipal refuse. This route will be cheaper than disposal to sewer and less restricted than land filling (Figure 5–27).

5.9.1.3 Anaerobic Filtration

The difficulties associated with settling and recycling biomass in CSTR reactors led to an alternative approach to reactor design based on filters. Bacteria were immobilised within a support medium to prolong solids retention time. Unlike aerobic filters but resembling BAF systems, anaerobic filters are both upflow and downflow. In both types of filter, the media are fully submerged. Downflow reactors have countercurrent gas and liquid flow; mixing is therefore intense and enhanced by recycle. Downflow reactors usually use orientated or tubular packings.

Figure 5–27 An anaerobic contact stirred tank reactor showing degassing tower, heat exchanger for cooling, and settlement tank for sludge return.

Upflow filters use both orientated and random packing; recycle is also used, but the upward flow velocities are lower compared with downflow filters. Upflow filters are usually cleaned with periodic flushes or desludging to dislodge accumulated biomass.

Packing Materials. The packing materials are designed to retain biomass within the reactor. In the early upflow anaerobic reactors, the media were stone and gravels. These had low voidage and blocked with solids and biomass quite quickly (Young & McCarty, 1969). Plastic rashig and pall rings were then tried; this prolonged the time intervals between blockage but did not overcome the problem totally, and increased flow rates were applied using recycle to overcome the problem. Blockages still occasionally occurred, and downflow filters were then tried. These downflow filters worked at very high liquid velocities. Blockage and channelling has still been reported in some downflow filters but only when the waste contained fat, insoluble organics, or other solid materials (Environment Canada, 1986). Present design practice anticipates some accumulation of solids and includes provision for desludging from the base of the reactor and from the media. The proportion of reactor containing packaging has also been reduced and is now typically 60% to 75% (Oleszkiewz, Hall, & Oziemblo, 1986; Oleszkiewz & Thadari, 1988). Solids may also be dislodged from the pack by temporary increases in upflow velocity induced by effluent and gas recycle (Camilleri, 1987; Ehlinger, Audic, Vernier, & Fraup, 1987). Once the solids have been removed, then performance is recovered.

Most downflow filters have used the modular sheet and block media, whilst most upflow filters have used the random ring type. Stone is no longer used except within low rate rural digesters; the voidage available for biomass accumulation is insufficient. The type of plastic material does not affect reactor performance, as most surfaces can be colonised by bacteria. Nylon, polypropylene, and polyvinylchloride have all been used, as well as ceramic blocks and rings (Table 5–31).

The media used must be mechanically stable and able to tolerate some long-term compression. The function of the media is to retain biomass within the reactor, but the predominant mechanisms are different in each type of anaerobic filter. In upflow anaerobic filters, most of the active biomass is held as flocculated, granulated solids in the interstices of the medium. This type of accumulated solids is also the source of clogging problems (Young & Dahab, 1983). Fine, loose solids are also present, held in place by the upward flow velocity. The specific surface area available within upflow anaerobic filters is not as important as in downflow filters. In downflow anaerobic filters, there are little suspended solids and sludge within the medium. The important biomass is as a thin film firmly attached to and close to the contours of the media. Media surface area and orientation are more important than in upflow systems (Kennedy & Droste, 1986). Cross-flow ordered

Table 5–31 Some Properties of Typical Media

Media Description and Name	*Material*	*Weight (kg/m³)*	*Specific Surface Area (m²/m³)*	*Voidage (%)*
Flocor R random corrugated tube (50 mm × 50 mm)	PVC	90	250	95
ETA pack pall rings (75 mm × 75 mm)	Polypropylene	85	160	97
Furnace slag porous stone	Expanded ceramic with significant iron content	975	100	55
Flocor E regular sheet modules (1200 mm × 600 mm)	PVC	90	100	94
Cloisonyle, vertical vaned tubes	PVC	200	80	95

media have been shown to perform slightly better than regular or tubular media and significantly better than random media, even with larger surface areas (Adams & McKinney, 1989).

There have been a number of studies on the ideal characteristics of media for downflow filters. Rough surfaces have been shown to have more rapid start-up, and some ceramic materials that leached trace nutrients (such as red tile clay) performed better than those that did not leach materials (Murray & van den Berg, 1981). In practice, however, cross-flow modular plastic media in combination with simple prefabricated reactor design usually offer the best value for money. There have been fewer investigations on the ideal characteristics of media for upflow filters. Larger media (>100 mm) would resist clogging better than the smaller materials (25 to 50 mm), but they may be more difficult to clean. More research is necessary on this topic.

Hydrodynamics and Mixing. There is a significant amount of operating experience on the performance of anaerobic filters. Further tests are required to provide models with a more fundamental basis, but there is much empirical information that indicates that reactor hydrodynamics are the most important feature of design. Some comprehensive surveys (Young, 1991; Young & Yang, 1989; Young & Young, 1988) suggest that retention time is the most useful single indicator of performance when other conditions are constant. Typically, poor performance arises from dead zones and short circuiting within the reactor. The degree of mixing is an important influence on the retention time: The greater the mixing, the less will be the amount of short circuiting and dead space. Upflow velocity and gas production control the degree of mixing. New upflow anaerobic filters behave as plug flow reactors, but active filters are mixed by the gas production. Early filter designs at low loading rates and without recycle did show plug flow behaviour (Dahab & Young, 1982; Wheatley & Cassell, 1985). Higher gas productions (>2 volumes gas/volume of reactor per day) show mixed behaviour (Young & Young, 1988). Downflow filters because of the countercurrent gas and liquid flow are nearly always completely mixed (Hall, 1983; Samson, van den Berg, & Kennedy, 1985). The amount of dead space or degree of clogging is measured by the standard retention time techniques (Levenspiel, 1972; Wheatley, Johnson, & Winstanley, 1988).

Recycle. Effluent recycle provides better mixing and lateral distribution, and can assist with keeping the media clean. Recycle also provides extra alkalinity and nutrients, and will dilute strong wastes. Providing the most appropriate recycle flow rate depends on reactor shape, gas production, and media, but it is usual to provide between 2:1 and 6:1. Reported upflow velocities vary widely, with satisfactory performance reported between 0.2 m/hr and 7 m/hr. In practice, upflow velocities need to be set to avoid washout of the biomass, which will depend on the degree of mixing and type of biomass; some versatility in pumping rates should be provided.

Distribution. Uniform distribution is also necessary to obtain optimum retention and performance. Typically, distributors are like cartwheels, with the number of arms depending on the diameter of the vessel. The distribution holes point downwards into the bed, with the diameter of the holes or number of holes increasing away from the inlet manifold. The recycle is introduced through a larger, totally separate but similar, system slightly above the feed distributor. The distributor arms should protrude through the reactor walls to allow some possibility of rodding.

Applications. The simplicity and robustness of the anaerobic filter make it ideally suited to the treatment of soluble food effluents. Loading rates are much higher than those for CSTRs. Typical operating conditions of an anaerobic filter are as follows:

- organic load: 2 to 10 kg COD/m^3 per day
- retention time: 10 to 50 hours
- COD removal: 70% to 80%
- critical solids concentration in the feed: 1000 mg/L

Despite these apparent advantages, the anaerobic filter is not as popular as the CSTR or the upflow anaerobic sludge blanket reactor. About 15% of the industrial anaerobic treatment plants in Europe are filters, with about equal numbers of downflow and upflow operation (Wheatley & Coombes, 1991). The major worry is excessive biomass accumulation and blockage of the packed bed. At the moment, there are 32 anaerobic filters in Europe with about equal numbers of downflow and upflow operation. Most are in France and all but two treat wastes from the agro-food industry. Nine have been built for the treatment of distillery wastes. Detailed information has been published on two (Camilleri, 1988; Racault, 1990); there are no reports of blockage. Young (1991) has published a list of upflow anaerobic filters in the United States and Canada. Most are for the treatment of food and distillery wastes, although three have been used for easily biodegradable chemical wastes (Witt, Humphreys, & Roberts, 1979). Organic loads were 0.2 to 16kg COD/m^3/day with hydraulic retention times of 12 to 96 hours. There are fewer downflow filters in the United States and Canada (Kennedy & Droste, 1991), but detailed information has been published on two. The performance and design of the downflow filter at Bacardi has been widely publicised (Szendry, 1983), and details on a dairy plant have also been reported (Environment Canada, 1986). The dairy filter used ceramic modular media and did block with fat after 2 years of operation. An extensive survey of laboratory and pilot studies on various types of waste has also been published (Bonastre & Paris, 1989).

5.9.1.4 The Upflow Sludge Blanket Reactor

The problems due to clogging of the filter media in anaerobic filters, particularly the early mineral packings, resulted in investigations into methods of eliminating the amount of media in the reactor. It was found that a high-density granular sludge could be produced and retained in the reactor despite the gassing and upflow velocity of the waste. The development of this type of reactor has been very successful for a wide range of food industry wastes and domestic sewages. There are about 200 full-scale operating plants world-wide, and it is the most popular type of high-rate anaerobic reactor, largely owing to the research and development efforts of Lettinga, Velsen, van Zeeuw, and Hobma (1979) and two Dutch companies marketing the process.

Biomass Granulation. Sludge granulation is a complex and unreliable process. To avoid problems, it is the usual commercial practice to use a large inoculum of granular sludge from an already operating plant. Inoculums of granular sludge between 10% and 15% of the reactor volume are used (Weiland & Rozzi, 1991). If an ideal start-up inoculum is available, full process efficiency may be obtained within 1 month, but if a suitable inoculum is not available, the start-up could take

2 to 6 months (Alibhai & Forster, 1986; Souza, 1986). Start-up procedures used are similar to the other types of anaerobic reactor, that is, stepped increases in load of 0.05 to 0.1 kg COD/kgVSS/day assuming the pH and concentrations of carboxylic acids remain satisfactory. The hydrodynamics of the reactor, created by the distribution and reactor shape, are important in retaining the correct form of biomass. To maintain process efficiency, biomass settling velocities of 10 m/h are required. Measured velocities up to 90 m/h have been reported (Lettinga & Hulshoff-Pol, 1986). Granulation is a natural process and is the result of a combination of microbial morphology, nature of the substrate, and accumulation of inorganic salts (Alibhai & Forster, 1986; Dolfing, 1986). The type of wastewater to be treated has an important role in the formation of granules. A satisfactory sludge blanket has been formed from yeast processing, sugar beet, and potato wastes but not with distillery, corn starch, abattoir, or dairy effluents (Stronach, Rudd, & Lester, 1986). Key elements in the feed substrate for successful granule formation are iron, calcium, phosphorus, ammonia, magnesium, aluminium, and silicon (Kissel, Grotenhuis, & Zehnder, 1988). The process is similar to scale formation. These materials are available in only some wastewaters. A large population of filamentous organisms (*Methanothrix sp.*) are also necessary and encouraged by a high concentration of soluble substrate, such as acetic acid. Generally, granular sludge will eventually develop from mostly soluble wastes (Lettinga & Hulshoff-Pol, 1991).

Hydrodynamics and Mixing. The shape of the upflow anaerobic sludge blanket reactor, the upward flow velocity, and the baffles inside the reactor are used to promote gas-solid separation inside the reactor. The most important feature governing reactor height is the treatability of the waste and the efficiency of granule formation. If the waste is soluble and a granular sludge has formed, upflow velocities of 3 m/hr can be achieved. Reactor heights can then be up to 10 m (Lettinga & Hulshoff-Pol, 1991). If the waste contains significant amounts of suspended COD, or during start-up when granulation is incomplete, the biomass is present as flocculent rather than granular sludge. Under these conditions, upflow velocities need to be more modest (e.g., 1 to 1.5 m/hr). Practically all wastes are heterogeneous, and a good compromise reactor height is 3 to 5 m (Figure 5–28).

The distribution system also exerts an important influence on the upflow velocity and can be used to promote good contact between sludge and wastewater. The large plan areas of reactors enables a large number of small distribution points (usually one every 1 to 2.5 m). A particular problem with upflow anaerobic sludge blanket reactors with a large number of small distribution points is clogging of the nozzles. Each influent system should be easy to clean and monitor.

The effect of gas production is more complex. Higher than optimum gas production will cause gulping and open paths through the blanket, while a lower gas

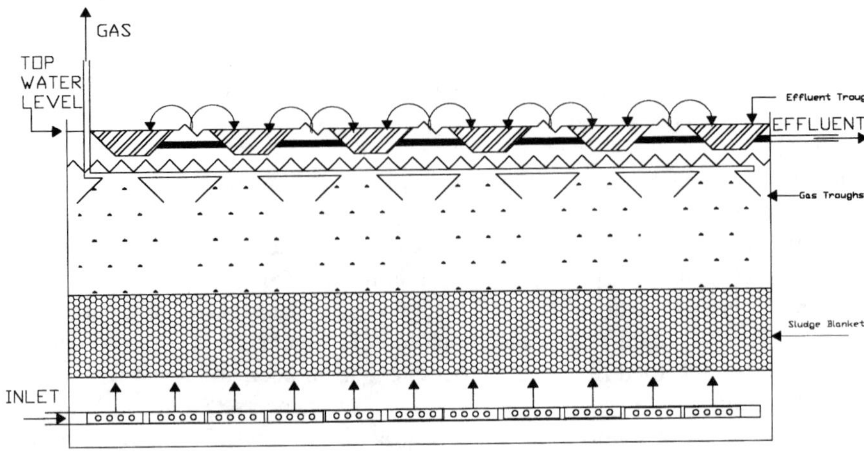

Figure 5–28 Typical upflow anaerobic sludge blanket baffle and feed arrangement.

production will not encourage enough mixing within the blanket to prevent short circuiting through the dense sludge granules. It has been shown (Bolle, van Bruegel, van Eyberger, Kossen, & Zoetmeyer, 1986) that at typical gas and liquid velocities (1 to 1.5m/h), a sludge blanket height of half the reactor height is optimal. Lettinga and Hulshoff-Pol (1991) have provided a design guide for the appropriate numbers of distributors according to the gas production and organic loading rate.

Various designs of baffles inside the upflow anaerobic sludge blanket reactor are used to promote gas-solid separation and therefore to retain the granules. The baffles provide extra surface area, which reduces upward flow velocity and promotes biomass flocculation. The baffles are usually made of reinforced plastic so as to be lightweight and corrosion resistant. Solids that collect in the cones should settle back into the reactor. Lettinga and Hulshoff-Pol (1991) have given a design guide based on waste characteristics. The gas-solids separator usually occupies 25% to 30% of the reactor height, the baffles are set with an angle of 45° to 60° and with an overlap of the plates of 200 mm to 300 mm. There may be special problems with wastewaters that produce foam. Many effluents that contain oil and fat, such as dairy wastes, will foam when vigorously aerated in activated sludge or in a gas-stirred reactor. Precautions against these foams blocking gas lines and outlet pipes are necessary. It is usually sufficient to install spray nozzles at the top of the reactor and above the baffles, but in very tall reactors it may also be necessary to add a chemical antifoam to avoid clogging the upflow anaerobic sludge blanket gas separator. In any event, periodic maintenance of both the top and bottom of the reactor to remove scale, scum, and grit will be necessary.

Applications. The upflow anaerobic sludge blanket reactor is the most widely used type of anaerobic reactor, and there is a great deal of design and operating experience. Usual design loadings of a upflow anaerobic sludge blanket reactor at mesophilic temperatures are as follows:

- load: 2 to 15 kg COD/m^3 per day
- retention time: 10 to 50 hours
- COD removal 70% to 90%

It has been widely applied to food processing wastes, and there are about 10 in the United Kingdom. The common problems have been the cost of alkalinity and losses of solids. Wastes with a high proportion of insoluble COD (>15%) can give rise to poor granulation. Lettinga and Hulshoff-Pol (1991) give further details on recommended design loadings. These vary according to the amount and solubility of the solids in the waste. High concentrations of suspended solids cause problems by reducing the activity of the retained sludge or lead to scum layers if the solids are fat. Prolonged and continuous entrapment of expanded flocculent sludge can lead to a spontaneous and sudden washout of the sludge bed. In order to avoid solids build-up and periodic loss of solids, washing of the sludge by increases in flow is recommended by the contractors (Lettinga & Hulshoff-Pol, 1991). If the waste does contain significant amounts of degradable suspended solids, a two-stage system with separate solids hydrolysis would normally be used in the design of an upflow anaerobic sludge blanket system.

5.9.1.5 The Expanded Bed Reactors

The difficulties of biomass separation in completely mixed reactors and the loss of granulation or blockage in plug flow reactors can be overcome by a combination of the two types. This is the basis of the fluidised bed. Fluidisation of carrier particles within a reactor occurs at a critical upflow velocity that depends on the density of the particles and other losses within the vessel. This fluidisation point can be defined by standard equations for pressure loss (Levenspiel, 1972). It is usually 10 to 30 m/hr. The performance of fluidised beds depends on the carrier materials and the hydrodynamics of the reactor.

Media Characteristics. Sand has frequently been used as media for fluidised beds, but other natural materials such as anthracite and gravels have also been used. To reduce fluidisation velocities, expanded lighter natural materials have been tried, for example, activated carbon, baked clays, and porous ceramics. Synthetic materials can also be fabricated into porous structures with a low density, and plastic carriers have included resin beads, reticulated foams, PVC, and other types of synthetic beads. Diffusion into porous particles may eventually become limiting, and the carrier may also break down with use. Much of the effectiveness of a fluidised bed can be attributed to the smaller carrier particles, which may be used without the risk of clogging. Start-up has been shown to be much quicker with smaller particles (Heijnen et al., 1985), which was attributed to the lower liquid shear velocities with smaller particles (0.35 compared to 0.7 mm). Particle sizes are usually between 0.3 mm and 1 mm; smaller sizes are difficult to control and retain within the reactor, larger sizes are difficult to fluidise. Surface roughness, specific surface area, strength, and stability will affect media performance as in anaerobic filters. The ideal physico-chemical characteristics of fluidised bed particles have been investigated (Iza, 1991; Switzenbaum, 1983). The particles must be uniform in size and shape to obtain an even fluidisation or bed expansion throughout the reactor height.

Hydrodynamics and Mixing. The support particles are expanded and mixed by gas and effluent recycle. Sufficient recycle is required to cause some lateral mixing of the particles and avoid channelling. Bed expansion is typically 20% to 25% of the reactor volume. Obtaining uniform bed expansion is relatively easy while the carrier particles are homogeneous but becomes more difficult as they change shape and density during colonisation.

Distribution. The design of the distribution system, like the other high-rate digesters, is a critical feature of the mixing characteristics. Many of the details of full-scale distributors are included in proprietary designs or covered by confidentiality agreements and not well understood. Small diameter reactors (1 to 2 m) can utilise an inverted cone with a downward pipe at the apex of the cone. Larger diameter reactors have used perforated plates, nozzles, and square annular or star-shaped pipework.

Reactor Aspect Ratio. To improve distribution and reduce the volume for recycle, fluidised beds are usually tall tanks with a 5:1 or 6:1 height-to-diameter ratio. A second, larger diameter zone is usually incorporated within the fluidised bed for clarification. This is an increase in the diameter of the top 20% to 25% of vessel (Figure 5–29). Foaming can be a problem with certain fat- and protein-containing wastes and if the fluidisation energy required is high. This may be counteracted by an antifoam agent and/or an upper grid to avoid particle loss. Designs in the United States usually use a separate carrier or particle separator.

Applications. Biological fluidised bed technology has been extensively studied at laboratory and pilot scale, but as yet there are few examples at full scale. The

Figure 5–29 An anaerobic fluidised bed reactor.

system has an important potential advantage in that very large concentrations of biomass can be achieved within the reactor without problems of blockage and channelling (40 to 100 kg VSS/m^3) (Heijnen et al., 1985). Potential performance is, therefore, also high. Typical operating conditions for a mesophilic fluidised bed are as follows:

- load: 2 to 50 kg COD/m^3 per day
- retention time: 0.5 to 24 hours
- COD removal: 70% to 80%

Fluidised and expended beds are very high-rate processes and are therefore difficult to scale up and operate (Sixt & Sahm, 1987). More experience at full scale is still required, but ultimately they should be a success for food industry wastewaters.

5.9.2 Materials of Construction

Most sewage works bioreactors and digesters are built from concrete. Capital costs were high, but the plant was expected to be operational for more than 50 years. This was appropriate at a time when standards and technology were changing slowly and inflation was low.

One of the effects of the high capital costs of concrete digesters was to restrict the application of sludge digestion to the largest sewage works. There are 75 sewage works in the United Kingdom that serve populations greater than 100,000 (equivalent to a 2,000 m^3 digester), and 95% of these use anaerobic treatment in concrete digesters. These large digesters are tanks of small height-to-diameter ratio and have flat or very slightly sloped bottoms. For large digesters (over 2,000 m^3 in size), concrete is still the cheapest method of construction, but with 80% of the UK population already served by sewage treatment, there will be few digesters of this size built in the United Kingdom in the future. A number of concrete digesters are being built in Europe. Continental digesters are much taller than most UK concrete digesters, and this tall, thin shape is intended to make the digester contents much easier to mix by vertical motion. A steep, conical bottom (45°) is also added to keep grit and scale in suspension for continuous removal. This type of tall digester needs to be well insulated because of the large surface area above ground, and it is also very expensive to build; two or three times more than flat-bottomed digesters. This has led to a change in profile to an egg shape, which can be constructed from successive ring segments to reduce the amount of shuttering and scaffolding. This is reported to reduce construction costs by 30% (Greeman, 1990). Two digesters at Dortmund, for example, are 12,000 m^3 in volume and 24 m in diameter at the widest point. They were designed to treat screened, consolidated, primary and secondary sewage sludges at a retention time of 30 days. The digesters are mechanically mixed, and sludge was drawn up a central draft tube and splashed over the surface of the digester liquid. The cost of construction was about £160/m^3; quoted UK costs for concrete digesters were £70/m^3. These figures were doubled by the costs of ancillary equipment.

Although some of the earlier digesters were made of concrete, it is generally too expensive and restricting for most industrial use.

5.9.2.1 Enamelled Steel

At the smaller sizes, prefabricated tanks made from bolted together enamelled-steel plates, are 70% cheaper than concrete tanks. The glass-enamel coating and holes for piping, as well as the holding bolts, can be made during fabrication of the plates. This reduces on-site work and associated delays. Such tanks have been used for many years as grain silos, storage tanks for farm slurries and other liquids, etc., so there is significant experience of their construction and life span. The tank is built on site from the shaped plates. The edges of the plates are sealed with mastic as they are bolted into the rings that form the tank. The bottom ring is grouted into a flat concrete base. The roof is of the same enamelled steel plates or plastic. Separate fixed or floating gas-holder roofs can be fabricated. The flat floors can accumulate grit, but manholes near the base will allow easy access and degritting. The tanks require insulation; this can be internal or, more generally, external. Polyurethane or polyisocyanurate foams in 50 mm to 100 mm layers can be used, or glass-wool (external only). All external insulations need a cladding or painting to waterproof the materials and to prevent pests from attacking the foam. The largest digester practical sizes are 1,000 m^3, the limit being set by the maximum size of the steel-plate roof, but larger sizes could be built by reinforcing the roof, which may be subject to significant flexing with the varying gas production. Multiple tank systems provide for greater capacity and have advantages over large single tanks in servicing and dealing with breakdowns. Increases in waste flow can be dealt with by adding extra tanks. Guaranteed plant life is 20 years, compared with 40 for concrete.

The prefabricated stirred-tank digester has been very successfully applied to sewage, industrial, and farm wastes.

5.9.2.2 Steel Tanks

Welded-steel tanks have been used for larger industrial use (greater than 2,000 m^3 in volume). They are about 10% cheaper than smaller enamelled tanks and 20% to 30% cheaper at the larger sizes, where multiple enamelled tanks would be needed. It is usual to coat the inside of mild-steel tanks with epoxy-rubber. Such coatings inside digesters can become detached or broken, and liquid will get behind them. There are doubts as to whether coating is needed. Corrosion of the inside surface is now not thought to be a problem; the inside surfaces are found to be covered with a layer of black sulphide. Mild-steel digesters need insulation.

Stainless-steel has been proposed for digesters, but its cost is too high for all except small or pilot-plant tanks. Aluminium and galvanising are unsuitable because of the high chloride concentration in the warm, digesting sludge. Jointing of pipes of different metals or alloys can lead to corrosion.

5.9.2.3 Plastics

Glass reinforced plastic (GRP) has been used for small digesters. There are difficulties with larger sizes because of its low strength, and the material is too flexible for large tanks without extra support. This can be overcome to a certain extent by excavation to sink the tank and backfilling, or the tank can be made into a horizontal tubular digester partly buried in the ground. GRP tanks are fabricated off-site and transported as a whole. This leads to an upper size limit for complete tanks based on ease of transport, but larger tanks can be made up from moulded rings bolted or glued together on-site. To provide sufficient self-supporting strength vertical digesters have to be tall and narrow. Digesters up to 500 cubic metres have been built in this way, but GRP digesters are no cheaper than enamelled-steel digesters at sizes above 200 m^3. The costs of moulds for plastic digesters also means that "one-off" tanks are hardly viable.

5.9.2.4 Gas-Holders

If the digester has a rigid roof, a separate gas-holder is needed. The most usual type is the floating dome in a tank of water. The gas holder is stabilised as it rises and falls by guide rails and rollers. The bottom of the gas tank is sealed in the water and the tank is stabilised by the guide rails. The gas tank may need counter-balancing by weights to produce the gas pressure required. Most gas appliances work better with a constant supply pressure. These gas-holders are usually made of mild steel, but some small ones in GRP are available. A separate water-sealed gas-holder or floating-roof digester have the advantage of acting as a safety valve if gas pressure is increased by failure in some other part of the system, as excess gas will bubble out through the liquid seal if the gas production exceeds storage capacity.

Plastic or rubberised-fabric digester tops are used as gas-holders mainly in continental Europe. Most plastic or rubber tops are used on small digesters. The top may be sealed by mechanical fixings to the edge of the treated effluent trough, or the bottom edge may drop below the sludge level to give a water seal. Butyl-rubber or plastic bags are also used as separate gas-holders. The bag or digester top is not usually inflated to stretch the fabric but to take up slack in a loose “bag”. A uniform pressure can be obtained by having weights directly on top of the bag. A number of designs were reviewed by Dohne (1980). Bags need separate safety valves to avoid overpressure, and care has to be taken to see that fabrics are not torn or otherwise damaged. In hot weather, dark-coloured gas bags can get very warm, and so gas pressure variations can be considerable.

All these systems store gas at low (50 to 150 mm or 2 to 6 inches WG) pressure, and the gas does not need to be specially cleaned before storage. The water in a water-sealed holder will scrub some impurities from the gas, so it is best changed each year. The capacity of the gas-holder depends very much on the use made of the digester gas. If the gas is running a boiler or an engine virtually non-stop and this uses the available gas, then a gas-holder capacity of only a few cubic metres will act as a gas ballast volume for a large digester. If the gas usage is intermittent—for example, just working day use—the gas produced at night might be stored for use during the day and the gas-holder will have to be sized on the nightly gas production. Gas storage is expensive and should be as small as possible.

The medium-pressure gas lines need mechanical safety valves; low-pressure lines can rely on blow-off from water-sealed gas holders or U-tube water seals on the digester or even from condensate pots in the gas lines. The U-tube water seals can freeze in very cold weather, so oil may be used in some cases, or a strong glycol solution. Gas pipes should be lagged to reduce the possibilities of water freezing out of the wet gas and blocking pipes or condensate traps. The bigger digesters usually have an automatic flare that burns off large excesses of unused gas and is started when the gas-holder is filled. Gas-holders should have pressure indicators to give some early warning of problems.

5.10 SLUDGE RESIDUES FROM TREATMENT

Effluent treatment processes generate solids; on average 70% to 80% of the original organic carbon is converted to dry solids. In a vegetable processing plant, for example, about 30% of the original COD might be removed by primary settlement to form a sludge, and about 50% to 60% of the remainder converted into dry solids by the activated sludge process. These sludges are putrescible, highly polluting, malodorous, and offensive. They can also be hazardous because they adsorb and accumulate pathogens and toxins. Raw sludges are saturated with bound water and not easily dealt with by simple chemical or physical processes. Disposal of these sludges may represent half the cost of the total effluent treatment.

Sludge production per volume of effluent treated is increasing with the popularity of intensive aerobic processing (see Section 5.8) and the greater use of additional chemicals to attain more stringent standards. Further treatment of coastal discharges will produce additional sludge. Traditionally, sludges generated from food effluents have been easy to recycle to land; they contain few materials that can be thought of as hazardous compared to sewage. It seems likely, however, that this task will become more difficult for the food industry as the water companies try to increase their use of these land outlets. In England and Wales, there are about 40M wet tonnes or 1M dry sewage solids to be disposed of each year: 16% of the sludge produced is tipped to landfill sites, 63% is used in agriculture, and 17% is incinerated (Figure 5–30).

Incineration is more common in Europe, but planning restrictions, the residual ash, and the efficiency of flue scrubbing are common difficulties (Hudson & Young 1991).

5.10.1 Sludge Thickening

Fresh or raw sludge is still largely water, some of which is biochemically bound to polymers or is intracellular, but at least half the water is held only by the sludge structure. In most cases, removal of this loosely held water can significantly improve the economics of sludge treatment and disposal. The most common method of achieving this is by gravity or consolidation in purpose-built tanks. Gravity thickening can double the solids content, and for most food wastes this is often sufficient treatment prior to land disposal or for mixing with other sludges. Sludge thickening is not a simple process but is caused by the sludge compressing under

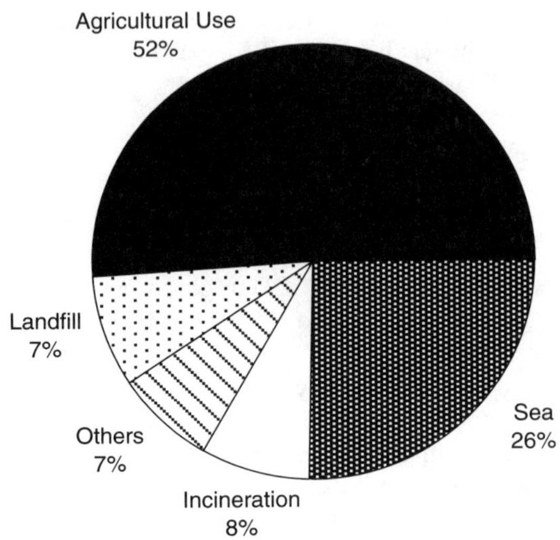

Figure 5–30 Sludge disposal routes, England and Wales.

its own weight. Water is thus squeezed out from between the solid particles. Sludges vary in their consolidation behaviour according to three characteristics:

1. The surface properties and charge, which will influence the degree of coagulation and interparticle attraction.
2. The origin of the solids, that is, whether the solids are surplus biomass from biological treatment or primary solids from settlement. Microbial biomass contains a higher proportion of bound water than primary solids because of intracellular water and polymer.
3. The age of the sludge can change these characteristics and lead to gas production as the sludge begins to break down anaerobically.

5.10.2 Design of Thickeners

In principle, because of the variability of sludges, the design of sludge thickeners can be improved by test work and modelling (Hoyland, Dee, & Day, 1989). In practice, for the average installation and for food processing wastes, prefabricated designs are available. Consolidation requires the interstitial held water to be displaced and for channels to be formed to allow the water to escape to the surface. Gentle mixing aids this process by promoting better flocculation and releasing gas and water. This type of mixing is carried out by a picket-fence-type agitator. Picket fences are a series of 50 mm to 60 mm diameter rods with about 150 mm to 200 mm spacings extending throughout the depth of the tank. It is important that the fence breaks the surface of the liquid to avoid the formation of a hard, dry crust. The picket fence rotates slowly, at 1 rpm to 2 rpm or 3 m to 4 m per minute (Figure 5–31).

A typical retention time for a gravity thickener is 2 days. Solids can be retained for longer without too much deterioration due to flotation, but there is little additional thickening. If the retention time is greater than 5 days, in warm conditions there are often significant problems due to flotation. The shape of the thickening tank is important because the taller the tank, the greater the solids thickness at the base; but, conversely, the rate of thickening is lower than in a shallow tank because the path length of the escaping water is greater. Typically, tanks are between 4m and 5m high and fitted with a number of draw-off points.

5.10.3 Sludge Treatment

Currently, the most common form of treatment is anaerobic digestion (Figure 5–32). Further sludge conditioning, either by heating or chemical conditioning, is

Figure 5–31 Standard prefabricated gravity sludge thickener with picket fence mixer.

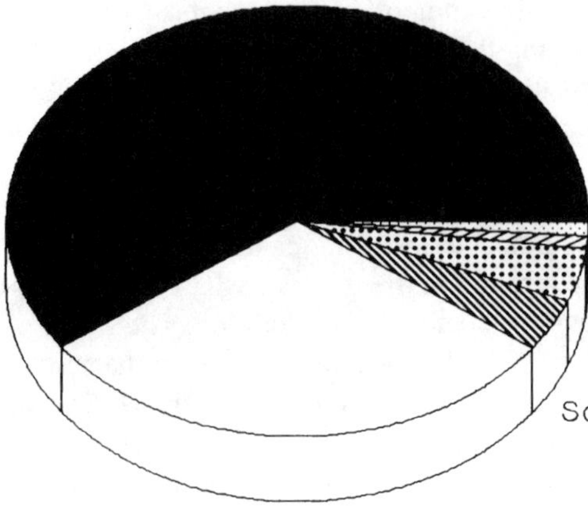

Figure 5–32 Treatment before disposal (%).

possible but rarely economic compared to anaerobic treatment and agricultural recycling. Polyelectrolytes, iron, aluminium salts, and lime are added to improve flocculation prior to mechanical dewatering. The most common dewatering processes are belt presses and centrifuges. Plate and frame presses and vacuum filters are less common. A further difficulty with these techniques, apart from the chemical and power costs, is control. The frequent changes in characteristics of biological sludges means that continual adjustments to the chemical dosing regime used are necessary. The most successful types of waste for further conditioning are those with a high fibre content (such as vegetable and paper wastes) when solids contents up to 50% are achievable; otherwise, 25% to 30% is more likely.

5.10.4 Sludge Disposal

5.10.4.1 Incineration

Incineration is not currently a common method of sludge disposal in either Europe or the United States (Table 5–32) because of its cost compared with the other routes. Traditionally, therefore, it has been used where there are particular difficulties with residual metals or organics from industrial effluents. Modern fluidised bed incinerators are cheaper to build and operate compared with the older multiple hearth type and would be more popular but for the constraint of public confidence. Concern over emissions means that planning applications will require a public enquiry. Incineration is one of the few disposal routes that is predictable and versatile. It is not affected by weather and not dependent on others (e.g., acceptance by farmers and retailers). It can be turned up or turned off; this flexibility makes it popular with water managers.

In a multiple-hearth incinerator, the sludge passes down through the furnace from top to bottom via a series of stages. The upper stages dry the sludge, which is then autothermic by the time it reaches the lower combustion zones. Volatile organics are released in the upper drying regions, and the air needs to be recirculated through the high-temperature combustion zone at 800° C to 900° C. Fluidised bed incinerators, recently introduced from Germany, are more efficient because of the much better mixing within the furnace. The sludge is fed directly into the hot combustion zone and needs to be pre-dried to at least 30% to 40% dry solids to be autothermic.

The main residues of burning sludge are carbon dioxide, nitrous oxides, sulphur dioxides, and trioxides, as well as the halogens and halides. All except the oxides

Table 5–32 Sludge Disposal Routes in Selected Countries

	000s (Tonnes Dry Solids per Year)				
	Total Sludge	*Agricultural Use*	*Landfill*	*Incineration*	*Sea*
United Kingdom	1,500	615	350	66	450
Belgium	29	8	15	6	0
Denmark	131	57	39	35	0
France	850	234	446	170	0
Germany	2,200	698	1,290	196	0
Greece	15	0	15	0	0
Ireland	23	7	4	0	12
Italy	800	270	440	90	0
Luxembourg	15	12	3	0	0
Netherlands	200	127	55	7	11
Spain	280	173	28	0	79
Sweden	180	108	72	0	0
Switzerland	250	113	80	57	0
United States	4,500	1,395	1,080	1,215	810
Canada	500	(No information available)			

Adapted from Matthews (1992) and Vincent and Critchley (1984).

of nitrogen can be removed by wet scrubbing with caustic soda. The nitrogen compounds are reduced by a catalytic after-burner.

One of the main concerns with incineration is the formation of dioxins, furans, and other residues from reactions at the lower temperatures (200° C to 400° C). These temperatures will occur in flue gas equipment, heat recovery, and so on, and the products of these reformation reactions may be more persistent and toxic than the original waste.

About 30% of the solids remain after incineration as ash, which is currently disposed of to landfill as a hazardous waste. In Japan, where there is little available land, there are a number of high-temperature vitrification incinerators. Additional fuel is required to achieve these temperatures (>1,000° C), and the final residue is a melted slag, which has then been used for making jewellery, paving slabs, and bricks (Tebbutt, 1992). Another novel approach is reductive hydrogenation of sludge at 400° C to 450° C, which gives a liquid oil or char (Campbell, 1989).

5.10.4.2 Agricultural Use

Agricultural use is the most popular sludge disposal route in the United Kingdom at present: About 65% of the sludge discharged to land is used as a soil conditioner. In the United States and other large European countries such as Germany, France, and Italy, landfilling is more popular. Treated sewage sludge provides both plant nutrients and a soil conditioner from the lignin or humus content. In the United Kingdom, there have been national guides (Institute of Water Pollution Control, 1978) for the use of sludge on land since 1975 (Exhibit 5–4). Many of these recommendations have been included in the more recent EC directive (see Section 5.2). Repeated experiments over the years have shown that 85%

Exhibit 5–4 Main Provisions of the 1989 Code of Practice for Agricultural Use of Sewage Sludge (DoE, 1989)

1. To encourage recycling of sludge and an understanding of the benefits
2. Indicate the treatments available; all sludge must be treated to avoid pathogens and odour unless sub-surface injected
3. Provide standard methods for the analysis of sludges and soils
4. A list of the most likely sources of problems, particularly metals and organics; to note the maximum application rates
5. Advice on the planting, grazing, and harvesting constraints for particular crops
6. To note the type of records that need to be kept for auditing purposes
7. To give guidance on odour, run-off, storage transport, and application

of the nitrogen and 50% of the phosphate can be taken up by growing plants from digested sludge (Table 5–33). A recent review of the benefits and hazards, using modern analytical methods, is given by Noguchi and Ho (1992).

Only small quantities of potassium and magnesium are present in sludge, and these may have to be supplemented. Sludge is also a very good source of inert organic matter. About 30% of digested sewage sludge is fibre, which provides structure and voidage to the soil for ventilation. Much of the nitrogen in digested sludge is in the form of ammonia, which, because of its solubility, can be lost during sludge dewatering. Thus, it is the water content of the sludge that governs its rate of application. Leaching losses in wet weather can be 75% of the nitrogen content. This run-off, or drainage through cracks in the soil in summer, can be a source of significant pollution.

Application rates, therefore, are varied according to conditions. Sub-surface injection is the best method of application; it relies on a special small chisel-shaped plough that lifts the soil and injects the sludge via a feeder tube 25 mm to 50 mm below the surface. A light compactor presses the lifted soil back into position after the plough has passed. Application rates of 1.5 to 1.8 wet tonnes per hectare are possible with sub-surface injection and good ground conditions. Sub-surface injection avoids any problems because of odour, aerosols, or pathogens on the soil surface. If the sludge is odour-free—and most food waste sludges are pathogen-free—then the sludge may also be distributed by standard manure tankers. Small standard manure tankers distribute the sludge via a simple fishtail nozzle or spinning disc. Sludge is rarely odour-free, and rain or manure guns are not acceptable except in very remote areas. There are also problems with aerosol formation and spray drift onto neighbouring areas. The advantage of high-pressure spray application is that it avoids all heavy transport on fields and therefore there is no risk of soil compaction.

A disadvantage of sewage sludge is that it may contain appreciable concentrations of potentially toxic materials. General pasteurisation at 70° C has recently been suggested (Pierrepont, 1998) to provide a further barrier against infection. Currently, a minimum of digestion is required for crops used for animal feeds or

Table 5–33 Nutrient Content of Various Types of Sludge

Type of Sludge	*Nitrogen (% as N)*	*Phosphate (% as P_2O_5)*	*Potassium (% as K_2O)*
Digested liquid sludge	0.2	0.2	0.010
Raw sludge (4% dry solids)	0.15	0.15	0.012
Sludge cake (20%–40% dry solids)	1–1.4	1–2.0	0.09

cereals. The best documented toxins in sludge are the heavy metals (Table 5–34). Metals added to soil are firmly bound by clay and organic matter and may remain in the soil indefinitely. The metals can be solubilized by acid conditions, and the pH of the soil needs to be 7 or above. The concentration of toxic metals in sludges has decreased markedly over the past 20 years, both as a result of more stringent conditions imposed by water companies and also because of the decline in heavy industry. Cadmium, lead, mercury, and chromium are highly toxic and need careful monitoring in soils and sludges. Cadmium accumulates in the leaves of plants, and additions have to be restricted to 5 kg/ha/a. Lead and chromium are less toxic, as they are not taken into the edible parts of the plant (Noguchi & Ho, 1992). It is not these metals, however, that are likely to limit agricultural use, because they are now present at low concentration and strictly controlled. Zinc and copper present a bigger risk. Sewage sludge contains high concentrations of zinc (from galvanising and cosmetics) and copper (from water pipes), and lesser quantities of chromium and nickel from stainless steel cooking utensils. There may also be some lead from car exhausts.

Fortunately, zinc is not highly toxic, and if the soil pH is above 6.5, 20 kg/ha may be added annually. If the application is to grassland where it is not intended to grow arable crops for many years, or if the soil is calcareous with a pH above 7.0, then 37 kg/ha/a may be added. Other metals that affect plant growth are copper (which is twice as toxic as zinc), nickel (which is eight times as toxic as zinc), and boron (which is 150 times as toxic).

Little is known about persistent trace organics in sludge because they are so variable. They depend on the nature of any industrial wastewater and, if they are present, steps should be taken to prevent them from entering the system.

In the long term, with changes in practice or with very clean sludges, the background metals and organics in soils are unlikely to be seriously affected by sludge application. The major limitation on the agricultural use of sludge is likely to be from excessive applications of nitrogen adding to the nitrate burden of ground and surface waters. Reuse of sewage and other sludges is the best environmental option for these materials, but research will need to continue to maintain the competitive advantage of sludge compared to artificial soil conditioners.

Table 5–34 Metals in Domestic Sewage Sludge (mg/kg dry solids)

Type of Sludge	*Dry Solids (%)*	*Cd*	*Cr*	*Cu*	*Fe*	*Pb*	*Ni*	*Zn*
Raw sludge Ex. 1	1.9	20	90	315	13,814	300	15	871
Raw sludge Ex. 2	4.9	20	42	169	6,651	206	15	568
Digested sludge	2.7	20	403	605	19,355	454	81	1,411

5.10.4.3 Non-Agricultural Use

Land Restoration. There is scope for additional non-agricultural uses of sludges. The nutrients and organic matter in sludge have been shown to help the restoration of derelict and contaminated land (Hall, 1993). The biological breakdown of recalcitrant and toxic residues in old industrial and mining sites may be accelerated by co-metabolism and enzyme induction. Applications of sludge bring about higher rates of general metabolic activity.

Forestry. Larger applications of sludge to forestry growth are possible than for simple agricultural use. Forests are usually grown on poor soil; in the United Kingdom, for example, the Forestry Commission uses 1.8 tonnes of fertiliser per hectare (Hall, 1993). Surveys have also shown that up to 20% of the Forestry Commission plantations may be within economical transport distances. Using sewage sludge for forestry is very popular in the United States, where transport distances greater than 100km have been shown to be worthwhile after special predrying (Hall, 1993).

Landfill. This is a popular disposal route in the other large industrial countries of Europe (about 40% of the total), but in the United Kingdom it is used as a reserve for difficulties with agricultural use; for example, because of weather or unforeseen contamination. The future of landfilling is uncertain because of impending changes in the legislation (EC proposed directive 91/C190/01). A disadvantage of landfilling of sludge is the high water content and consequent production of leachate, although the water, nutrients, and inoculum of active bacteria increase the rate of degradation of domestic refuse. A number of the large water companies have invested in landfill sites to protect their continued use of this route as a standby facility (Hudson & Young, 1991).

Compost. At present there are only pilot schemes for composting sludges in the United Kingdom and Europe. In the United States it is a common process. There is renewed interest in the United Kingdom because of the public recognition of limited supplies of horticultural peat and the environmental damage caused by its extraction (640,000 tonnes per annum).

Composting is an aerobic process, and it relies on the structural support of the solids to maintain openness and ventilation during the process. Straw, wood chips, shavings, bark, and nut husks are commonly used to provide this bulking material. Straw, for example, will absorb three times its own weight of water and can be co-composted with liquid wastes such as sludges and farm slurry. The benefits of composting are similar to those of anaerobic digestion; these are the controlled decomposition and stabilisation of organic waste, odour control, and the removal of pathogens. Composting is a successful process if the transport and operating costs can be overcome. In the United Kingdom, the most common use of composting is for mushrooming. Various demonstration schemes have been conducted on the large-scale composting of animal and domestic wastes with straw, but there are insufficient financial incentives to make it competitive with peat and other soil conditioners (also a better growing medium).

The best use of compost is as a soil conditioner. The humus and fibre content of compost make it suitable as a supplement to clay and sandy soils. Compost is not a good fertiliser; combined N:P:K are less than 3%, and it is too bulky to transport. It has so far not been used for agriculture in Europe, but in Japan and the United States there are special projects to promote its use to avoid dependence on chemical fertilisers (United States Environmental Protection Agency, 1991). Premium uses, such as in landscaping, horticulture, and land reclamation, are the most economic uses of compost. Horticultural use will be impossible for sewage sludge or domestic refuse because of the risks, although slight, of contamination with metals and organics. Food industry wastes are contamination-free and could be used for horticulture; this could be a valuable outlet in the future.

Composting is a batch process, and there is a succession from mesophilic to thermophilic ecology. The heat generated during the rapid growth of the bacteria and actinomycetes breaking down the smaller organic molecules raises the temperature to 40° C to 50° C when a thermophilic population becomes established. The thermophilic fungi are efficient at breaking down cellulose. Temperatures frequently reach 70° C when the activity of most organisms becomes inhibited. These high temperatures are beneficial for the composting of animal and domestic wastes, since it assists the removal of pathogens.

Another consequence of the heating is that the wastes dry as composting takes place. This may limit activity once the moisture content falls below 25%. The moisture content and organic content of the starting materials govern the drying process.

Different techniques of composting have developed for different types of waste. The simplest technique is the open, static pile. In this process, solid waste is collected in a pyramid about 2m to 3m at the base and 1m high. The pile is periodically aerated by a perforated pipe at the base of the heap. Static pile composting can be carried out in the open, but it is more controllable if it is roofed or enclosed. An alternative method is to arrange the waste in long rows and aerate the stack by periodic mixing using special turning machines. This is known as the windrow process. There may be problems with dust and odour during turning. The aerial spores of the fungi growing in the compost, particularly *Aspergillus,* can be a potential hazard to the operators. Bioreactor composting systems have been used for domestic refuse in an attempt to reduce the land required for pile and row composting. Bioreactors are easier to control but expensive, and they cannot be served from the value of the product. The are two basic types: an inclined rotating tunnel or a standard stirred tank (Anderson & Smith, 1987). Residence time in the reactor is 5 to 10 days, but this has to be followed by a maturation period of several weeks outside the reactor.

5.11 CONCLUSION

Greater environmental awareness is a major issue facing the food industry. Many people perceive industry as harmful and polluting (Liardet, 1991). Surveys also show that about 30% of the general public think that the environment and pollution are one of the most important problems that government has to deal with (Department of the Environment, 1992). The legislative response has been three key principles: *integrated pollution control,* which is designed to lead to a fundamental rethinking of production processes and encourage clean technology; *the duty of care,* which is aimed at ensuring that waste is disposed of in the best environmental way; and *waste directives,* which impose limits on the amounts of a range of materials released into the environment. Encouragement is provided by accreditation via ISO 14000.

Overall, there are opportunities for reducing wastage and the costs of effluent disposal by a better understanding of the waste-generating processes. Simple pretreatments such as screening, balancing, settlement, and filtration will in most food industry cases prove cost-effective compared with the discharge of untreated material. Most complete effluent treatment systems are biological and require skilled operation.

Waste treatment is the largest practical use of micro-organisms. The diversity, volume, and complexity of wastes mean that there are few alternatives to this type of treatment. The size of the application, coupled with popular demand for a better environment, has continuously stimulated the development of more efficient, reliable processes.

A disadvantage of increasing microbial efficiency for waste treatment is biomass yield. This can result in the conversion of one problem (polluted water) into another (surplus sludge). A second problem is that more intense processes inevitably require better control, and reliable sensors for wastewater are not often available. New sensors and control models are an important area of future research.

Traditional processes (although they need space) have a substantial case history, and most of the problems have previously been encountered. Additional technologies, such as membranes, and physico-chemical oxidation, such as photolysis and electro-magnetic radiation, will in future aid existing technologies to meet the new standards more consistently.

REFERENCES

Adams, C.D., & McKinney, R.E. (1989). Anaerobic trickling filters: A new treatment potential. In *Proceedings of the 44th Purdue Industrial Waste Conference* (pp. 1–12). Michigan: Ann Arbor Science.

Alibhai, K.R.K., and Forster, C.F. (1986). An examination of the granulation process in UASB reactors. *Environ. Technol. Lett., 7*, 193–200.

American Society of Civil Engineers (1984). *A Standard Measurement of Oxygen Transfer in Clean Water.* New York: ASCE.

Anderson, J.G., & Smith, J.E. (1987). Composting. In J.M. Sidwick & R.S. Holton (Eds.), *Biotechnology of Waste Treatment and Exploitation* (pp. 301–325). Chichester: Ellis Horwood.

Baumann, P.G., & Huibregtse, G.L. (1981). Evaluation and comparison of digester mixing systems. *J Water Pollution Control Fed., 54*, 1194–1203.

Bishop, P.L., & Kinner, N.E. (1986). Anaerobic fixed-film processes. In H.J. Rehm & G. Reed (Eds.), *Biotechnology: A Comprehensive Treatise* (pp. 113–176). Weinheim: VCH.

Bolle, W.L., van Breugel, J., van Eyberger, G.C., Kossen, N.W.F., & Zoetmeyer, R.J. (1986). Modelling the liquid flow in upflow anaerobic sludge blanket reactors. *Biotechnol. Bioeng., 28*, 1615–1620.

Bonastre, N., & Paris, J.M. (1989). Survey of laboratory pilot and industrial anaerobic filter installations. *Proc. Biochem., 241*, 15–20.

Brade, C.E., & Noone, G.P. (1981). Anaerobic sludge digestion—need it be expensive? I. Making more of existing resources. *Water Pollution Control, 80*, 70–90.

Brooking, J., Buckingham, C., & Fuggle, R. (1990). Constraints on effluent plant design for a brewery effluent treatment and waste disposal. *Int. Chem. Eng. Symp., 116*, 109–126.

Brown, M.J., & Lester, J.N. (1979). Metal removal in activated sludge: The role of bacterial extra cellular polymer. *Water Res., 13*, 817–837.

Butcher, G.J. (1988). Experiences with anaerobic digestion of wheat starch processing waste. *Int. Biodeterioration, 25*, 71–77.

Butler, G.A. (1984). Anaerobic digestion of starch process effluents at Tenstar Products. In J. Coombs (Ed.), *Anaerobic Treatment of Industrial Waste.* Reading, UK: BABA.

Bruntland, G.H. (1987). *Our Common Future.* World Commission on Environment and Development. Oxford: Oxford University Press.

Camilleri, C. (1987). Operating results from a fixed film anaerobic digester for pollution abatement and methane production from industrial wastes. In G. Grassi, B. Delmon, J.F. Molle, & H. Zibetta (Eds.), *Biomass for Energy and Industry, 4th EC Conference* (pp. 1338–1342). London: Elsevier Applied Science.

Camilleri, C. (1988). Anaerobic digestion of food processing wastewater: Industrial performance of fixed film technologies for methane recovery and pollution abatement. In A. Tilche & A. Rozzi (Eds.), *5th Symposium on Anaerobic Digestion, Bologna* (pp. 473–476). Monduzzi, Bologna.

Campbell, H.W. (1989). A status report on Canada's oil from sludge technology. In A.H. Dirkzwager & P. L'Hermite, *Sewage Treatment and Use* (pp. 281–290). London: Elsevier.

Chambers, B., & Tomlinson, E.J. (Eds.). (1982). *Bulking of Activated Sludge: Preventive and Remedial Methods.* Chichester: Ellis Horwood for the Water Research Centre.

Chartered Institution of Water and Environmental Management (CIWEM). (1997). *Handbooks of UK Wastewater Practice Activated Sludge Treatment.* London: CIWEM, 202 pp.

Cooper, P.F., Drew, E.A., Bailey, D.A., & Thomas, E.V. (1977). Recent advances in sewage effluent denitrification. *Water Pollution Control, 76*, 287–300.

Cooper, P.F., & Downing, A.L. (1998). Milestones in the development of the activated sludge processes over the past eighty years. *J. Inst. Wat. Environmental Management, 12*(5), 303–313.

Corbitt, R.A. (1989). *Standard Handbook of Environmental Engineering.* New York: McGraw-Hill.

Dahab, M.F. & Young, J.C. (1982). Retention and distribution of biological solids in fixed-bed anaerobic filters. *Proceedings of the 1st International Conference on Fixed Film Biological Processes* (pp. 1337–1351). Kings Island, Ohio.

Department of the Environment. (1979). *Methods for the Examination of Waters and Associated Materials Analysis.* London: HMSO.

Department of the Environment. (1982). The bacteriological examination of drinking water supplies. *Reports on Public Health and Medical Subjects No. 71.* London: HMSO.

Department of the Environment (1989). *Code of Practice for the Agricultural Use of Sewage Sludge.* London: HMSO, 20 pp.

Department of the Environment. (1992). *The UK Environment Statistics.* London: HMSO Government Statistical Service.

Department of Scientific and Industrial Research. (1960). *Water Pollution Research: The Report of the Water Pollution Research Laboratory, 1955–1959.* London: HMSO.

Dohne, E. (1980). Bio gas storage and utilization. In D.A. Stafford, B.I. Wheatley, & D.A. Hughes (Eds.), *Anaerobic digestion, 1st International Symposium* (pp. 429–445). London: Elsevier Applied Science.

Dolfing, J. (1986). Granulation in UASB reactors. *Water Sci. Technol., 18*, 15–25.

Eaton, A.D., Clesceri, L.S., & Greenberg, A.E. (Eds.). (1995). *Standard Methods for the Examination of Water and Wastewater*, 19th Edition. American Public Health Association, American Water Works Association and the Water and Environment Federation.

Ehlinger, F., Audic, J.M., Vernier, D., & Fraup, G.M. (1987). The influence of the carbon source on microbiological clogging in an anaerobic filter. *Water Sci. Technol., 19*, 261–273.

Eikelboom, D.H. (1975). Filamentous organisms observed in activated sludge. *Water Res., 9*, 365–388.

Environment Canada. (1986). *Anaerobic Treatment of Dairy Effluent* (Report EPS 3/FP/1). Ottawa: Environment Canada.

Foot, R.J. (1992). The effects of process control parameters on the composition and stability of activated sludge. *J. Inst. Water Environ. Mgmt., 2*, 215–228.

Gloyna, E.F., Comstock, R.F., & Renn, C.E. (1952). Rotary tubes as experimental trickling filters. *Sewage Ind. Waste, 24*, 1355–1357.

Gorsuch, T.T. (1986). *Food Processing Research: A Report to the Priorities Board Ministry of Agriculture Fisheries and Food.* London: HMSO, 38 pp.

Greeman, A. (1990). Good eggs: a new concrete structural design for wastewater treatment. *New Civil Engineer, 10*, 14–17.

Hall, E.R. (1983). Biomass retention and mixing characteristics in fixed film and suspended growth anaerobic reactors. *Water Sci. Technol., 15*, 371–396.

Hall, J.E. (1993). Recent developments in sewage sludge disposal and use. *Chem. Ind., 6*, 188–191.

Hawkes, H.A. (1963). *The Ecology of Waste Water Treatment* (1st ed.). Oxford: Pergamon Press.

Hawkes, H.A., & Shephard, M.R.N. (1972). The effect of dosing frequency on the seasonal fluctuations and vertical distribution of solids and grazing fauna in sewage percolating filters. *Water Res., 6*, 721–730.

Heijnen, J.J., Enger, W.A., Mulder, A., Lourens, P.A., Keijzers, A.A., & Hoeks, F.W.J. (1985). Awendung der anaeroben wirbel schichttechnik in der biologischen abwasseinigung. *Wasser/ abwasser, 126*, 115–130.

Hellawell, J.M. (1978). *Biological Surveillance of Rivers: A Biological Monitoring Handbook.* Water Research Center Medmerham, 331 pp.

Henze, M., & Harremoes, P. (1983). Anaerobic treatment of waste water in fixed film reactors: A review. *Water Sci. Technol., 15*, 1–90.

Hobson, P.N., & Wheatley, A.D. (1993). *Anaerobic Digestion: Modern Theory and Practice.* London: Elsevier, 288 pp.

Horan, N. (1990). *Biological Wastewater Treatment Systems and Operation.* Chichester: Wiley, 310 pp.

Horan, N. Bu'ali, A.M., & Eccles, C.R. (1988). Isolation, identification and characterization of filamentous and floc forming organisms from activating sludge. *Environ. Technol., 9*, 449–457.

Housley, J.N., & Zoutberg, G.R. (1994). Application of the biothane wastewater treatment process in the soft drinks industry. *J. Inst. Wat. Environmental Management, 8*(3), 239–245.

Hoyland, G., Dee, A., & Day, J.J. (1989). Optimum design of sewage sludge consolidation tanks. *J. Inst. Wat. Environ. Mgmt., 3*(5), 505–5161.

Huber, L., & Metzner, G. (1986). In H.-J. Rehm & G. Read (Eds.), *Examples of Industrial Waste Water Treatment in Biotechnology: A Comprehensive Treatise* (pp. 269–305). Weinheim: VCH.

Hudson, J., & Young, G. (1991). Strategic role of anaerobic digestion in sludge management in Yorkshire. In G.E. Richards & P.B. Mistry (Eds.), *Anaerobic Digestion Seminar.* Oxford: Harwell Press.

Institute of Water Pollution Control. (1978). Sewage sludge 3. Utilisation and disposal. *Manuals of British Practice.* Maidstone, Kent: IWPC.

Institute of Water Pollution Control. (1979). Sewage sludge 1. Production, preliminary treatment and digestion. *Manuals of British Practice.* Maidstone, Kent: IWPC.

Iza, J. (1991). Fluidised bed reactors for anaerobic treatment. *Water Sci. Technol., 24*, 109–132.

Jones, G.A., & Franklin, B.C. (1985). The prevention of filamentous bulking of activated sludge by operation means at Halifax Sewage Treatment Works. *Water Pollution Control, 84*, 329–344.

Kennedy, K.J., & Droste, R.L. (1986). Anaerobic fixed film reactors treating carbohydrate wastewater. *Water Res., 20*, 685–695.

Kennedy, K.J., & Droste, R.L. (1991). Anaerobic wastewater treatment in down flow stationary fixed film reactors. *Water Sci. Technol., 24*, 157–177.

Kissel, J.C., Grotenhuis, J.T.C., & Zehnder, A.J.B. (1988). Computer simulation of granule growth in an upflow anaerobic sludge blanket reactor (UASBR). In A. Tilche & A. Rozzi (Eds.), *5th International Symposium on Anaerobic Digestion* (pp. 207–210). Bologna: Monduzzi.

La Grega, M.D., Buckingham, P.L., & Evans, J.C. (1994). *Hazardous Waste Management.* New York: McGraw Hill.

Lettinga, G., & Hulshoff-Pol, L.W. (1986). Advanced reactor design, operation and economy. *Water Sci. Technol., 18*, 12, 41–53.

Lettinga, G., & Hulshoff-Pol, L.W. (1991). UASB—Process design for various types of wastewaters. *Water Sci. Technol., 24*, 87–107.

Lettinga, G., Velsen, A.F.M., van Zeeuw, W.J., & Hobma, S.W. (1979). The application of anaerobic digestion to industrial pollution treatment. In D.A. Stafford, B.I. Wheatley, & D.A. Hughes (Eds.), *Anaerobic Digestion 1st International Symposium.* London: Elsevier Applied Science.

Levenspiel, O. (1972). *Chemical Reaction Engineering* (pp. 253–325). Chichester: Wiley.

Liardet, G. (1991). Public opinion and the chemical industry. *Chem. Ind., 4*, 118–123.

Lilly, W., Brown, H., Crabtree, H., Upton, J., & Thomas, V. (1991). The production of high quality effluents in sewage treatment by using the Biocarbone process. *J. Inst. Water Environ. Mgmt., 5*(2), 123–133.

Matthews, P.J. (1992). Sewage sludge disposal in the UK: A new challenge for the next twenty years. *J. Inst. Water Environ. Mgmt., 6*, 551–559.

Micklewright, A.T. (1986). A review of the practice of trade effluent control and charging in the North West. *Water Pollution Control, 85*, 324–336.

Ministry of Housing and Local Government. (1954). *Report of an Informal Working Party on the Treatment and Disposal of Sewage Sludge.* London: HMSO.

Morgan, H. (1980). The development of an anaerobic process for the treatment of wheat starch factory effluent. In A. Herzka & R.G. Booth (Eds.), *Food Industry Wastes: Disposal and Recovery.* Barking: Elsevier, 246 pp.

Murray, W.D., & van den Berg, L. (1981). Effects of nickel cobalt and molybdenum on the performance of methanogenic fixed film reactors. *Appl. Environ. Microbiol., 42*, 502–505.

Nahle, C. (1991). The contact process for the anaerobic treatment of wastewater technology, design and experiences. *Water Sci. Technol., 24*, 179–191.

Noguchi, H., & Ho, H. (1992). Long term experiments of applying sewage sludge fertilizer to agricultural land. *J. Inst. Wat. & Env. Manag., 6*(5), 576–582.

Noone, G.P., & Brade, C.E. (1985). Anaerobic sludge digestion—Need it be expensive? III. Integrated and low cost digestion. *J. Inst. Water Pollution Control, 84*, 309–328.

Office of Water Services. 1996. *1996–97 Report on Tariff Structures and Charges of the Water Companies.* Birmingham, UK: Office of Water Services.

Oleszkiewz, J.A., Hall, E.R., & Oziemblo, J.Z. (1986). Performance of laboratory scale anaerobic hybrid reactors with varying depths of media. *Environ. Technol. Lett., 7*, 445–452.

Oleszkiewz, J.A., & Thadari, V.J. (1988). Effects of biofilter media on performance of anaerobic hybrid reactors. *Environ. Technol. Lett., 9*, 89–100.

Oliva, E., Jacquart, J.C., & Prevot, C. (1990). Treatment of brewery wastewater by methanization in fluidized bed reactors. *Water Sci. Technol., 22*, 483–490.

O'Shaughnessy, F.R. (1931). Some considerations in the oxidation of sewage. *Proceedings of the Association of Managers of Sewage Disposal Works* (pp. 74–92).

Pauss, A., Depaepe, D., & Nyns, E.J. (1990). Review of AD plants in Europe. In G.E. Richards (Ed.), *Euroenviron Anaerobic Digestion Workshops.* Oxford: Harwell.

Pierrepont, P. (1998, December). The ADAS Matrix: the food and water industry in agreement. *Wastes Management*, 28–29.

Pike, E.B. (1975). The aerobic bacteria. In C.R. Curds & H.A. Hawkes (Eds.), *Ecological Aspects of Used-Water Treatment*, Vol. 1 (pp. 1–63). London: Academic Press.

Pike, E.B., Carlton-Smith, C.H., Evans, R.H., & Harrington, D.W. (1982). Performance of rotating biological contractors under field conditions. *Water Pollution Control, 81*, 10–27.

Pike, E.B., & Carrington, E.G. (1972). Recent developments in the study of bacteria in the activated sludge process. *Water Pollution Control, 71*, 583–605.

Pike, E.B., & Carrington, E.G. (1979). The fate of enteric bacteria and pathogens during sewage treatment. In A. James & L.M. Evison (Eds.), *Biological Indicators of Water Quality, a Symposium* (pp. 20-1–20-32). Chichester: Wiley.

Postgate, J. (1989, January). Microbial happy families. *New Sci.,* 40–44.

Racault, Y. (1990). Treatment of distillery wastewater using an anaerobic downflow stationary fixed film reactor: Performance of a large scale plant in operation for four years. *Wat. Sci. Technol., 22* (1/2), 361–372.

Royal Commission on Sewage Disposal. (1908). 5th Report. London: HMSO.

Rundle, H., & Whyley, J. (1981). Comparison of gas recirculation of systems for mixing in anaerobic digestion. *Water Pollution Control, 80*, 463–480.

Samson, R., van den Berg, L., & Kennedy, K.J. (1985). Mixing characteristics and start up of anaerobic stationary fixed film reactor. *J. Biotechnol., 2*, 95–106.

Shore, M., Broughton, N.W., & Bumstead, N. (1984). Anaerobic treatment of waste waters in the sugar beet industry. *Water Pollution Control, 83*, 499–506.

Sixt, H., & Sahm, H. (1987). Biomethanation. In J.M. Sidwick & R.S. Holdom (Eds.), *The Biotechnology of Waste Treatment and Exploitation* (pp. 149–172). Chichester: Ellis Horwood.

Smith, L.J. (1986). The treatment of wastewaters from malting, brewing and distilling. *Water Sci. Technol., 18*, 127–135.

Smith, M.O., Ferrall, J., Smith, A.J.T., Winstanley, C.I., & Wheatley, A.D. (1988). The economics of effluent treatment: A case study at Bovril. *Int. Biodeterioration, 25*, 97–105.

Smith, N. (1984). Anaerobic effluent treatment in British Sugar. In J. Coombs (Eds.), *Anaerobic Treatment of Industrial Waste.* Reading, UK: BABA.

Souza, M.E. (1986). Criteria for the utilization design and operation of UASB reactors. *Water Sci. Technol., 18*, 55–69.

Stephenson, T., Mann, A., & Upton, J. (1993). The small footprint wastewater treatment process. *Chemistry and Industry, 14*, 533–536.

Stronach, S.M., Rudd, T., & Lester, J.N. (1986). *Anaerobic Digestion Process in Industrial Waste Treatment.* Berlin: Springer-Verlag, 184 pp.

Swanwick, J.D., Shurben, D.G., & Jackson, S. (1969). A survey of the performance of sewage sludge digestion in Great Britain. *Water Pollution Control, 68*, 639–661.

Switzenbaum, M. (1983). A comparison of the anaerobic filter and anaerobic expanded/fluidised bed process. *Water Sci. Technol., 15*, 399–413.

Szendry, L.M. (1983). Scale up and operation of the Bacardi Corporation anaerobic filter. In *Proceedings of the 3rd International Symposium on Anaerobic Digestion* (pp. 365–377). Boston.

Tebbutt, T.H.Y. (1992). Japanese sludge treatment utilization and disposal. *J. Inst. Water Environ. Mgmt., 6*, 628–632.

Tilche, A., & Vierra, S.M.M. (1991). Reactor design of anaerobic filters and sludge bed reactors. *Water Sci. Technol., 24*, 193–206.

Upton, J.E., & Churchley, J.H. (1993, February). Process options for biological nutrient removal in activated sludge plants. Paper presented to Institution of Water and Environment Management at Leicester.

United States Environmental Protection Agency. (1991). *Technical Bulletin 430 981-011:* Composting processes to stabilize and disinfect municipal slewage sludge. Washington, D.C.: USEPA, Office of Water Programs.

Vincent, A.J., & Critchley, R.F. (1984). A review of sewage sludge treatment and disposal in Europe. In A.M. Bruce (Ed.), *Sewage Sludge Stabilisation and Disinfection* (pp. 550–571). Chichester: Ellis Horwood.

von Gunten, H.R., & Lienert, C. (1993). Water and ground water quality in Switzerland. *Nature, 364*, 220.

Water Data Unit. (1979). *Water Data 1978*. Reading: Department of the Environment, Water Data Unit.

Weiland, P., & Rozzi, A. (1991). The start up, operation and monitoring of high rate anaerobic treatment systems. *Water Sci. Technol., 24*, 257–277.

Wheatley, A.D. (1984). Biotechnology and effluent treatment. In G.E. Russell (Ed.), *Biotechnology and Genetic Engineering Reviews*, vol. 1. Newcastle-upon-Tyne: Intercept Press.

Wheatley, A.D. (1985). Waste water treatment and by-product recovery. In J.M. Sidwick (Ed.), *Topics in Waste Water Treatment* (pp. 68–105). Oxford: Blackwell.

Wheatley, A.D. (1990). The treatment of industrial effluent. *Critical Reports in Applied Chemistry 31*, 172–220.

Wheatley, A.D., & Cassell, L. (1985). Effluent treatment by anaerobic biofiltration. *Water Pollution Control, 84*, 10–20.

Wheatley, A.D., & Coombs, J. (1991). *Anaerobic Digestion: A Technical and Marketing Report.* Newbury: CPL Scientific, 330 pp.

Wheatley, A.D., Johnson, K.A., & Winstanley, C.I. (1988). The reliability of anaerobic digestion for the treatment of food processing effluents. In *Advances in Water Pollution Control Anaerobic Digestion* (pp. 135–146). Oxford: Pergamon.

Wheatley, A.D., Fisher, M.B., & Grobicki, A.M.W. (1997). Applications of anaerobic digestion for the treatment of industrial wastewaters in Europe. *J Inst. Wat. Environmental Management, 11*(1), 39–46.

Wilson, F., Hamoda, M.F., Islam, H., & Buranasin, P. (1988). The treatment of high strength vegetable picking waste using the RBC process. *Env. Tech. Lett., 9*(11), 1201–1212.

Wishart, J.M., & Wilkinson, R. (1941). Purification of settled sewage in percolating filters in series with periodic change in the order of the filter. *J. Proc. Inst. Sewage Purification, 1*, 15–38.

Witt, E.R., Humphreys, W.J., & Roberts, T.E. (1979). Full scale anaerobic filter treats high strength wastes. *Proceedings 34th Purdue Industrial Waste Treatment Conf.* (pp. 229–234).

World Health Organisation. (1973). Reuse of effluents: methods of wastewater treatment and health safeguards. *Technical Report Number 517*. Geneva: WHO.

Young, H.C., & Young, J.C. (1988). Hydraulic characteristics of upflow anaerobic filters. *J Environ. Eng., 114*, 621–638.

Young, J.C. (1991). Factors affecting the design and performance of upflow anaerobic filters. *Water Sci. Technol., 24*, 133–155.

Young, J.C., & Dahab, M.F. (1983). Effect of media design on the performance of fixed bed anaerobic filters. *Water Sci. Technol., 15*, 369–383.

Young, J.C., & McCarty, P.L. (1969). The anaerobic filter for waste treatment. *J. Water Pollution Control Fed., 41*, R160–R173.

Young, J.C., & Yang, B.S. (1989). Design considerations for full-scale anaerobic filters. *J Water Pollution Control Fed., 61*, 1576–1587.

CHAPTER 6

Cooling and Temperature-Controlled Storage and Distribution Systems

Clive V.J. Dellino, Tony Palmer, and George Hazle

6.1 INTRODUCTION

This chapter is concerned with the effects on the environment of current and future techniques used to chill, freeze, store, and transport food products. Simple cooling processes, such as the circulation of ambient air through warm products, have little or no effect on the environment; this chapter ignores them and concentrates on those processes that require the application of refrigeration.

6.2 REFRIGERATION

Refrigeration can be simply defined as any process that removes heat from where it is not wanted, and rejects it to a place where it can be tolerated. In its simplest form, this process is typified by a domestic refrigerator, where heat is extracted from the storage volume and dissipated via a "radiator" coil on the back into the room air.

As heat will only flow from a warm source to a cooler one, the job of the refrigeration unit is to collect heat by providing a cold surface to attract it, gather it up, and boost its temperature to a level that will allow it to leave the system. By far the most common "heat sink" is ambient air, and most refrigeration systems use this convenient dumping ground, one way or another.

To move heat about in this way—by providing low temperature (at the cooler) and a higher temperature than, say, the ambient air—most systems use a volatile fluid (the refrigerant) that can be made to "boil" at low pressure, thus picking up heat in the same way as a kettle takes heat from flame, and can be condensed back to a liquid by giving up heat somewhere else.

To do this, the refrigerant is pumped around the system by a compressor, and the work done by this compressor has to be paid for. Payment is made by the cost of the energy required at the compressor. The greater the temperature lift between the cooled area and the heat sink, the more horsepower is required at the compressor shaft (Figure 6–1).

Most of the energy used during this process is converted into heat, and it, too, is ultimately rejected to the heat sink along with the heat taken up at the cooler. Most of the heat rejected from refrigeration plant condensers is low-grade heat, that is to say, it is available at relatively low temperatures, around 30°–35° C, and not much use for most heating requirements.

There are some exceptions to this but, in the main, condenser heat is most commonly just discharged into the air, or sometimes into surface water (Figure 6–2).

6.2.1 Cryogenic Refrigeration

In addition to the vapour compression systems described in the previous paragraphs, the food industry also uses "total loss" systems of cooling such as liquid nitrogen and solid carbon dioxide. In these cases, ready-made "cold" is purchased in the form of a liquefied gas or a solidified gas that will cool or freeze a product by taking from it the heat it requires to evaporate back to a gas. The gas then produced is never recycled; it just goes into the atmosphere. Whilst nitrogen is the main constituent of the air we breathe, CO_2 is not so welcome in our environment.

The energy required to produce these cryogenic substances, and to transport them, should be taken into account when considering the total impact of their use on our environment. In most applications, the total fossil fuel consumption to freeze with liquid nitrogen, compared with freezing with a conventional vapour compression system, is very much greater. There will probably always be instances where the food industry will use cryogenics: in laboratory work, in the

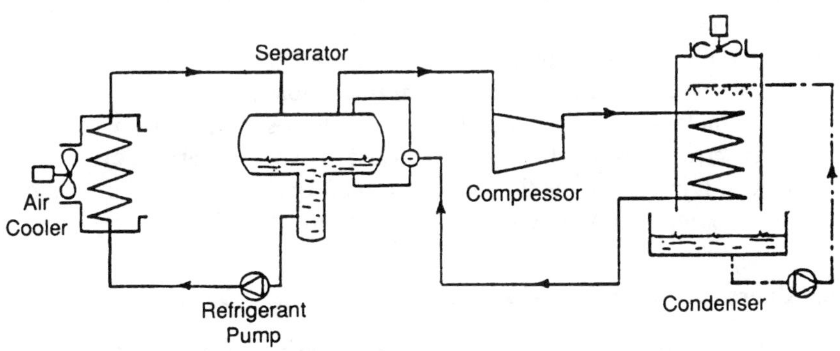

Figure 6–1 Simplified refrigeration plant circuit diagram.

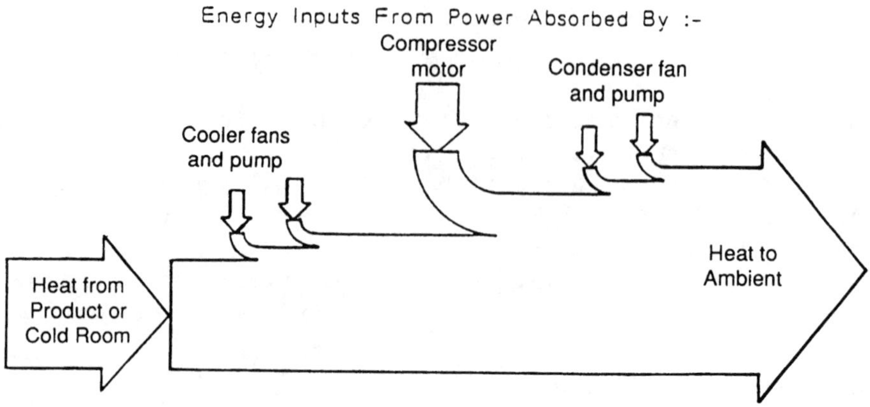

Figure 6–2 Heat extracted versus heat rejected by refrigeration plant.

transport of small quantities of frozen products, and in highly specialised freezing processes. But there are sound environmental and economic reasons for questioning their use in the average factory production line.

6.2.2 Energy Use

From the foregoing notes, it is clear that industry will continue to cause concerns to anyone taking a responsible attitude towards the environment, as long as refrigeration plays a part in the food production and distribution chain. Industry must be seen to be among those with the responsible attitude, and one obvious area for re-evaluation is our attitude to energy use.

The environmental effects of inefficient energy use are twofold: we may pollute our local environment and also put avoidable loads on power stations (and cryogenic plants) that, in their turn, are discharging pollutants.

As far as cooling and temperature-controlled storage are concerned, we use some liquid fuels (for example, in refrigerated vehicles) but, in the main, we burn up electrical energy that has been generated elsewhere, with varying effects on the environment. It is in the reduction of electrical energy use that industry can make a significant contribution to improving the environment.

6.2.3 Energy Costs and the Food Industry

The rise in the cost of electrical energy to industrial consumers is indicated in Figure 6–3, which shows how the cost per unit (kWh) has risen over the period of

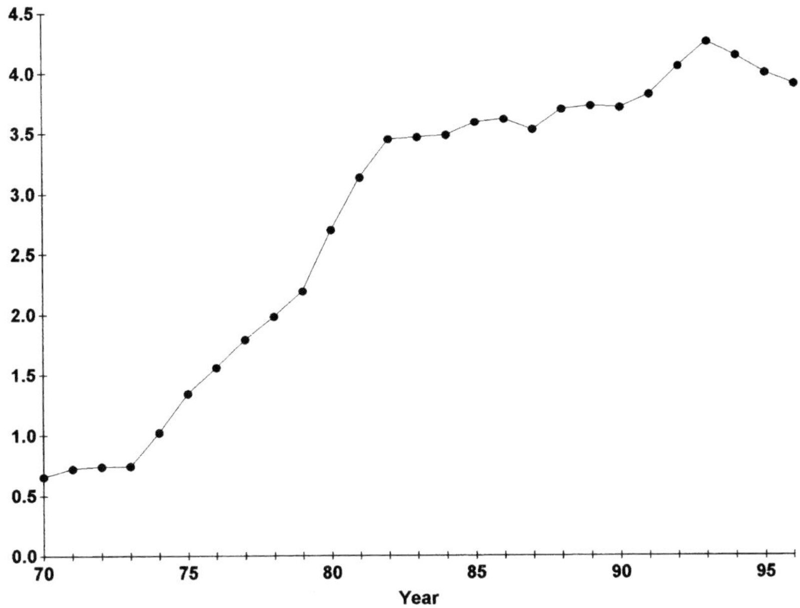

Figure 6–3 Electricity costs to industrial users (average figures, per kWhour).

maximum growth of the UK frozen foods industry. The costs shown are average figures and do not include special tariffs or standing charges.

It would be expected that the dramatic increase in such a basic cost would have concentrated the minds of cost-conscious managers; however, this has not been so. There seem to be two reasons for this. First, their electricity bills have been one of their smaller accounts and, second, other prices have risen at much the same rate, thus keeping relative power costs in much the same position.

Producers of frozen foods have power costs (per ton of product) that are similar to their packaging costs, way down the table when compared with raw material and labour costs. Additional investment in automation and more efficient use of raw material has proved to be more attractive than improved energy efficiency.

Figure 6–3 illustrates how electrical energy has increased in price in the United Kingdom over a period of 26 years but gives no indication of its relative cost to users. If these prices are indexed to 1990 and further related to the price of UK-produced goods (i.e., including exports but excluding imports) over the same period, a significant picture emerges. The top line in Figure 6–4 shows how, when using the gross domestic product (GDP) deflator, the relative cost of electrical energy in the United Kingdom has fallen steadily for the past 13 of those 26 years.

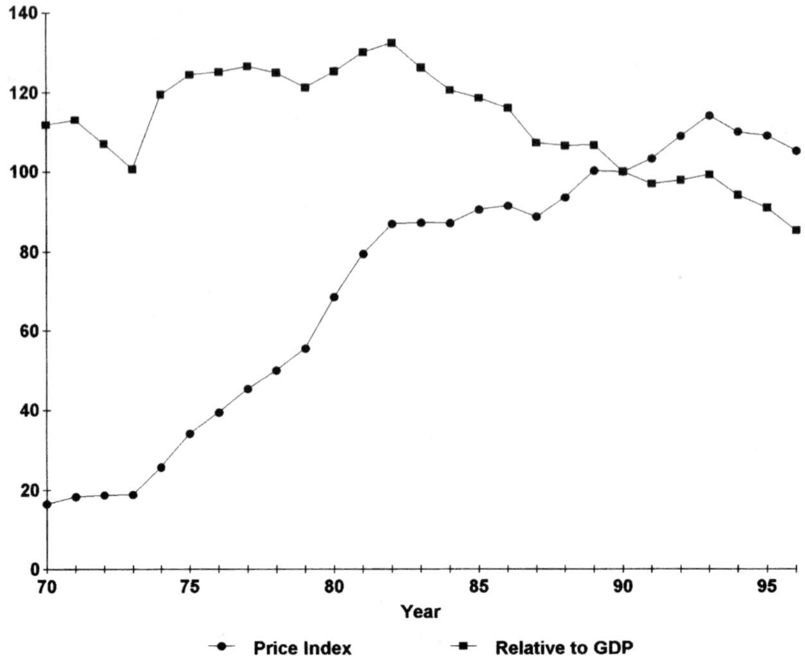

Figure 6–4 Electricity price indices.

There are, obviously, financial benefits from energy-saving practices, and one area of our industry has had a greater incentive to implement these practices. The public cold store operators have electricity bills that are much more significant in their operating costs, and it is interesting to see how their improved operating practices have reduced their energy costs per ton of product stored.

If we consider the period of maximum growth in their activities—from, say, 1965 to 1985—the one consistent factor has been the increase in store size.

The Year Book of the National Cold Storage Federation for 1965 shows that, of the major operators, the average cold store capacity was 200,000 ft^3 (5,663 m^3). By 1985, however, similar statistics show the average storage module to have risen to 1,000,000 ft^3 (28,318 m^3). Over this period, mechanical handling, plant design, materials of construction, and store layouts have developed rapidly, and an "optimum" store evolved that has not changed very much since.

Figure 6–5 shows the effect of these design changes on the amount of energy consumed per ton of product stored, and how a cost-conscious sector of the industry has (albeit as a secondary effect) reduced environmental stress to some 35% of its old level.

There are many ways in which the environmental impact of the use of refrigeration can be minimised with regard to electrical energy, and much has been published on the subject. Good Practice Guides Nos. 42 and 44 are a useful starting point for further reading on this subject. They are published by the Energy Efficiency Office of the Department of Energy.

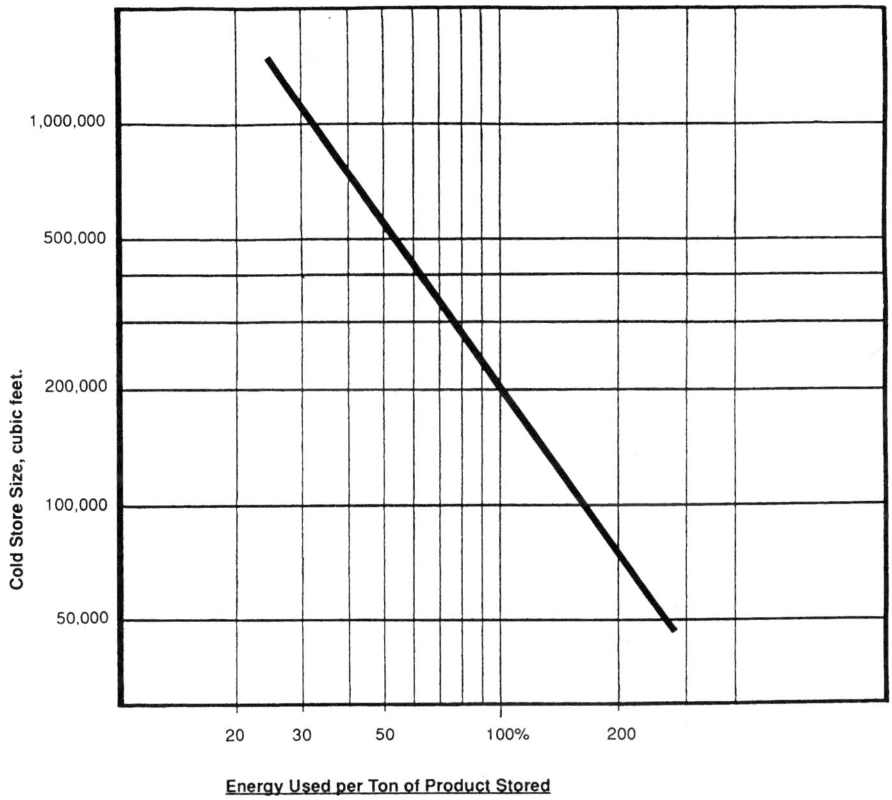

Figure 6–5 This graph illustrates the reduction energy use per ton of product stored as public cold stores grew in size. In 1965, the National Cold Storage Federation Year Book showed the average store to be some 200,000 ft^3 capacity, and this is taken as datum. By the late 1980s (and little changed today), a typical cold store for frozen goods (–25 to –29° C) was closer to 1,000,000 ft^3.

Fuel economy has been considered first in this chapter because it will always be something that directly and indirectly affects our environment and should therefore be of permanent concern. A more high-profile but, hopefully, transient problem is the effect on the environment of the substances we use.

6.3 GLOBAL WARMING AND THE OZONE LAYER

Some of the gases we use as refrigerants have been identified as being among those that, when released into the atmosphere, can alter the natural balance of the global environment. The two areas of concern are quite separate and, although the remedy is the same (do not release them!), it may be helpful to define the perceived dangers in each case. The processes are complicated, and well defined elsewhere, but the basic concerns can be set out as follows.

6.3.1 Global Warming

Some of the energy received from the sun gets bounced back into space from the earth's surface. The earth's atmosphere has always contained the so-called "greenhouse gases" that have acted like a well-chosen blanket, keeping enough heat in the bed for us to be comfortable but allowing the remainder to escape. Of the man-made greenhouse gases, CO_2 is the most in evidence but, fortunately, it is not a very good material to make blankets from. The refrigerant R12, however, is some 7,300 times more effective and, if we want to stay comfortable, our future actions should be obvious.

6.3.2 The Ozone Layer

Further out from our blanket of gases, in the stratosphere, is another natural protective layer. This one is more like sun-tan oil than a blanket, as it protects us from the harmful effects of ultraviolet (UV) light from the sun. We know that small amounts of UV radiation can be harmful to plants, plankton, and people. It can damage DNA, suppress our immune systems, and cause cataracts and cancer. It is hardly surprising that international action is being taken to limit the depletion of this layer of ozone by man-made chemicals.

Unfortunately, many of our refrigerants are in this group. The destruction of the ozone is brought about by the presence of free chlorine in the stratosphere layer; because they are so stable, the chlorofluorocarbons (CFCs) manage to get all the way up there before they break down and release free chlorine. It can take up to 10 years for them to get there and the worst of them can remain active for another 100 years. It is hard to predict the effects of what we have already released. The Global Atmosphere Division of the UK Department of the Environment says " . . . it will not be until the second half of the next century that any significant improvement in the ozone layer is observed." That prediction is based on the revised Montreal Protocol (see Section 6.3.3).

6.3.3 What Can We Do About It?

The food chain, from farmers via processors to retailers, is a high-profile target for the consumer and environmental watchdogs, and their viewpoints and findings are guaranteed to get wide media attention. After all, everyone eats!

Our ultimate clients, the consumers, are already showing awareness of environmental issues and will continue to do so. It would seem important that the food industry is seen by the consumers as being on their side in this respect. They will expect us to be pro-active in publicised areas of concern, such as global warming and the depletion of the ozone layer. Not only should we take steps that improve our impact on the environment, we must be clearly seen to be doing so.

Refrigeration is one of the technologies that has made considerable use of the CFCs and hydrochlorofluorocarbons (HCFCs) that are now seen as being among the prime dangers to the ozone layer and global warming. These chemicals are now being phased out of production under the terms of the Montreal Protocol, signed in 1987. It is, perhaps, worthwhile to point out that the full name of this agreement is the Montreal Protocol on Substances that Deplete the Ozone Layer; it is not concerned with anything else. This is important to consider when we look for alternatives to these substances. It is not beyond possibility that, should research show the need, other similar protocols could be signed to limit global warming, discharge of CO_2, or the use of other man-made chemicals.

So, where can we look for acceptable refrigerants for the future? There are alternatives to the CFCs and HCFCs on the market, and more are to be expected from the chemical industry. The "old" refrigerants are still available, and other established volatiles—such as hydrocarbons—are now being actively promoted as viable options. Each has its pros and cons and, in terms of potential environmental impact (if taken into general use across our industry), they should be compared and considered.

In using any alternative to the CFCs and HCFCs, we must take into account the secondary environmental effects of their use by us on our sites. For example, a new refrigerant may be seen as totally benign in our plant room; but if it requires more horse-power to drive the compressors, there will be more CO_2 discharged at the power stations. Similarly, even if the power consumed is unchanged, the production process required to produce our benign refrigerant may, in itself, be found unacceptably energy expensive. In both these examples, a future "energy tax" would change the equations to a considerable degree.

In making our choices of replacement substances, we should be pro-active and not forget that the environmental pressure groups will have the same information we have and will be looking critically at how we tackle this problem.

6.3.4 CFCs and HCFCs

These chemicals have been used in refrigerated facilities in two main areas: as blowing agents in the production of insulating foam and as refrigerants in enclosed systems. Used in the production of foams, they have allowed insulation thicknesses to be reduced, thus saving space and/or weight. Alternative production methods for the manufacture of insulation have always been available, and reverting to them is not too difficult in the case of cold stores, as an extra few centimetres of wall thickness or a bit more weight on the foundations is hardly noticed.

This is not true of refrigerated transport, where every litre of transit space is important and every kilogram of unladen weight has to be paid for twice.

Apart from transport, therefore, it could be expected that the use of CFC blown foam will be avoided by a greater use of some of the original man-made insulants, such as expanded polystyrene.

It is in the matter of replacements for the refrigerants that the picture is less clear for the static refrigeration installation. The new refrigerants are, at this time, still under development and long-term testing for health and safety. The preliminary results are encouraging in some respects, but some questions still remain. Much has been written and published on this subject; to repeat it in this chapter would be wasteful and, in view of the continuing developments, could even be misleading.

6.3.5 Replacement Refrigerants

In hunting for new compounds to replace the refrigerants we use, the chemical industry is formulating new ones that have better characteristics regarding their impact on the ozone layer and their contribution to global warming.

The yardsticks by which they are measured have been given the names ozone depletion potential (ODP) and global warming potential (GWP), and these are useful in making comparisons between them. Table 6–1 shows the ODPs, GWPs, and approximate stratospheric lifetimes of the current refrigerants and the replacements. The GWPs are based on a 50-year time horizon and the ODPs on CFC 11 = 1.0.

The table clearly shows the high risk values of the CFC refrigerants, the lower risks of the HCFCs, and the limited risks associated with the hydrofluorocarbons (HFCs). The Montreal Protocol will phase out production of the CFCs on a short time scale and the HCFCs on a longer one.

Table 6–1 Refrigerant Characteristics

	ODP	*GWP (CO_2 = 1.0)*	*Stratospheric Lifetime (Years)*
CFC 11	1.0	4100	55.0
CFC 12	1.0	7400	116.0
CFC 113	1.07	4700	110.0
CFC 114	0.8	6700	220.0
CFC 115	0.5	6200	550.0
HCFC 22	0.055	2600	15.2
HCFC 123	0.02	150	1.6
HCFC 124	0.022	760	6.6
HCFC 141b	0.11	980	7.8
HCFC 142b	0.065	2800	19.1
HFC 125	0.0	4600	32.6
HFC 134a	0.0	3400	14.6
HFC 143a	0.0	5000	44.3
HFC 152a	0.0	460	1.5
HFC 404A	0.0	4760	14.6–48.3
HFC 407C	0.0	3401	5.6–32.6
HFC 410A	0.0	3350	5.6–32.6

Missing from the table is R502, which is an azeotropic mixture of 49% HCFC 22 and 51% CFC 115 (i.e., one good and one bad). R502 will, therefore, disappear with the CFCs.

Table 6–1 compares only some of the more common, older man-made refrigerants with some of the newer ones being offered as replacements by the chemical industry. The hydrocarbons (such as propane) are not listed (although they have zero ODPs, minimal GWPs, and are proven refrigerants) because their future use and acceptability in the food industry may turn on factors other than environmental ones. They are flammable and, therefore, their use would be influenced by many existing regulations, laws, by-laws, standards, and codes of practice.

Also missing from the table are "not in kind" refrigerants (NIKs) such as CO_2, water vapour, air, and ammonia. With one exception (ammonia), the introduction of NIKs as replacements for the environmentally unacceptable refrigerants would involve serious cost implications. CO_2 is a high-pressure refrigerant requiring heavy plant and pipework; water vapour systems (and air systems) involve the transfer of large volumes of vapour (or air). So, to use CO_2, water vapour, or air would mean considerable investment in heavy or large plants that use uncommon techniques. However, just as cold store designers can revert to the use of expanded polystyrene for insulating a store, so can they reconsider ammonia to refrigerate it.

6.3.6 Ammonia as a Replacement Refrigerant

Before discussing its future potential in our industry, it may be helpful to review the past role of ammonia (NH_3) and how its use has progressed alongside the man-made refrigerants (for simplicity, let us just call them CFCs).

Before the arrival of the CFCs, the most common refrigerants in use were NH_3, CO_2, SO_2, and methyl chloride. CO_2 required very high operating pressures and was confined, in the main, to those applications where an inert gas was essential, such as refrigerated shipping. All the others had "nasty" characteristics of some sort: they were poisonous, explosive, irritant, inflammable, and sometimes combined these hazards. At the same time, plants were somewhat fundamental and leaky!

Under pressure from their enormous air-conditioning market, the American chemical industry came up with the CFCs, which at the time were seen to be the complete answer. At first, the CFCs penetrated the market at the consumer level: household refrigerators, windowsill air-conditioners, display cabinets, and so on. They soon moved into commercial applications: back-room storage at shops, central air-conditioning plant, refrigerated transport, and distribution stores serving the (then) infant frozen foods market.

These new refrigerants were expensive, and any industry that used tonnage quantities had problems. Industrial users—such as dockside cold stores, ice-making plants, meat handlers, and fishing ports—had every incentive to settle for NH_3 as their primary refrigerant. It was cheap, easily available, well known, and (of the alternatives) the least problematic. Most important, it was the most efficient in terms of horsepower input for refrigerating output. This created a divide that is still with us today.

As the frozen foods industry has grown over the past 30 years, so has the division grown between those who are comfortable with NH_3 and those who are not. This applies equally to the users of refrigeration and the suppliers of plant. Those companies that had their roots in the old industrial market have different attitudes to NH_3 from those who are part of the later retail developments. The big producers of frozen foods still rely, in the main, on NH_3 plants, as do the major cold store operators, who have similar histories.

The large CFC installations are, almost without exception, owned by companies that have moved to the industrial scale from operations that were commercial or even retail in their original concepts. Naturally enough, those companies who cut their teeth on CFC charged installations are wary of NH_3; it does have disadvantages.

In today's climate, it is vital to recognise that NH_3 plants have changed dramatically from those that were around when the first CFCs came on the scene. The designers and contractors who supplied heavy NH_3 plants in the past soon learned how to build large CFC plants when they were required. The high cost of these refrigerants (and the difficulty of sensing a leak, when compared to NH_3) meant that systems had to be gas-tight to a degree that had not been justified before.

New techniques—such as weld-in-line valves, hermetically sealed pumps, capped valves, welded pipework, and sophisticated shaft seals—were used on industrial scale CFC plants, together with gas leak detectors. It did not take long for the "ammonia men" to realise that they no longer needed to go about their business with eyes streaming and noses running. The same gas-tight techniques are now applied to NH_3 plants. Further developments of small screw compressors and low charge plants have increased the potential application of NH_3 in areas where it has not been previously considered.

6.3.7 Ammonia and the Environment

Owners and operators of refrigeration plants, generally, are very conscious of the effect of release of refrigerant into the atmosphere, regardless of which refrigerant. However, leaks do occur; always accidentally, sometimes from worn components, sometimes during service and maintenance operations. The harmful effects of CFC emissions are now recognised, but what about NH_3 release?

As a naturally occurring substance, it has no effect on the earth's environment. Being twice as light as air, the gas disperses quickly and decomposes in the atmosphere in a few days. A liquid spill evaporates almost immediately (its boiling point at atmospheric pressure is –30° C), so it does not find its way into subsurface water.

Its disadvantages are that it can explode, burn, choke, or blind, and it smells. If we are going to give serious thought to why it is in such common use and look at extending its use in industry, we must put these disadvantages in perspective.

Ammonia can be flammable in air. For this to be so, the concentration in the air has to be between 16% and 25%, but a developing leak can be "smelled" at concentrations of 1/20,000 of this, and the flashpoint would be well over 600° C. In practical terms, the risk of fire or explosion is fairly small.

NH_3 has a suffocating effect and the smell may cause panic. The highest level for safe human exposure is about 25 parts per million (ppm) in air, but the fatal dose is calculated to be 30,000 ppm. There are no recurring toxic effects after human inhalation. NH_3 is highly soluble in water; eye irritation can be severe when leaks occur. The known risks in NH_3 have caused it to be classed as a "substance hazardous to health" but not a poison. By adopting techniques and practices that:

- keep it in the system
- keep potential leak points away from people
- automatically monitor potential leak points

- keep total quantities in the system to a minimum
- rapidly ventilate high risk areas (such as plant rooms)

most major users of refrigeration have been able to continue to use this highly efficient refrigerant. Not only is it by far the cheapest refrigerant to buy, its coefficient of performance (COP) (which is simply a measure of horsepower required to provide a certain cooling capacity) is better under most low temperature conditions than the alternatives.

Recognising that there will always be some areas—corner shops, vehicles, etc.—where it will be impossible to introduce NH_3, there is every reason to look for a much wider use of this efficient, cheap, and environmentally acceptable refrigerant.

6.3.8 Conclusion

The environmental impact of the cooling, freezing and temperature controlled storage sectors of our industry can be minimised by:

- energy efficient operation
- heat recovery systems
- responsible purchase of materials of construction
- wider use of environmentally benign refrigerants

But, above all, we need constant awareness of the overall effects of our activities and the potential future risks that we build into our operations by purchasing plant and equipment that could be seen as less than responsible.

6.4 REFRIGERATED TRANSPORT SYSTEMS

Primary and secondary distribution of perishable cargos under controlled temperature conditions grew significantly during the 1970s, 1980s, and 1990s, largely under the influence of market forces. The major food retailers, led by the big-name supermarket chains, have (under the cost pressure of competition) striven to extend "sell by" dates and times by retarding product deterioration. In many cases, this has been forced on them by the increasing scale of their distribution operations, which require product to be carried over longer distances, in vehicles of greater capacity, from larger and fewer depots, and delivered on a just-in-time basis.

6.4.1 The Food Hygiene (Amendment) Regulations 1990

Introduced on 1 April 1991, this regulation lists the foods that must be kept at or below 8° C (or at or above 63° C in the case of foods held hot after a period after cooking) throughout the food chain, with some time-limited exemptions. On 1 April 1993, the chill temperature requirement was reduced from 8° C to 5° C for many of these foods. In addition, directives for specific products laid down various chill temperatures from 2° C to 8° C at certain stages in the food distribution chain.

6.4.2 Primary Distribution

Primary distribution is defined in the United Kingdom as those journeys that, regardless of number of drops, do not include a direct retail outlet or catering establishment amongst the deliveries to be made. Local distribution is defined in the United Kingdom as direct retail outlets and catering establishment deliveries.

The Quick Frozen Food Regulations stipulate that products must be stored and transported at a produce temperature of –18° C or colder throughout primary distribution. A 3° C temperature tolerance is permitted in primary distribution. In local distribution, quick frozen product temperatures must be at –18° C, with a temperature tolerance allowance of 3° C (10 January 1997).

6.4.3 Temperature Monitoring

On 1 August 1994, a Quick Frozen Foodstuffs Amendment Regulation came into effect, making the fitting of approved temperature monitoring equipment mandatory in storage and transport for quick frozen foods by the manufacturer, temperature-controlled store, transport, local distributor, and retailer. They must record air temperature frequently and regularly, date these records, and keep them for at least 1 year, and make them available to any authorised person.

6.4.4 Food Safety (Temperature Control) Regulation 1995

This regulation came into force on 15 September 1995 and introduced some major changes to the previous legislation:

1. The 5° C temperature limit for chilled foods was withdrawn.
2. A single temperature limit of 8° C was introduced.

To some extent, the temperature limit of 8° C is considered to be too high by sections of the chilled foods industry and, for many products, the major supermarkets and distributors show no signs of abandoning their 5° C maximum upper limit.

6.4.5 Chill Control Requirements

Temperatures specified refer to that of the food and not the air in the storage facility. This applies to foods including raw materials and ingredients during transport, preparation, processing, storage, and display for sale at retail and catering outlets.

6.4.6 Types of Food That Require Chilled Storage

In any legal proceedings, the regulations state: "Where food has not been kept at or below 8° C, the burden of proof would be on the Food Authority to demonstrate that it was likely to support the growth of pathogenic micro-organisms or formation of toxins."

Typical foods are seen as soft or semi-hardened cheeses ripened by mould and/or bacteria; dairy-based desserts, except where the pH of those products would prevent the growth of pathogenic micro-organisms or the formation of toxins (such as yoghurt with a pH below 4.5) or where other preservatives have been used to render temperature control unnecessary; cooked products containing meat, fish, eggs (or substitutes for them), milk, hard or soft cheese, cereals, rice, pulses, and vegetables; ready-to-eat products such as sandwiches containing fillings or toppings prepared with the foods mentioned above; smoked or cured fish; smoked or cured ready-to-eat meat that is not ambient shelf-stable; ready-to-eat prepared foods such as vegetables; vegetable salads containing fruit or prepared salads such as coleslaw containing other products; prepared products such as mayonnaise; and uncooked or partly cooked pastry and dough products such as pizzas, sausage rolls, or fresh pasta containing fish (or substitutes) or vegetables. Fresh pasta should normally be temperature controlled, whether or not it contains these products.

6.4.7 Exemptions from Chill Control Requirements

Exemptions include products that present no risk to health when stored at ambient temperatures; dehydrated or canned foods; food that must be ripened or matured at ambient temperatures (does not apply once the ripening or maturation is complete); mail-order food (subject to the general requirements that the product be supplied by the mail-order business at a temperature that is not likely to give rise to a risk to health).

A further, less predictable, fillip to those parts of the transport industry concerned with moving foodstuffs arose through the 1980s as the big supermarkets focussed their attention increasingly on fresh produce, that is, fruit and vegetables. Although it accounts for a relatively small percentage of turnover, it has become symbolic of product freshness and, by association, overall quality of a much wider range of products.

Freshness, both actual and perceived, has become paramount. Shoppers' expectations have been raised to the point where the small independent greengrocer, whose produce has traditionally been collected in the early morning from central markets using simple non-temperature-controlled vans or covered dropside vehicles, is almost extinct.

Buyers of fresh produce for retailers now commonly deal directly with growers, both in the United Kingdom and further afield. In the case of green vegetables and indigenous fruit, they endeavour to establish a "cool chain" extending sometimes from the field or orchard all the way to the supermarket shelf.

In the United States, the USDA Food and Safety and Inspection Service is responsible for food safety. Storage temperature depends on product and manufacturer; typical temperatures for a public warehouse are 0 to –5° F; cooler is +28 to +38° F. Trucks must maintain the temperature of the foodstuffs as they come out of the warehouse, and the receiver will check the temperature at delivery.

6.4.8 Environmental Impacts

The move towards refrigerated transport for products traditionally carried at ambient temperature has increased initial vehicle cost, and refrigeration equipment adds to unladen weight, thus diminishing payload capacity, so that heavier duty vehicles or bigger fleets are implied for hauling the same tonne-mileage.

Whatever the measures taken to compensate for loss of overall vehicle carrying capacity (brought about through refrigerating hitherto ambient, or "dry freight," fleets), the environmental impact is significant. There has unavoidably been an increase in exhaust emissions, from vehicle tail-pipes and from the small engines that drive most mechanical fridge units.

Emissions from vehicle engines—almost entirely diesels in a food industry transport context—are subject to progressively stricter legislative control. It is over 20 years since regulations concerned with truck and van diesel tail-pipe emissions first came under legislators' scrutiny.

British Standards BS AU 141a came into force in 1971, but it covered only visible emissions, that is, the black smoke that in former days characterised the diesel engine as a species. The aim of the legislation was to avoid the unsightly appearance of lorries belching smoke as they ground their way up long hills. There was little thought to deleterious effects on health or on the environment.

It should be pointed out that in the 1970s, a much higher percentage of companies distributing food were the manufacturers themselves or, alternatively, the retailers rather than contracted distribution specialists. Vehicle operators were only too conscious of the impression given by their vehicles to actual and would-be customers who noticed them on the street. And of course they and the contractors working for them remain so in the 1990s.

Accordingly, food-carrying vehicles were and are kept clean in the widest sense, including their exhaust emissions. The fact that 20 years ago most trucks were underpowered and needed to spend a high percentage of their running time developing full power only served to aggravate the smoke problem, both at the time and in the longer term. This, in turn, made it difficult to maintain the conditions needed for clean combustion beyond the first few thousand miles.

By the late 1980s, vehicle emission concerns had grown beyond the mere negative aesthetics of a dirty, smoking exhaust. Immediate effects on the health of plants, animals, and, of course, humans, as well as the longer-term greenhouse effect and global warming implications, came within the ambit of the legislators' concerns and enactments.

Visible emissions, now referred to as particulates—comprising mainly solid or liquid carbon or hydrocarbon particles, together with three different categories of gaseous (invisible but polluting) emissions—are now subject to limiting European Community legislation, under Directives 88/77/EEC and 91/542/EEC.

Controls on gaseous diesel exhaust emissions, oxides of nitrogen (NO_x) carbon monoxide (CO), and hydrocarbons (HC) were introduced in 1990. In 1992–1993 (in two stages), tighter gaseous emission limits—and for the first time a particulates limit—were applied, in so-called Euro 1 legislation. A further reduction in permitted levels of exhaust gas and particulate pollutants came into force, again in two stages, for Euro 2 legislation during 1995–1996.

In the United States, vehicle emissions are also subject to ever-tightening control. Code of Federal Regulations (CFR) 85 deals with limits for engines. Limits on NO_x have been cut by half since 1985; particulates have been cut from 0.6 g/brake horsepower hour to 0.1 since 1988 (Table 6–2).

All new trucks being put into service on the public road are now subject to emission controls. Those serving the food industry are affected no more nor less than any other vehicles in this respect.

There is, in any case, a measure of environmental controversy surrounding the emission requirements. The permitted NO_x and particulate levels in particular come into conflict with each other in the sense that it is technically difficult to satisfy both without compromising fuel economy. While engine researchers are striving to reduce it to a minimum, some fuel penalty seems inevitable on the truck

Table 6–2 Limits for 1993 and Later Model Year Heavy-Duty Diesel Engines

	g/Brake Horsepower-h	*g/MJ*
Hydrocarbons (petroleum diesel) and organic material hydrocarbons (methanol diesel)	1.3	0.48
Carbon monoxide	15.5	5.77
NO_x	50.0	1.9
Particulate	0.10	0.037

diesels of the late 1990s. If more fuel is burned, more carbon dioxide (CO_2) is released, making a possibly unwelcome contribution to global warming; CO_2 has not yet been added to the list of proscribed diesel exhaust pollutants.

On a refrigerated truck distributing food, a second, albeit smaller, diesel engine is likely to be found driving the compressor of the refrigeration equipment. As the law stands at present, that auxiliary power unit can emit unlimited noxious fumes into the atmosphere. It escapes completely the legislation controlling emissions from the vehicle's propulsive engine.

Because refrigeration unit engines are small, their output of pollutant gases and particulates is correspondingly limited. But EC discussions on the control of emissions from engines other then those powering vehicles on the public highway are proceeding, with off-road construction equipment, generator sets, and large items of engine-driven industrial plant the main areas of attention. A tentative target date of the year 2000 has been suggested for implementation of such regulations.

It is probable that small "donkey engines" of the kind employed in vehicle refrigeration units will be brought into the same category for emissions legislation purposes. Whether permitted pollutant levels will be related, as they are on vehicle diesels, to engine output (in horsepower or kilowatts) remains to be seen.

The cost in basic engine research and development—in other words, in new technology—of reducing pollutant levels from small, relatively cheap industrial diesels of the kind used to drive mechanical refrigeration units will be disproportionately large. It might even be unacceptable, implying a need for exhaust after-treatment kits that can, in particular, filter out the particulates, although Thermo King claims to meet Euro 2 engine exhaust standards already.

A likely alternative development will be for mechanical fridge unit producers to switch to a different kind of engine, able to run on a more environmentally friendly fuel. Liquefied petroleum gas (LPG) was at one time considered the alternative fuel for auxiliary engines, as the many LPG-powered fork-lift trucks in warehouses and transport depots bear testimony.

However, attention is now turning to natural gas, of the kind piped into most households for heating and cooking, as an engine fuel. It is inherently cheaper to produce than LPG, needing no refinery processing, with all that implies by way of additional energy input. As such, in the context of the global energy equation, it is less wasteful of fossil fuel resources and so more environmentally acceptable.

When natural gas is burned in an engine's combustion chamber, it produces far lower levels of proscribed pollutants. For practical storage and handling reasons, natural gas (chemical formula CH_4) has, for on-board use (powering the vehicle itself or a piece of auxiliary equipment such as an engine-driven fridge unit), to be at a considerably higher pressure than in one's domestic mains supply.

It can be fully compressed to a liquid state, in which form it is commonly referred to as LNG. Because storage pressure is high, the volume occupied for a given usable quantity of fuel, and hence the tank capacity needed, is small. But those tanks have to be correspondingly heavy and expensive to withstand the pressure, with sophisticated and costly valve arrangements.

Natural gas in less concentrated form, that is, at about one-third the pressure and still in a gaseous state (known as CNG), is the increasingly favoured alternative to diesel fuel in environmentally sensitive applications. CNG-fuelled auxiliary engines for on-board vehicle refrigeration equipment are a distinct possibility in the medium to long term.

Unfortunately, for a given quantity of stored energy that can be equated to fridge unit engine running time, CNT tanks occupy something like 3.7 times the cubic space needed for diesel fuel. However, on most refrigerated vehicles, especially the heavier articulated outfits, there is ample space below floor level.

A high proportion of mechanical refrigeration units on temperature-controlled vehicles now incorporate an overnight mains electric stand-by drive. It is established transport industry practice for trucks to be preloaded with, say, Thursday's consignments soon after their return to base late on Wednesday afternoon, in readiness for an early departure.

Clearly, the emissions and noise created by the auxiliary fridge unit engines on 20 or 30 reefer vehicles lined up in a depot yard and running all night are unacceptable, not least to local residents. Plugging into the mains allows the fridge unit compressor to be switched from engine to mains electric drive. Noise is thereby reduced considerably and pollutant exhaust emissions from the auxiliary engines are avoided altogether. Environmentalists will point out that the mains electricity has to be generated in a coal-, oil-, or gas-fired power station, which itself pushes noxious gases and particulates into the atmosphere.

The electric versus engine power environmental argument in any case hinges on economic issues, and the economics are blurred by the differential taxes levied (or not) on various liquid, gaseous, and solid fuels and by variations in electricity costs (a) for large versus small industrial consumers and (b) at night, when overall demand for electrical power is low.

Mechanically driven vehicle refrigeration equipment and its accompanying auxiliary engine emissions, along with the environmentally damaging refrigerants described in the first part of this chapter, can be avoided altogether using fundamentally different forms of refrigeration, although, again, the global energy versus emissions balance sheet must not be ignored.

Cheap night-time electricity can be used to power passive (and silent) eutectic vehicle refrigeration systems, whose function can be likened to a domestic night storage heater in reverse.

A conventional refrigeration process is applied, using the same kind of on-board compressor, evaporator, and fluid refrigerants as those in a so-called mechanical system. But instead of the evaporator cooling a flow of air being simultaneously blown into the loadspace, it cools a eutectic (low melting point) brine mixture contained in hollow beams or plates mounted inside the vehicle body. Before the vehicle leaves the depot in the morning, the system is unplugged. "Cold" is then progressively released from the beams into (or, as the technically pedantic would put it, heat is progressively extracted from) the loadspace.

The principle of gradual heat extraction means that eutectic systems can be seen as natural successors to the solid carbon dioxide and even ice-blocks placed directly in the loadspace of, respectively, ice-cream and fish-carrying vehicles in years gone by.

On the debit side, eutectic plate systems are heavy, and they lack the operational flexibility needed to cope on large vehicles with prolonged door openings in warm ambient conditions. Eutectic plates, although capable of holding a set box temperature for many hours, cannot achieve the rapid pull-down in temperature required, for health and hygiene reasons, once the doors have been closed before the vehicle gets on its way to the next drop.

Fast pull-down is an attribute of the other alternative to mechanical refrigeration: that is, direct liquid nitrogen cooling, exemplified by BOC's Polarstream system. The liquefied gas is carried on board, under pressure, and released in a fine spray that immediately evaporates. The process requires heat, which is extracted from the surrounding air in the vehicle loadspace. Cooling of the air is rapid, thereby arresting further heat gain into the product, which, by definition, has of course been pre-cooled in a cold/chill store.

Nitrogen is an inert harmless gas; it makes up 80% of the air we breathe. For BOC and competitors like Air Products it is effectively a by-product of the production of oxygen. The energy input and consequent environmental cost involved in its manufacture is therefore difficult to quantify objectively.

Liquid nitrogen has obvious practical limitations. Where a vehicle is making multi-drops, so that there are frequent door openings, the system can run out of nitrogen. The on-board pressure tanks, like the CNG fuel tanks referred to earlier, take up significant space. The number of liquid nitrogen replenishment points around the country, enabling a vehicle to recharge its refrigeration system away from base, is undeniably limited.

For many refrigerated vehicle operators, away from their base depots for days or even weeks at a time, there is currently no viable alternative to mechanically driven compressor-evaporator refrigeration. Although the compressor might be driven by an auxiliary engine or by a land-based overnight electric stand-by motor, the process of refrigeration is the same.

A suitable substance, that is, a refrigerant, is needed to circulate through the system. The chlorofluorocarbons (CFCs) used for many years are, because of their chlorine content, now universally acknowledged to be damaging to the earth's ozone layer. They also contribute to the global warming phenomenon.

A total ban on CFCs by members of the Montreal Protocol is now in place and includes the EC and the United States. CFCs have already been replaced in most refrigeration equipment (static and vehicle-mounted) by a family of hydrochlorofluorocarbons that, in simple terms, have some of their damaging chlorine replaced by hydrogen. The range of HCFCs now available to equipment producers are much more environmentally acceptable than CFCs, although in absolute terms they vary considerably in their ozone depletion and global warming potential. They are also more expensive, by a factor of between 2 and 5.

Rival refrigeration equipment makers have striven to "brand" the new generation of refrigerants with what amounts to proprietary (rather than generic) designations, such as R22 and R69L. Claims of superior environmental characteristics are made for each new contender, under the influence of what the manufacturers see (although they might deny it) as a marketing (i.e., profit) opportunity, with all that implies in product cost.

Undoubtedly, the new HCFCs are more expensive to produce than CFCs, but many environmentalists and food industry observers would like to see more collaboration between erstwhile refrigeration equipment and chemical company competitors in developing an industry standard for refrigerants. Such a move would help contain the cost penalties now anticipated, while at the same time contributing to the greater good of the environment.

Such considerations can be extrapolated into the next generation of refrigerants now being developed, the chlorine-free hydrofluorocarbons (HFCs), whose ozone depletion potential is nil, although their possible effect on global warming remains a concern.

In a transport and distribution context, the harmful effects of the older (and, in most cases, cheaper) refrigerants on the earth's ozone layer are given an extra edge by the sort of operational hazards that could lead to an escape of the offending gas into the atmosphere.

Leaks, most obviously from pipe joints, are inevitably a greater risk on a vehicle than in a cold/chill store fridge installation. There are vibrations to contend with, not just from the vehicle and the fridge unit engine, but from all but the smoothest road surface. Repairs and maintenance also have to be carried out under less controlled conditions; sometimes away from home, possibly in a non-specialist contractor's workshop.

There is also the often unspoken, but real, danger of the refrigeration equipment suffering physical damage in an accident. A front bulkhead (or trailer front wall) mounted unit is in a vulnerable position in the event of a frontal impact. On those multiple-box installations, where the compressor and evaporator are several metres apart, the connecting refrigerant pipe-runs are often even more exposed to accident damage.

It should be remembered that the dramatic growth in vehicle refrigeration in the late 1980s centred on chilled distribution, largely motivated by the need for retailers to project that image of freshness. Volumes of frozen product (meat, fish, and packaged vegetables) moved by road have grown less markedly. Nevertheless, almost without exception, vehicles are specified for deep frozen food carrying. They have thicker insulation and higher capacity (in heat extraction terms) refrigeration equipment than is needed for chilled (down to 0° C) operation only. The reasons are economic.

Refrigerated vehicle buyers, in assessing total (i.e., whole life) vehicle running costs, are all too aware of residual values. With a frozen food carrying capability, preferably verified by the international ATP (Accord Transport Périssables) class C certification, a vehicle's secondhand value is greatly enhanced. Used reefers are invariably bought by small entrepreneurial carriers who need to be able to carry every category of perishable foodstuff: chilled, deep frozen, and ambient.

The need to build in second-owner appeal, which often prompts the choice of other truck features such as sleeper cabs, translates into a mandatory deep-frozen capability, which can only be readily achieved using the modern ozone-depleting refrigerants or their new, more environment friendly, and much more expensive substitutes.

If new vehicle refrigeration equipment could be specified for its specific purpose, then all those vehicles put on the road to distribute dairy products, fruit, vegetables, and other chilled cabinet merchandise could be equipped at much lower cost, while also incurring less payload penalty, with more conservatively rated refrigeration units containing less refrigerant.

In reality, many food carriers feel obliged to specify refrigeration equipment and bodywork that will give them the operational flexibility to carry either chilled or frozen produce, or even both, in a multi-compartment loadspace. This applies especially to the third-party transport sector, embracing food hauliers and contract distributors.

Ozone-depleting CFCs and HCFCs have traditionally been employed for a quite different purpose related to refrigeration, namely in the manufacture of foam slab or *in situ* foamed insulation, principally polyurethane. Alternative, more environmentally acceptable foaming agents are readily available. Indeed, water-blow urethanes are now in production, but their insulation properties per centimetre of material thickness are poorer. So too are those of older foam materials like expanded polystyrene and foamed PVC. Expanded PVC has been favoured for some time where the insulation material is required to contribute significantly to the structural strength of the body shell. This applies especially in the case of a chassis-less refrigerated semi-trailer.

Compensation has to be made for inferior insulation properties in a cellular material employed in a vehicle body structure. The most obvious ploy is to increase thickness. In a cold store, an additional 10 mm of wall thickness is easily accommodated. On a vehicle, space is at much more of a premium.

Legal limits are set on vehicle dimensions, most pertinently on width. For many years, until the mid-1980s, a 2.5-m width limit applied through most of Europe. It meant that two metric (1,200 x 1,000 mm^2) pallets could be accommodated alongside each other, with their long sides across the vehicle, only by making the walls far thinner than a refrigeration engineer would recommend. In the United States, the width limit is 2.6 m (CFR 658).

Excessive heat gain through the sides of the body was compensated by increases in insulation thickness elsewhere, notably in the floor. However, this was not a thermally ideal solution, because the heat path through those thin side walls close to floor level led to "hot spots" in the loadspaces. This was made worse when loads were jammed tight against the walls of the body/trailer, restricting the circulation of cold air from the fridge unit evaporator.

Compliance with ATP class C standards was, in many instances, marginal, although use of the most efficient CFC-blown insulation material wherever possible minimised these thermal performance shortcomings. Vigorous lobbying in Brussels and at the national level by the Transfrigouroute reefer operators' organisation won a notable regulatory concession from the transport legislators. The permitted width of trucks carrying loads at controlled temperatures was increased, in stages, to the current 2.6 m limit. That extra 100 mm has seen the effective demise of the early 1980s "thin-wall" reefers, which were unloved for a variety of reasons, not least their questionable structural integrity.

Continuing research is needed to develop new foaming agents if we are to avoid a re-run of the thin-wall saga, with wall thicknesses having to be increased again, prompting refrigerated vehicle operators and manufacturers to agitate for another vehicle width concession.

Vehicles fitted with streamlined bodywork and air management equipment show a significant reduction in fuel consumption when used on long-haul, motorway, and deliveries. Also, the use of speed limiters and driver training programmes can further improve fuel economy without extending journey times. These actions, when taken together as part of an environmental policy, can reduce fuel consumption by up to 20%.

Only by integrating transport policy into a broader environmental commitment can today's food distribution specialist make a true contribution to saving the environment.

CHAPTER 7

Packaging Options

Susan E. Selke

7.1 INTRODUCTION

The goal of this chapter is to examine packaging options, especially for food products, in light of their environmental impacts. Although the entire life cycle of packages, from acquisition of raw materials through eventual disposal of the package, must be examined for a complete evaluation of environmental impact, impact associated with disposal is covered in a later chapter, so it will not be included here. Thus, we examine environmental impact resulting from the acquisition of raw materials, processing those raw materials into packaging materials and then to packages, and distributing the packages (filled or unfilled) to their users.

First, some preliminaries are needed. Packages exist because they perform certain desired functions. An understanding of these functions is necessary in order to determine what packaging changes can or cannot be successfully adopted. In addition, it is necessary to have some basic background about packaging materials in order to evaluate the environmental impact of those materials.

7.2 FUNCTIONS OF PACKAGING

Before evaluating the environmental impact of various packaging options, we should consider our reasons for using packages in the first place. After all, the package with the least environmental impact of all is the package that does not exist. Lists of the functions served by packages of various types can be very long. For simplicity, these will be classified in three major categories: protection, communication, and utility.

7.2.1 Protection

The protection function of packaging includes all those package attributes that act to protect the product from harm or damage from the environment surrounding it, as well as those attributes that act to protect the environment from harm or damage from the product contained in the package.

A basic protection function is containment itself. This allows the product to be moved from one place to another as a unit and is obviously absolutely essential for products that are liquids, gases, or fine particles. It is hard to imagine how we might be able to have milk and cornflakes for breakfast, for instance, without both the milk and the cereal being contained in some type of package so we can get them from the cow and the cereal factory to our breakfast table.

For most products, some additional degree of protection beyond simple containment is also required. In our milk and cornflakes example, we wish to protect both foods from exposure to dust and dirt. We further wish to protect the milk from micro-organisms, light, excessive heat, and so on. We want to protect the cornflakes from exposure to moisture, which will make them lose their crunch, and from crushing, which will reduce them to a powder. This additional protection may be provided by the package itself, or it may be provided by modifications of the environment the product and package inhabit, such as refrigeration. For example, we could prevent moisture gain in our cornflakes by maintaining the cereal in a dry atmosphere at all times, but we are more likely to use containment in a package, which provides a barrier to the passage of water vapour. On the other hand, we are likely to use refrigeration to retard the spoilage of the milk, though we may alternatively use aseptic packaging techniques.

Protection from exposure to micro-organisms and protection against excessive water gain are two common functions of packaging. Others include (but are certainly not limited to) protection against exposure to oxygen, shocks, vibrations, electrostatic discharges, ultraviolet light; as well as protection against loss of carbon dioxide, flavour compounds, and organic vapours.

If our product is toxic or hazardous in some way, we are likely to use the package to provide some degree of protection against exposure to the product. For example, child-resistant packaging is designed to protect small children from harm resulting from exposure to various types of products, including medications and toxic household products. The stringent regulations governing shipment of hazardous materials include restrictions relating to the packaging of these goods in ways that will minimise exposure of people and other parts of the environment to these hazardous products, both during normal distribution and in the event of accidents. An additional aspect of protection is package components that are designed to deter tampering and/or to alert the potential user that product tampering may have occurred. Protection, including containment, can be regarded as the most basic function of packaging.

7.2.2 Communication

Another set of package functions have communication to the potential user as their primary purpose. Many packages perform a wide variety of communication functions. The most basic communication function is to answer the question "What is the product?" Often the identity of the product is conveyed through written words on the package. Other times it may be conveyed simply by transparency of the package so that the product itself can be seen and identified.

A variety of messages are legally required to be conveyed for many products, including such things as how much of the product is in the package and who is the product manufacturer. Such legally required messages are often specified down to the minimum size print that may be used and the place they must appear on the package. Other package communications are included because the seller wishes to convey certain messages to the potential consumer. These may include, for example, opening instructions for a blister package (a package that confines the product between a plastic bubble and a paperboard backing), or recipes and appetising pictures on a box of spaghetti.

A significant communication goal for many packages is to convince the potential purchaser to pick that particular brand of product over the competitors on the shelf in the store. In addition to the package itself and its messages, the space the product occupies on the shelf also has communication value. It simply is easier for the consumer to see and be attracted to a product if it occupies more visible space.

As illustrated by some of the examples above, the communication aspects of the package are not necessarily printed words. Pictures, colours, shape, and even package transparency can all be used to communicate to potential users. This is one reason why manufacturers of consumer goods are often reluctant to make major changes in package design; they risk losing the brand identity that is a major factor in their sales. Therefore, when significant changes are desired, they may be phased in gradually to facilitate their customers' ability to continue to easily recognise the product.

7.2.3 Utility

The utility features of the package are those that provide some degree of usefulness to people or entities handling or using the product or package. These can be as varied as a dispensing closure on a bottle of hand lotion or a circle cut out of the backing of a blister package to permit hanging the product on a display. In the first case, the utility is provided to the user of the product. In the second case, the utility is provided to the retailer. For some products, such as many aerosols, the utility provided by the package is essential to the proper functioning of the product. In other cases, the utility simply provides added convenience of use or handling.

7.2.4 Combination of Functions

Package features do not always fit nicely into one of the three categories discussed above. Many package features perform more than one function. For example, a bar code on a product has both communication and utility functions. The oversize backing card on a tiny watch battery provides both communication and protection (in this case, primarily protection against shoplifting).

7.3 PACKAGE SELECTION

There are almost an infinite number of combinations of package materials and package designs that could be considered for packaging a product. Most of these choices, for any given product, are impractical for one reason or another. Many will not do an adequate job of performing the functions that are necessary or desirable for the product. Others will be eliminated on cost grounds. The overall goal of package design is generally the selection of the package that will offer the greatest economic benefit to the producer of the product.

Thus, the package designer is trying to select an optimum package, balancing the cost of the package against the benefits (functions) the package will provide. A general rule is that the more the package does, the more it will cost.

In the area of protection, this can often be measured in a straightforward way. The more protection provided by the package, the more material is used, the more the package costs. On the other hand, the less the package does, the more the product is likely to be damaged. An economic relationship can be calculated showing overall package and product system costs to be high at very high package costs and also high at very low package costs, because of high damage rates. The overall system cost will be lowest at an intermediate package cost, where product protection is sufficient and package costs are not excessive. The value of the product is a very significant variable in determining how much damage can be tolerated.

In the areas of communication and utility, determination of the optimum package is usually more complex. Here, the determining variables often are relationships between the packaged product and the product user's purchasing decisions. These decisions tend to be highly variable from one consumer or group of consumers to another, and can be extremely difficult to determine with any degree of certainty. Test marketing becomes extremely important, and a high degree of product failure in the marketplace is commonplace.

It is impossible to cover, in the course of a single chapter, all the myriad of package designs that are available or even the considerably smaller, but still large, number that are commonplace. Fortunately, the environmental impact of package alternatives is determined more by the basic package material or materials than by the package type or style. Thus, our discussion is organised around the five basic package materials: wood, paper, metal, glass, and plastics.

7.4 WOOD

Wood in packaging is used primarily in pallets, with additional uses in crates, boxes, and so on. Wood is a renewable resource, though utilising it can lead to decrease in soil fertility, erosion, loss of habitat for wildlife, and other environmental impacts. Using wood also requires energy, for both harvesting and processing the wood.

The environmental impact of using wood packaging is, in large part, a subset of the environmental impact of using paper packaging. References to the impact of paper packaging are considerably more available in the literature than are references dealing with the impact of wood itself.

Franklin Associates (1992) examine the energy required to produce wood packaging from raw materials, calculating an average value of 12.23 MJ (megajoules)/kg (5,269 Btu/lb), not including the energy content of the wood itself.

Boustead and Hancock (1981) provide an analysis of the energy required to produce a typical pallet, weighing 25.40 kg and containing 17.91 kg of softwood, 6.33 kg of hardwood, and 1.16 kg of steel nails and wire. Total energy required was 1,106.7 MJ and includes the feedstock energy of the wood. This total energy is broken down into 150.48 MJ for production of 18.81 kg of softwood, 3.19 MJ for sea transport of the softwood, and 6.25 MJ for road transport of the softwood. Production of 6.65 kg of hardwood consumes 53.20 MJ. Boustead and Hancock found published values for production of raw wood from standing timber ranging from 3.33 MJ/kg to 12.59 MJ/kg, and chose 8 MJ/kg as a representative average. The wood feedstock energy is 437.91 MJ, based on a 12% moisture content. Sawing and makeup of the pallet consume 418.96 MJ; 36.19 MJ are used for production of the steel products and 0.52 MJ for their delivery. The difference between the wood weights required for production and the weight in the product represents an average of 5% waste. All energy values were based on the typical situation in the United Kingdom at the time of the analysis.

7.5 PAPER AND PAPERBOARD

Paper and paperboard are widely used in packaging, in forms ranging from corrugated boxes to glassine margarine wraps. In fact, it is estimated that about half of all packaging is paper and paperboard. The environmental impact associated with paper packaging includes impact resulting from harvesting and distribution of wood, pulping processes, paper and paperboard manufacture, converting operations, filling and sealing, and distribution.

7.5.1 Raw Materials

The raw materials used to make paper and paperboard packaging are generally softwoods, though hardwoods are used in making corrugated medium (the fluted inner layer in corrugated board). As mentioned above, wood is, in principle, a renewable resource. Recycled paper is also an important fibre source.

7.5.2 Energy

Energy requirements for making paper and paperboard packaging are a function of transportation distances for raw materials, pulping processes used, papermaking machinery employed, thickness, and other variables. Energy required for conversion of the paper material into the finished product likewise varies depending on the package type being produced and so on. Calculations of energy requirements generally include the energy value of the wood raw material, as well as energy used during manufacture. As a rule, the energy associated with production of the packaging material is much greater than that associated with its conversion to packages.

One of the early reports of comparative energy requirements for packaging systems was published by the Midwest Research Institute in 1974 (Hunt & Welch, 1974). They reported energy requirements of 7.6 MJ (7,515 Btu) for production of a paper gallon milk container, 0.65 MJ (612 Btu) for production of a paper gallon-size produce bag, 0.34 MJ (324 Btu) for production of a 9-oz paper vending cup, and 0.89 MJ (847 Btu) for production of a pulp meat tray.

Gaines (1981) listed production energy requirements for various grades of packaging paper, ranging from 58.8 MJ/kg (25,300 Btu/lb) for natural linerboard to 77.6 MJ/kg (33,400 Btu/lb) for bleached board. Use of recycled fibre yielded an energy savings of about 46.4 MJ/kg (20,000 Btu/lb). Corrugating medium, with a recycled content of 30%, had a production energy cost of 37.3 MJ/kg (16,050 Btu/lb). This energy requirement included the combustion energy of the raw material itself. Gaines further estimated that about 75% of the energy input is provided by wood and wood by-products, with the remainder being purchased energy. About 30% of the purchased energy was electricity, and most of the rest was oil and natural gas.

Bousted and Hancock (1981) reported energy requirements for pulp production ranging from 29 to 31 MJ/kg. They chose 28.59 MJ/kg as a representative average, and selected 2% fibre loss as representative. Based on these values, they calculated production of 1 kg base paper from standing timber required 83.91 MJ, production of 1 kg of coated paper from standing timber required 99.95 MJ, and production of 1 kg of paperboard from standing timber required 102.85 MJ of energy. The energy requirement for producing paper sacks from paper was found to be insignificant compared with the energy required for producing the paper itself.

Franklin Associates (1992) report values for manufacture and fabrication of paper products that exclude the energy value of the wood raw material. They report an average value of 50.19 MJ/kg (21,624 Btu/lb) for manufacture and fabrication of unbleached kraft paper and paperboard products, 64.18 MJ/kg (27,650 Btu/lb) for bleached kraft paper and paperboard products, and 46.95 MJ/kg (20,227 Btu/lb) for manufacture of corrugated board products. Energy for manufacture of wax-coated paper products is reported as 70.92 MJ/kg (30,555 Btu/lb) and for low-density polyethylene (LDPE)-coated paper products as 76.98 MJ/kg (33,166 Btu/lb), including the energy content of the petroleum feedstock but excluding that of the paper. Cellophane manufacture and fabrication was reported to require 135.23 MJ/kg (58,266 Btu/lb); moulded pulp products from virgin fibre, 56.42 MJ/kg (24,310 Btu/lb); and moulded pulp products from recycled fibre, 46.55 MJ/kg (20,058 Btu/lb).

Fenton (1991) reported published values for energy requirements for kraft paper bags ranging from 47.83 to 61.87 MJ/kg. Van Eijk, Nieuwenhuis, Post, and DeZeeuw (1992) reported an energy requirement of 50.5 MJ/kg for production of polyethylene-coated kraft paperboard for the manufacture of cups (in the Netherlands).

The Tellus Institute (1992) reported energy requirements ranging from 55.3 MJ/kg (47.57 MM Btu/ton) for corrugating medium to 89.8 MJ/kg (77.29 MM Btu/ton) for bleached kraft paperboard. These values also included the energy value of the incorporated wood. For recycled grades, energy requirements ranged from 40.1 MJ/kg (34.48 MM Btu/lb) for unbleached coated boxboard to 41.2 MJ/kg (35.46 MM Btu/lb) for recycled linerboard. The study includes analysis of energy requirements for each step of the production process.

Gaines and Stodolsky (1993) report energy requirements for production of 100% recycled paperboard as 23.1 MJ/kg (19.9 x 10^6 Btu/ton).

Franklin Associates (1990) reported life cycle energy requirements for 9-inch paperboard plates (1.04 MJ [984 Btu]), 4-inch paperboard hinged containers (0.97 MJ [922 Btu]), and 16-ounce wax and LDPE-coated paperboard cups (0.93 MJ and 0.76 MJ [878 and 718 Btu]). Of the total energy required, 95%–96% was for the containers themselves, 3.6–4.7% was consumed by secondary packaging, and disposal energy was insignificant.

7.5.3 Other Environmental Impacts

Many studies, especially earlier ones, comparing environmental effects of various packaging materials sum emissions across a category and simply report tonnage of air emissions and tonnage of water emissions. They then use these values to compare materials. This methodology cannot be justified, because there are major differences between pollutants of various chemical types in toxicity and other environmental impacts that are obscured by aggregation of data. Therefore, such aggregated values are not reported here.

Effluents emitted during the manufacture of paper and paperboard packaging arise mostly during the paper manufacturing stage. The types and amounts of effluents emitted are highly process-specific, depending heavily on the particular pulping process employed, along with other process variables. The Tellus Institute study (1992) provides a detailed description of the various stages of papermaking and the effluents associated with each.

Harvesting and transportation of the wood raw material can result in soil erosion if logging roads are improperly constructed, and in emission of particulates into the atmosphere if harvesting residues are burned. Washing of logs before debarking, a common operation, is estimated to result in 0.5 g to 4.0 g of biological oxygen demand (BOD) per kg of wood (1–8 lb/ton), and in 2.5 g to 27 g of total suspended solids (TSS) per kg of wood (5–55 lb/ton). The suspended solids are primarily silt. Debarking contributes 7.5 g to 10 g of BOD and 25 g to 50 g of TSS per kg of wood debarked (15–20 lb/ton BOD, 50–100 lb/ton TSS) if wet drum debarking is used. Hydraulic debarking results in lower emissions, 0.5–5.0 g BOD and 3.0–27.5 g TSS per kg (1–10 lb/ton BOD, 6–55 lb/ton TSS). (Tellus Institute, 1992).

Kraft pulping uses sodium hydroxide and sodium sulfide (added as sodium sulfate) as pulping chemicals. It is the most widely used pulping process for packaging paper and paperboard. The Tellus Institute study (1992) estimates that a third to a half of the BOD and TSS from a kraft mill arise from spills and wash-ups related to equipment breakdown, routine maintenance, power failures, shutdowns and start-ups, and product grade changes. Air emissions include highly objectionable odours caused by emission of reduced sulfur compounds. The primary sulfur emissions are hydrogen sulfide, methyl mercaptan, dimethyl sulfide, and dimethyl disulfide. Recovery of pulping chemicals is routinely practiced, and in part involves combustion of the "black liquor" effluent from the pulping digester, after it has been concentrated. The black liquor contains a substantial amount of organic matter, and its combustion provides process heat, as well as aiding in the conversion of the pulping chemicals back to desired forms.

The other major pulping process used for packaging paper is neutral sulfite semi-chemical (NSSC) pulping, which is the major process used in making corrugating medium. These plants emit particulates of sodium sulfate and sodium carbonate if spent liquor recovery systems are not used. They typically operate with a high degree of water recycling, so the effluent flow is low, but the BOD load in the effluent is high.

Controlled air emissions for kraft and NSSC pulping processes are summarised in Table 7–1. Water emissions from paper manufacturing processes are listed in Table 7–2.

When white paper or paperboard is desired, the pulp must be bleached to remove the brown colour that naturally results from both kraft and NSSC pulping. Bleaching operations historically typically employed chlorine, chlorine dioxide, and caustic. Energy use and emissions associated with these processes are reported in the Tellus Institute study (1992); they include chlorine gas, chlorine dioxide gas, chloroform, and sometimes sulfur dioxide. Bleach plants are reported to be the largest source of water pollution in a bleached board mill, contributing up to 40% of the BOD, 25% of the suspended solids, 70% of the colour, and virtually all of the chlorinated organics in the waste stream. Bleaching chemicals are generally not recovered. However, significant changes in bleaching operations are under way, usually involving changing from chlorine bleaching to bleaching with peroxides, and resulting primarily from the need to reduce emissions of tetrachlorodibenzo-p-dioxins and tetrachlorodibenzofurans associated with chlorine bleaching. Therefore, the reported emissions are out of date, but current values are not yet available.

Table 7–1 Controlled and Uncontrolled Air Emissions from Pulping Processes (g/Bone Dry kg Unbleached Pulp)

Emission Type	*Kraft (Uncontrolled)*	*Kraft (Controlled)*	*Neutral Sulfite Semi-Chemical (NSSC) (Uncontrolled)*
Total suspended particulates	151.8	11.5	78.9
SO_x	2.89	2.36	11.6
NO_x	3.06	8.93×10^{-5}	0.95
Volatile organic compounds	1.39	3.61×10^{-7}	0.157
CO	6.19	6.14	2.49×10^{-2}
Lead	1.06×10^{-5}	2.38×10^{-8}	2.51×10^{-4}
Methane	8.54×10^{-5}	1.93×10^{-7}	1.60×10^{-3}
Hydrogen sulfide	4.45	1.24	
Reduced sulfur compounds	3.44	0.61	
Sodium hydroxide	2.22	1.39×10^{-2}	

Note: Data derived from Tellus Institute, 1992; controlled emissions for NSSC not available.

Table 7–2 Controlled Water Emissions from Production of Kraft Linerboard, Corrugating Medium, and Kraft Paper (g/kg paper)

Emission Type	*Liner*	*Medium*	*Paper*
5-day biological oxygen demand	1.77	2.53	1.96
Suspended solids	2.32	4.09	3.83
Abietic acid	4.72×10^{-4}	1.05×10^{-3}	1.48×10^{-2}
Dehydroabietic acid	5.19×10^{-4}	3.77×10^{-4}	7.64×10^{-3}
Isopimaric acid	2.83×10^{-4}	1.88×10^{-4}	1.34×10^{-3}
Pimaric acid	4.72×10^{-5}	1.07×10^{-4}	2.87×10^{-3}
Oleic acid	1.79×10^{-3}	8.89×10^{-4}	3.41×10^{-3}
Linoleic acid	0.00	3.77×10^{-4}	0.00
Linolenic acid		9.44×10^{-4}	
Chlorodehydroabietic acid		6.46×10^{-4}	
Xylenes	0.00	5.39×10^{-5}	0.00
Benzene		5.39×10^{-5}	
Chloroform	0.00	0.00	0.00
Ethylbenzene		2.69×10^{-5}	
Methylene chloride	0.00	1.34×10^{-4}	3.59×10^{-4}
Naphthalene		0.00	
Isophorene	0.00		
Pentachlorophenol		2.69×10^{-5}	
Phenol	1.42×10^{-4}	3.77×10^{-4}	0.00
Bis(2-ethylhexyl) phthalate	1.42×10^{-4}	4.04×10^{-4}	8.94×10^{-5}
Butyl benzyl phthalate		0.00	
Di-n-butyl phthalate	4.72×10^{-5}	0.00	0.00
Toluene	0.00	5.39×10^{-5}	0.00
Trichloroethylene		0.00	
Chromium	3.31×10^{-4}	5.09×10^{-4}	1.07×10^{-3}
Copper	2.36×10^{-4}	6.74×10^{-4}	8.09×10^{-4}
Lead	2.36×10^{-4}	1.34×10^{-4}	1.43×10^{-3}
Mercury	1.18×10^{-6}	6.74×10^{-7}	2.24×10^{-6}
Nickel	2.36×10^{-4}	2.69×10^{-4}	4.48×10^{-4}
Zinc	3.16×10^{-3}	1.86×10^{-3}	7.24×10^{-3}

Note: Data derived from Tellus Institute, 1992.

After pulping, the fibres are subjected to a mechanical treatment known as refining before being made into paper. Various additives can be employed to impart desired properties to the paper or paperboard, and different types of pulp may be blended together. Common additives include sizing agents such as rosin and alum combinations; fillers such as clay, calcium carbonate, or titanium dioxide; and starch or synthetic polymers to enhance strength or performance. The actual formation of paper or paperboard from pulp generally employs either a fourdrinier machine or a cylinder machine. The pulp is used in a dilute form, generally only about 0.5% fibre and filler, with the rest being water. The "whitewater" that leaves the machine contains substantial amounts of fibre and is commonly reused. Slimicides, biocides, and defoamers are sometimes added to the water to inhibit growth of biological organisms in the water and to reduce foaming. Much of the energy used in papermaking is expended to dry the paper from its typical 65% water content on leaving the press, to its final water content of about 6%. Air emissions during this process are insignificant compared with those during pulping. However, water emissions of many types result, as summarised in Table 7–2 (Tellus Institute, 1992).

Paper or paperboard for packaging use may be coated with a variety of materials. Environmental effects are determined by the particular coating and coating application techniques used. Typical effluents from production of polyethylene-coated bleached kraft paper for paper cups are listed in Table 7–3 (van Eijk et al.,

Table 7–3 Effluents from Production of LDPE-Coated Bleached Kraft Paper for Production of 1 kg Cups (g/kg)

	Paper	*LDPE*	*Total*
Air emissions			
Particulates	3.27		3.27
SO_x	13.06	0.08	13.14
NO_x	5.98	0.08	6.06
CO	3.27	0.05	3.31
Hydrocarbons	7.87	0.61	8.48
Water emissions			
Biological oxygen demand	3.00		3.00
Chemical oxygen demand	25.3		25.3
Suspended matter	1.84		1.84
Oils and fats	0.038	0.02	0.06
Phenols		2.67×10^{-4}	2.67×10^{-4}
Dissolved matter	2.68	1.64	4.32
Organic substances		0.01	0.01
Fluoride	2.34×10^{-3}	6.67×10^{-5}	2.41×10^{-3}
Salts	44.9		44.9
Mercury	1.22×10^{-6}		1.22×10^{-6}
Chloride	18.7		18.7
Ammonia	1.08×10^{-3}		1.08×10^{-3}

Note: Data from van Eijk et al., 1992.

1992). Franklin Associates (1990) also present data for atmospheric and waterborne emissions for hinged paperboard containers, paper plates, and wax-coated and polyethylene-coated paper cups.

Solid waste resulting from papermaking is primarily residues from air pollution control devices and sludge from wastewater treatment (Tellus Institute, 1992).

Instead of using virgin fibre, paper and paperboard can be made in whole or in part from recycled fibres. These materials must be repulped and cleaned before use, and depending on the application, may also need to be de-inked and/or bleached. However, for most packaging applications, de-inking and bleaching are not required. Typically, about 5% to 10% of the input wastepaper ends up as waste for disposal. Typical emissions from production of paper and paperboard from recycled fibre are presented in the Tellus Institute (1992) study.

7.6 ALUMINIUM

Aluminium is widely used in packaging for beverage cans, especially for carbonated beverages such as soft drinks and beer. Other major uses include semi-rigid packaging such as foil trays and pie plates, and flexible packaging, in which aluminium foil is often combined with other materials such as paper and plastics. Aluminium is also used in paper and film. In flexible packaging applications, aluminium most often functions as a barrier layer.

7.6.1 Raw Materials

Aluminium is produced from bauxite ore, which consists of 40% to 60% aluminium oxide, and the remainder iron oxide, titanium oxide, and silicon oxide (Tellus Institute, 1992). The ore is typically mined by open pit methods and is then crushed, washed, screened, and dried, producing a relatively pure bauxite. The purified bauxite ore is converted to alumina by the Bayer process, which involves mixing the bauxite ore with lime, grinding it, and then mixing it with caustic soda. The resulting solution is pumped into large pressure tanks (digesters) where steam heat and pressure are applied, resulting in dissolving of the alumina in the caustic soda, forming sodium aluminate. Clarification and filtration of the liquid separates the aluminium solution from the impurities. Next the aluminium is recovered by precipitation as aluminium trihydrate. The crystals are filtered, washed, and then subjected to high temperatures to drive off the water, leaving pure alumina (Al_2O_3). Next, the alumina is subjected to electrolytic reduction to pure aluminium and oxygen, generally using the Hall-Heroult method. In this process, alumina is dissolved in a molten bath of a fluoride electrolyte that contains cryolite, fluorspar, and aluminium fluoride in an electrolytic cell consisting of a carbon cathode cell lining and a consumable anode made of petroleum coke and pitch. A high-amperage, low-voltage current reduces the aluminium to molten metal, which pools at the bottom of the cell. Oxygen reacts with carbon, sulfur, and other impurities at the anode, forming carbon dioxide, carbon monoxide, and sulfur dioxide. The molten aluminium is removed from the cells with a ladle or crucible and is then cast as ingots. Alloying materials can be added during this step (Tellus Institute, 1992).

Aluminium can stock is made from aluminium alloyed with magnesium and manganese. A typical can body stock contains 0.8% to 1.3% magnesium and 1.0% to 1.5% manganese, while a typical end stock contains 4.0% to 5.0% magnesium and 0.2% to 0.5% manganese (Boustead & Hancock, 1981).

Production of 1 kg of aluminium ingot requires mining approximately 0.16 kg of limestone, 2.75 kg of bauxite, and 0.14 kg of salt. In addition, 0.55 kg of crude oil is required for the manufacture of coke and pitch (calculated from Hunt & Welch, 1974).

Aluminium for packaging use can also be produced from recycled aluminium, and in fact this is widely done. The difference in alloys for body and end stock results sometimes in the need to remove magnesium from body stock manufactured from a very high percentage of recycled cans. This process is known as demagging, and uses fluxing agents such as chlorine, aluminium fluoride, or aluminium chloride to remove magnesium and other contaminants (Tellus Institute, 1992).

7.6.2 Energy

Mining and initial processing of bauxite consumes approximately 0.59 MJ of energy per kg of bauxite produced (0.51 MM Btu/ton). Lime production consumes 5.70 MJ energy per kg of lime produced (4.91 MM Btu/ton). The Bayer process for alumina production uses 26.1 MJ energy per kg of alumina (22.5 MM Btu/ton). Reduction of the alumina to aluminium consumes 241.6 MJ/kg aluminium (208 MM Btu/ton) (Tellus Institute, 1992).

Processing of collected beverage cans into hot metal for recycling consumes about 9.60 MJ/kg (8.26 MM Btu/ton) (Tellus Institute, 1992), and thus represents a very significant energy savings.

Boustead and Hancock (1981) calculated the total energy required to produce 1 kg of aluminium can body stock in the United Kingdom as 346 MJ. Gaines and Stodolsky (1993) calculated US energy consumption for production of aluminium beverage containers as 230.6 MJ/kg (198.7 x 10^6 Btu/ton) from virgin raw materials, and 52.1 MJ/kg (44.9 x 10^6 Btu/ton) if 100% recycled aluminium is used. Little (1984) calculated energy requirements for production of aluminium ingot from virgin raw materials as 204.5 MJ/kg, and from 100% recycled materials as 11.4 MJ/kg. For virgin aluminium, 69% of the energy was electricity, while for recycled aluminium, only 22.5% was electricity. Franklin Associates (1992) reported average energy requirements for manufacture of aluminium products as 241.7 MJ/kg (104,131 Btu/lb) from virgin aluminium and 57.2 MJ/kg (24,651 Btu/lb) from 100% postconsumer raw material.

7.6.3 Other Environmental Impacts

Bauxite mining and processing of bauxite produces about 0.75 g total suspended particulate air emissions per kg bauxite produced (1.5 lb/ton), with emission controls in place, and not including emissions associated with energy production. Lime production has controlled emissions of about 21.4 g TSP per kg lime, 1.4 g NO_x/kg, and 1.0 g CO/kg. Production of caustic results in air emissions of 1.2 g chlorine/kg caustic and water emissions of 7.04 x 10^{-3} g chlorine, 4.37 x 10^{-3} g copper, 2.14 x 10^{-3} lead, and 3.30 x 10^{-6} g nickel per kg caustic. Alumina production results in air emissions of 19.2 g TSP per kg alumina. Production of 1 kg aluminium from alumina results in air emissions of 5.9 g TSP and 3.70 g fluoride, and water emissions of 1.91 x 10^{-2} g fluoride, 2.41 x 10^{-5} g benzo(a)pyrene, 4.23 x 10^{-3} g aluminium, 1.34 x 10^{-3} g antimony, and 5.74 x 10^{-4} g nickel. In all of the above, emissions associated with energy production are neglected (Tellus Institute, 1992).

Petroleum coke production is associated with a variety of air and water emissions. A summary of total emissions, as well as breakdowns by subprocess, is available in the Tellus Institute report (1992).

Recycling of aluminium results in air emissions of 25.6 g total suspended particles (TSP) per kg aluminium, again not including emissions associated with process energy. Nearly all of these emissions are associated with demagging. Controlled water emissions for recycling processes were not available (Tellus Institute, 1992).

It should also be noted that during the electrolytic reduction of alumina, the carbon cathode lining of the cell becomes contaminated with fluorides, sodium, cyanide, and other hazardous compounds, resulting in its classification as a hazardous waste (Tellus Institute, 1992).

7.7 STEEL

The largest use of steel in packaging in the United States is in steel food cans. Steel is also used for beverage cans, drums, strapping, and other forms of packaging.

7.7.1 Raw Materials

The major raw materials required for the manufacture of steel are iron ore, limestone, and coal. Iron ore is most often produced from open pit mines. Much of the iron ore used is relatively low-grade ore, containing 23% to 25% iron. A total of 3.2 tons of ore are required to produce 1 ton of iron pellets. Processing of the ore includes crushing, grinding, and separation of the iron-containing particles from the rest of the material, using either magnetic separation or separation by density. The separated material is then pelletised by adding a binder and hardening the material in a furnace. The resultant pellets contain about 64% iron. Limestone is also produced in open pit mines. Treatment includes crushing and screening. In some cases, the limestone is further processed in a cement kiln to convert the limestone to lime. Coal can be produced in either open pit or underground mines. It is cleaned to lower its ash and sulfur content, and then subjected to destructive distillation to yield coke. The coke is used in the blast furnace as both a fuel and an oxygen-reducing agent. About 1.2 kg of mined coal is needed to produce 1 kg of clean coal, which then yields about 0.75 kg of coke. The blast furnace is fed with iron ore, coke, and lime, producing molten iron (pig iron) and slag. Next, the pig iron is converted to steel, generally, for packaging uses, in a basic oxygen furnace. This furnace is charged with a combination of steel scrap and molten pig iron, and oxygen is supplied to oxidise impurities. After an initial heating period, lime and fluorspar are added to combine with the impurities and form a slag layer. Alloying metals are added, if desired, and the steel is then delivered for casting or ingot production. Table 7–4 summarises the raw materials typically used for steel production (Tellus Institute, 1992).

Table 7–4 Raw Materials Used for Production of Steel in a Basic Oxygen Furnace (g/kg Steel)

Material	*Quantity*
Pig iron	829
Scrap	320
Limestone	5
Lime	75
Fluorspar	8
Ferromanganese	11
Ferrosilicon	1
Aluminium	0.5
Pellets	3
Mill scale	8
Refractories	6.5
Oxygen	6.0×10^{-5} m^3

Note: Data from Tellus Institute, 1992.

For most packaging applications, provision of a non-corrosive layer to keep the steel from rusting is required. This is generally either a layer of tin (tinplate) or a very thin layer of chromium (tin-free steel, chromium type). Steel scrap is routinely used in steel manufacture. Collected post-consumer steel cans are sometimes detinned before use, and sometimes used directly.

7.7.2 Energy

The Tellus Institute (1992) calculated the energy requirement for production of steel from the mix of ingredients listed in Table 7–4 as 22.8 MJ/kg (19.6 MMBtu/ton). Boustead and Hancock (1981) calculated the energy requirement to produce 1 kg of liquid steel in the United Kingdom from ore as 20.4 MJ, and that to produce 1 kg of hot rolled strip from ore as 30.67 MJ. Production of 1 kg of tinplate sheet from ore required 49.70 MJ, and production of 1 kg of tin-free steel sheet required 48.08 MJ. Production of 1 kg of steel strapping used 35.56 MJ. Franklin Associates (1992) calculated the average energy requirements for manufacture of coated steel products for packaging as 48.47 MJ/kg (20 885 Btu/lb). Gaines and Stodolsky (1993) calculate the energy requirement for production of steel cans from virgin raw materials as 65.2 MJ/kg (56.2 x 10^6 Btu/ton), and from recycled materials as 50.8 MJ/kg (43.8 x 10^6 Btu/ton).

Energy requirements to produce steel cans or other packaging materials from sheet are dependent on the design and size of the package. Gaines (1981) lists the energy requirement for the manufacture of 12-oz steel beverage containers with aluminium easy-open ends as 151 MJ/kg from virgin material and 98 MJ/kg from recycled material. Little (1991) lists the total life cycle energy requirement for 2-piece bimetallic 12-oz beverage cans as 102 MJ/kg, for 2-piece draw/redraw tinfree steel tuna cans as 108 MJ/kg, and for three-piece welded tinplate vegetable cans as 83.5 MJ/kg. These life cycle energy requirements include energy used in distribution and recycling as well as manufacture, raw materials, and so on.

7.7.3 Other Environmental Impacts

Air and water emissions associated with mining, concentration, and pelletising of iron ore are not available. Production of lime results in air emissions of 19.7 g TSP per kg lime produced, along with 1.3 g of NO_x and 0.9 g of CO. Production of coke from coal results in emissions of 1.48 g TSP/kg coke, 0.04 g benzene, 2.47 x 10^{-3} g POM, 0.22 g toluene, and 0.02 g manganese. These emissions do not include emissions associated with required energy or with coal mining. Coke production also results in emission of a long list of water pollutants. The Tellus Institute study (1992) contains a summary of the emissions associated with the overall steel production process, along with breakdowns by subprocess. Comparable values for steel production with 40% recycled content are also presented.

7.8 GLASS

Glass packages are primarily bottles and jars. Because of the inertness, transparency, barrier, and consumer appeal of glass, it is widely used, especially for food and beverage products.

7.8.1 Raw Materials

The primary raw materials for glass production are sand (or sandstone, silicon dioxide), soda ash (sodium carbonate), limestone (calcium carbonate), and feldspar (aluminium silicate). The carbonates are converted to oxides during glass formation, yielding a packaging glass with a typical composition of 70% silicon dioxide, 15% sodium oxide, 12% calcium oxide, 2% aluminium oxide, and 1% other ingredients. Cullet (recycled glass) is a necessary ingredient for efficient glass production.

The raw materials required to manufacture 1 kg of glass containers include 591 g sand, 181 g limestone, 84 g feldspar, 197 g soda ash, 100 g cullet, and 13.8 g of other materials (Tellus Institute, 1992).

Limestone is generally quarried in open pit mines. Processing consists of crushing and screening, and results in about 90 g solid waste per kg of limestone produced (Tellus Institute, 1992).

Soda ash can be produced from natural minerals or can be manufactured synthetically. All the soda ash used for glassmaking in the United States is natural, and the United States is a net exporter of soda ash. Natural soda ash is produced from trona ore, which contains 86% to 95% sodium sesquicarbonate. The ore is crushed, screened, and calcined to form sodium carbonate, which is then purified, crystallised, and dried (Tellus Institute, 1992). In the United Kingdom, all soda ash used is synthesised from limestone and salt (Boustead & Hancock, 1981) or imported.

In-house cullet typically constitutes 8.5% of the batch weight. The remainder of the cullet used comes from post-consumer sources. Current cullet use rates in the United States are about 25% to 30% for flint (clear) glass, 50% for amber glass, and 70% for green glass (Tellus Institute, 1992). One hundred percent cullet can be used for glass manufacture, although this is seldom done.

Processing of collected cullet for use includes size reduction and removal of contaminants. Colour separation is, as a rule, required before crushing of the glass. The beneficiation process for cullet is designed to remove contaminants, especially aluminium, plastic, and paper.

7.8.2 Energy

According to the Tellus Institute (1992), production of one kg of limestone requires 0.19 MJ of energy. Production of 1 kg of natural soda ash requires 9.98 MJ of energy. The total energy input for the manufacture of 1 kg of glass containers is estimated as 15.7 MJ (13.5 MM Btu/ton). Production of glass containers from 100% cullet is estimated to require 11.5 MJ/kg (9.92 MM Btu/ton).

Franklin Associates (1992) calculate the average energy requirement for production of glass containers as 17.01 MJ/kg (7,330 Btu/lb), using 21% post-consumer cullet and 11.9% in-house cullet.

Boustead and Hancock (1981) report the energy requirement for the production of synthetic soda ash as 14.23 MJ/kg soda ash. They report the average energy required to produce 1 kg of saleable glass containers from a mixture containing 20% in-house cullet, in the United Kingdom, as 22.2 MJ. Individual glass factories had energy requirements ranging from 18.03 MJ/kg to 29.39 MJ/kg. The effect of glass colour on energy requirements was considerably smaller than the average variation between plants.

Gaines (1981) reports energy requirements for 12-oz beverage containers as 20.2 MJ/kg for virgin glass and 15.2 MJ/kg for recycled glass. Gaines and Stodolsky (1993) report the energy requirements for glass containers at the maximum sustainable recycling rate as 15.3 MJ/kg (13.2 x 10^6 Btu/ton).

7.8.3 Other Environmental Impacts

The Tellus Institute study (1992) estimates emissions from glass manufacturing, using 10% cullet (in-house) and using 100% cullet, to be as shown in Table 7–5.

7.8.4 Reuse of Glass Containers

Glass bottles, especially for carbonated beverages, are one of very few types of consumer packages that have an extensive history of reuse. Although reuse of glass containers has become much less common, it still occurs to a significant degree in many places.

The environmental impacts of reuse of glass packaging can be divided into three categories. First, because bottles designed for reuse need to maintain strength for a longer time and through more abuse, they must be made stronger initially than one-way containers. Therefore, reusable glass bottles are heavier than non-reusable ones. This does not appreciably affect the environmental impact discussed on a weight basis but makes a considerable difference in per container impact. Gaines (1981) estimates that refillable bottles are, on average, 50% heavier than nonrefillable ones and therefore require 50% more energy to manufacture. It should also be remembered that increased size of the reusable bottles will, in general, lead to an increase in the size of any associated secondary and tertiary packaging as well.

Table 7–5 Controlled Emissions from Glass Manufacturing (g/kg Glass Containers)

Emission Type	*10% Cullet*	*100% Cullet*
Air		
Total suspended particulates	0.31	0.10
SO_x	2.09	0.96
NO_x	4.36	2.70
Volatile organic compounds	4.48	4.43
CO	0.33	0.15
Lead	8.34×10^{-4}	8.44×10^{-5}
Methane	0.015	.005
Water		
Oil	4.69×10^{-4}	4.69×10^{-4}
Total suspended solids	4.69×10^{-4}	4.69×10^{-4}

Note: Data derived from Tellus Institute, 1992.

The second category of impact is that associated with the return system. Effects associated with transportation of the containers from the use site to the refill location are, in principle, no different from effects associated with transportation of containers from the production site to the fill location. In general, transportation effects are heavily dependent on the weight and/or volume of the packages, and on the distance involved, as discussed in Section 7.10. Gaines (1981) states that the marginal energy cost for reuse is low, since the nonrefillable bottles must be transported to disposal or recycle anyway. This argument is valid only if the distances are comparable.

The remaining category of impact is that associated with the cleaning necessary to prepare the containers for refill. The cleaning cycle involves use of detergents and of significant quantities of water and energy. Detergents, as well as product residues and dirt, are discharged into the wastewater stream. Van Eijk et al. (1992) present an evaluation of the environmental impact of washing porcelain crockery in a dishwasher, a process that is similar (although certainly not identical) to that undergone by glass bottles. Gaines (1981) states that new glass bottles must also be washed, therefore decreasing the net effect of this aspect of the reuse cycle. However, Boustead and Hancock (1981) point out that returnable bottles need a hot detergent wash to remove labels and clean and sterilise the bottles, while nonreturnable bottles require only a rinse. They calculated the energy required to fill nonreturnable beer bottles as 64% to 84% of the energy required to fill returnable beer bottles, with the greatest difference for the smallest bottles. For soft drinks, the range was 57% for 6 ounce bottles to 87% for 1-litre bottles.

Gaines and Stodolsky (1993) report on a German study that found the energy requirements for a refillable beer bottle used 50 times were only 6% of that for single-use bottles.

7.9 PLASTICS

The use of plastics in packaging continues to increase rapidly. Plastic packaging often permits significant source reduction (use of less packaging, by weight or by volume), as well as significant cost advantages. Unlike other packaging materials, it is difficult to assess the environmental impact of plastics packaging as a group, because various specific plastics have their own characteristics, which can differ significantly from other members of the plastics packaging family. Therefore, in addition to some general discussion, we look in more detail at the major packaging plastics, including low- and high-density polyethylene (LDPE and HDPE), linear low-density polyethylene (LLDPE), polypropylene (PP), polystyrene (PS), polyvinyl chloride (PVC) and polyethylene terephthalate (PET).

7.9.1 Raw Materials

The basic raw materials for production of plastics are petroleum and natural gas. Natural gas yields ethane and propane for plastics production, while petroleum yields propane, butane, naphtha, and gas oil. Propane and butane, combined, are termed liquefied petroleum gas (LPG).

Natural gas is processed by removal of water and hydrocarbon condensate, removal of non-hydrocarbon gases, liquefaction, and distillation. The hydrocarbon condensate is heavy hydrocarbons with five or more carbon atoms. The principal non-hydrocarbon gas found in natural gas is hydrogen sulfide, although nitrogen, hydrogen, and helium are also found. Liquefaction and distillation are carried out to separate the hydrocarbons from each other. Up to 97% of the natural gas is methane. Heavier hydrocarbons are separated from methane by liquefaction. Distillation, or fractionation, of the liquefied gases separates ethane, propane, and butane from each other. Ethane is the major feedstock for manufacture of ethylene, and propane is used to produce propylene (Tellus Institute, 1992).

Processing of crude oil also involves separation of the oil mixture into a number of components. Because crude oil is chemically a more complex mixture than natural gas, its processing is also more complex. Preliminary processing of petroleum involves removal of water and suspended solids, and removal of dissolved salts in the oil (desalting). This is followed by a multi-stage distillation process that produces, among other products, propane, naphtha, and gas oil. Residuals from the distillation are sent through vacuum distillation and catalytic cracking, yielding naphtha and gas oil, among other products.

The four major petrochemical building blocks for production of packaging plastics are ethylene (used to make polyethylene, polystyrene, PET, and PVC), propylene (used to make polypropylene), benzene (used to make polystyrene), and paraxylene (used to make PET). Ethylene and propylene can be made from liquefied petroleum gases, naphtha, and gas oil, as well as from natural gas. Benzene and paraxylene are manufactured from naphtha and gas oil (a small amount of benzene is made from coal) (Tellus Institute, 1992).

7.9.2 Energy

Onshore natural gas extraction and processing is estimated to require 0.793 MJ per MJ of natural gas produced, if a mixture of 50% sweet gas and 50% sour gas is produced. Sour gas has a hydrogen sulfide content greater than 0.25 grains per 100 standard cubic feet, and requires treatment with an amine solution to absorb the hydrogen sulfide. Processing of sweet natural gas requires only 0.310 MJ per MJ of natural gas produced, while processing of sour gas requires 1.132 MJ per MJ of gas. The drilling process requires 0.072 MJ per MJ of natural gas. The total energy input to produce 1 kg of ethane from natural gas is 91.3 MJ (78.58 MM Btu/ton). A total of 89.7 MJ of energy is required to produce 1 kg of propane from natural gas (77.22 MM Btu/ton) (the fuel value of the product is included) (Tellus Institute, 1992).

Onshore extraction of petroleum requires 0.00148 MJ per MJ of crude oil produced. Production of propane from crude oil requires 51.8 MJ per kg of propane. Production of naphtha by atmospheric distillation requires 45.4 MJ per kg (39.04 MM Btu/ton). Production of naphtha by catalytic cracking requires 51.3 MJ per kg (44.15 MM Btu/ton) (Tellus Institute, 1992).

According to the Tellus Institute study (1992), production of ethylene from ethane requires 110.7 MJ/kg, after credit for by-products is taken and including energy content of the raw materials. For manufacture of ethylene from propane, 71.04 MJ per kg are required. If the raw material is naphtha, 61.32 MJ are required, and from heavy gas oil, 59.47 MJ. If a raw material mix of 25% from each source is assumed, an average energy cost of 75.64 MJ per kg of ethylene produced can be calculated, again including the energy value of the raw materials.

Production of propylene from ethane requires 110.7 MJ/kg, again including by-product credits and the energy content of the raw material. From propane, 71.08 MJ are required; from naphtha, 61.33 MJ; and from heavy gas oil, 59.47 MJ. If an average split of 2% ethane, 26% propane, 34% naphtha, and 48% heavy gas oil feedstock is assumed, the average energy required to produce 1 kg of propane is 63.95 MJ (Tellus Institute, 1992).

For production of benzene, the Tellus Institute (1992) calculated a split of 62% manufactured by catalytic reforming and 38% by toluene dealkylation. Overall energy requirements were 67.47 MJ/kg. For paraxylene formation by catalytic reforming, energy requirements were 93.02 MJ/kg.

7.9.3 Other Environmental Impacts

A variety of environmental impacts are associated with natural gas and petroleum drilling and processing. Drilling wastes include drilling muds, drill cuttings, and drilling fluids. The American Petroleum Institute estimates the average amount of drilling waste per well as 5,183 barrels (Tellus Institute, 1992). Pollutants in these wastes include a large number of compounds in small concentrations. Air pollutants during drilling operations arise primarily from the large diesel engines used to power the drill and include NO_x, CO, SO_x; volatile organic compounds (VOCs); and particulates. These emissions are generally uncontrolled.

Production wastes for oil and natural gas include water found in the oil or gas, along with water added during removal of the gas or oil from the well. As in drilling effluents, a large number of pollutants are found in small concentrations. Air emissions arise from fuel use and from oil and gas separators, gas dehydrators, and heater treaters (which remove water from crude oil). Other wastes include tank bottoms, separator sludges, cooling water, and other miscellaneous wastes (Tellus Institute, 1992).

Processing of oil and natural gas also results in a variety of emissions, mostly in small quantities. Worthy of note are average emissions of 1.23 g SO_x per MJ natural gas produced (Tellus Institute, 1992).

The Tellus Institute summary (1992) of air pollutant emission factors for ethylene production includes controlled emissions of 0.084 g benzene per kg of ethylene. The only other significant air pollutant quantified was 8.34 g/kg uncontrolled emissions of VOCs. A controlled emission value for VOCs was not available. For propylene production, air emissions of 0.50 g VOC/kg propylene were listed as uncontrolled emissions. For benzene production, 0.55 g VOC/kg benzene can be expected in uncontrolled emissions, along with 0.015 g toluene per kg of toluene used in benzene production. For production of mixed xylenes and paraxylene, 0.55 g VOC and 1.50 g xylene are emitted per kg of xylenes produced (Tellus Institute, 1992).

Water pollutant emissions for ethylene, propylene, benzene, and xylene include a wide variety of compounds and are summarised in the Tellus Institute study (1992).

7.9.4 High-Density Polyethylene

HDPE is the plastic resin most often used for bottles. Bottles and other containers are also the most common use of HDPE in packaging. However, a significant amount of HDPE is also used in film applications.

HDPE is manufactured from ethylene at moderate temperatures (93° C) and pressures (20–48 atm) (Tellus Institute, 1992), using a catalyst. The polymerisation processes can be either liquid phase solution process, liquid phase slurry process, or gas phase. HDPE manufactured with Phillips catalysts using a slurry process predominates, accounting for 58% of all HDPE manufactured. Ziegler catalysts and either a solution or slurry process account for 32% of production, and gas phase reactions account for 10% (Tellus Institute, 1992). For analysis of environmental impacts, the Tellus Institute study assumed 74% of HDPE produced by the slurry process, 16% by suspension, and 10% in gas phase reactions.

Production of HDPE by the slurry process involves feeding ethylene, hydrogen as a molecular weight regulator, a diluent such as hexane, and the catalyst into the reactor. Output from the reactor consists of polymer, diluent, off-bases, unreacted monomer, catalyst, and waxes. The unreacted monomer and diluent are recycled back into the process. The solution process is similar to the slurry process, but temperatures are higher so the polymer is dissolved rather than suspended in particle form. The gas phase polymerisation polymerises purified ethylene in the presence of a dry powder catalyst, a comonomer, and hydrogen as a molecular weight regulator. Unreacted monomer and comonomer are recycled (Tellus Institute, 1992).

Overall energy requirements for production of HDPE are 92.5 MJ/kg, including the energy value of the HDPE. Process energy is 43.1 MJ, with the rest being material energy. One kg of polyethylene requires 1.02 kg of ethylene as raw material. The polymerisation process itself requires 6.60 MJ steam and 8.71 MJ electricity per kg HDPE produced (Tellus Institute, 1992). An earlier estimate by Boustead and Hancock (1981) was that 107.76 MJ/kg are required to produce HDPE from crude oil, again including the energy value of the product.

Franklin Associates (1992) calculate an energy requirement of 87.4 MJ/kg (37,650 Btu/lb) for manufacture of HDPE resin, including the energy content of the raw materials. Fabrication of blow-moulded products of HDPE (and of other plastic resins except PET) is estimated to require 7.68 MJ/kg (3,311 Btu/lb). Fabrication of injection-moulded products (for any plastic) is estimated to require 20.66 MJ/kg (8,901 Btu/lb), and manufacture of film, 8.24 MJ/kg (3,552 Btu/lb). Thermoforming averages 6.92 MJ/kg (2,980 Btu/lb), except for foam polystyrene.

Air emissions associated with HDPE production from ethylene are 16.7 g VOC, uncontrolled, per kg HDPE. Controlled air emission factors were not available. Water emission factors, uncontrolled and controlled, are presented in Tables 7–6

Table 7–6 Uncontrolled Water Emissions from HDPE, LLDPE, and LDPE Polymerisation (g/kg Product)

Pollutant	*HDPE*	*LLDPE*	*LDPE*
5-day biological oxygen demand	5.29×10^{-4}	2.17×10^{-4}	2.14
Chemical oxygen demand	7.04×10^{-4}	2.90×10^{-4}	5.74
Total organic carbon	6.39×10^{-4}	2.62×10^{-4}	3.27
Oil and grease			11.1
Total suspended solids	2.35×10^{-3}	9.64×10^{-4}	4.17
Total dissolved solids	2.10×10^{-2}	8.64×10^{-3}	4.73
Acenaphthylene			139
Benzene	2.52×10^{-5}	1.03×10^{-5}	2.48×10^{-2}
Benzo[*a*]anthracene	1.64×10^{-6}	6.74×10^{-7}	
Dimethyl phthalate			8.24×10^{-3}
Di-n-butyl phthalate			1.11×10^{-3}
Ethylene dichloride			4.76×10^{-4}
Phenol	1.84×10^{-6}	7.54×10^{-7}	1.46×10^{-3}
Toluene	4.04×10^{-4}	1.66×10^{-4}	2.36×10^{-2}
Aluminium	1.29×10^{-2}	5.29×10^{-3}	
Chromium			5.09×10^{-2}
Copper			2.36×10^{-3}
Lead			8.74×10^{-4}
Mercury			2.70×10^{-5}
Titanium	3.54×10^{-2}	1.45×10^{-2}	

Note: Data from Tellus Institute, 1992.

and 7–7. A complete set of controlled water emissions was not available, so these data are incomplete (Tellus Institute, 1992).

7.9.5 Linear Low-Density Polyethylene

The primary use of LLDPE in packaging is in film applications, where it competes primarily against low-density polyethylene. HDPE and LLDPE are linear in molecular structure, while LDPE is a branched structure.

LDPE is produced by processes similar to those used to produce HDPE, but is a copolymer of ethylene and 1-butene, 1-hexene, or 1-octene. Process equipment designed to manufacture either HDPE or LLDPE can, in general, also be used to produce the other, giving manufacturers what is known as "swing capacity".

The two primary methods for manufacture of LLDPE are gas phase and solution polymerisation. In the Tellus Institute study (1992), it was estimated that

63% of LLDPE is manufactured by gas phase processing, and 37% by solution processing.

Energy requirements to produce LLDPE from ethylene and 1-butene by gas phase polymerisation total 83.02 MJ/kg LLDPE (71.46 MM Btu/ton), including 75.87 MJ for 0.99 kg of ethylene, 2.9 MJ for 0.04 kg of 1-butene, and 4.24 MJ of process energy. The estimate for the 1-butene is low, since it does not include energy required to manufacture the 1-butene, only the energy content of the butene itself. Energy values for solution phase processing were not available, so were assumed to be the same as for gas phase polymerisation (Tellus Institute, 1992). Franklin Associates (1992) calculate a requirement of 87.39 MJ/kg (37,650 Btu/lb) for LLDPE production.

Air pollutants from LLDPE production are primarily VOCs, amounting to 25.4 g/kg LLDPE of uncontrolled air emissions. Again, controlled emission amounts were not available (Tellus Institute, 1992).

Table 7–7 Controlled Water Emissions from HDPE, LLDPE, and LDPE Polymerisation (g/kg Product)

Pollutant	*HDPE*	*LLDPE*	*LDPE*
5-day biological oxygen demand			
Chemical oxygen demand			
Total organic carbon			
Oil and grease			
Total suspended solids			
Total dissolved solids			
Acenaphthylene			41.3
Benzene	6.64×10^{-7}	2.74×10^{-7}	7.44×10^{-3}
Benzo[*a*]anthracene	1.11×10^{-7}	4.57×10^{-8}	
Dimethyl phthalate			5.09×10^{-3}
Di-n-butyl phthalate			8.09×10^{-4}
Ethylene dichloride			2.16×10^{-4}
Phenol	2.00×10^{-6}	8.24×10^{-7}	8.79×10^{-4}
Toluene	9.09×10^{-6}	3.75×10^{-6}	7.19×10^{-3}
Aluminium	1.40×10^{-2}	5.79×10^{-3}	
Chromium			1.51×10^{-2}
Copper			7.09×10^{-4}
Lead			2.64×10^{-4}
Mercury			1.80×10^{-5}
Titanium	2.04×10^{-2}	8.39×10^{-3}	

Note: Data from Tellus Institute, 1992; treated wastewater emissions not available for all effluents.

Gas phase polymerisation of LLDPE results in no water pollutants. Solution polymerisation results in water pollutants in essentially the same types and amounts as solution phase HDPE production, except that differences in the proportional use of each polymerisation method must be accounted for (Tellus Institute, 1992). Amounts are shown in Tables 7–6 and 7–7.

7.9.6 Low-Density Polyethylene

The major packaging market for LDPE is film, where, as mentioned earlier, it competes with LLDPE. Coating is another major market. LDPE is the oldest member of the polyethylene family.

LDPE is produced by polymerisation of ethylene at relatively high temperatures (150–300° C) and pressures (1,000–3,400 atm) in an autoclave or tubular reactor. In addition to ethylene, a chain transfer agent such as hydrogen and a free radical initiator such as oxygen or a peroxide are added to the reaction vessel. When the product is removed, the unreacted ethylene is separated and recycled. By-products such as oils and waxes are commonly used as boiler fuel (Tellus Institute, 1992).

The total energy required to manufacture LDPE is 98.3 MJ per kilogram. Process energy amounts to 46.4 MJ, and the remainder is the energy content of the materials; 1.05 kg of ethylene is required to produce 1 kg of LDPE (Tellus Institute, 1992). Boustead and Hancock (1981) estimated the energy required to manufacture LDPE as 21.01 MJ/kg from ethylene or 104.89 MJ/kg from crude oil. To produce 1 kg polyethylene film from crude oil requires, by their calculations, 188.72 MJ. Fenton (1991) reported literature values of energy required to produce LDPE bags ranging from 74.44 to 116.27 MJ/kg. Franklin Associates (1992) calculated a requirement of 95.48 MJ/kg (41,136 Btu/lb) for production of LDPE resin.

The primary air emission from LDPE polymerisation is VOCs, in amounts of 8.73 g per kg LDPE in uncontrolled emissions. Uncontrolled benzene emissions are 1.16 g per kg LDPE. Data for controlled emissions were not available (Tellus Institute, 1992).

Wastewater from LDPE manufacturing is produced during removal of the oils and waxes from the unreacted ethylene and when the pelletised LDPE is water-cooled (Tellus Institute, 1992). Water emissions for LDPE polymerisation are also listed in Tables 7–6 and 7–7.

7.9.7 Polypropylene

PP is used in packaging in the production of closures, as well as in containers and film. Nearly all the polypropylene used in packaging is a stereochemically ordered structure known as isotactic polypropylene. Polymerisation processes generally produce some undesired, unordered atactic polypropylene as well, which does not have desired properties and must be separated from the product.

The polymerisation process for PP may involve solution, suspension, or gas phase reactions. For purposes of analysing environmental impacts, the Tellus Institute study (1992) assumed a mix of 75% suspension polymerisation and 25% gas phase polymerisation. Polymerisation processes can also be classified as high-yield or low-yield, based on the required ratio of catalyst to feedstock. Low-yield catalysts are present in the polymer in high concentrations and must be removed, while high-yield catalysts are present in low concentrations and do not require removal. High-yield catalysts are the preferable method.

In continuous suspension polymerisation with high-yield catalysts, liquid propylene, hydrogen as a molecular weight regulating agent, a polymer suspension medium, and a catalyst are fed into the reactor. Excess liquid propylene is used as a diluent. Polymerisation occurs at a temperature of about 60° C and pressures of 20 atm to 27 atm. After the reaction mixture exits the reactor, unreacted monomer and atactic polymer are removed, and the isotactic PP product is recovered. The process is the same for low-yield polymerisation, except that hexane or xylene is used as a diluent, and the catalyst and diluent must be removed from the product stream (Tellus Institute, 1992).

In gas phase polymerisation, the reaction vessel is fed with a mixture of propylene, hexane as a diluent, and catalysts. Unreacted monomer is recycled back into the reactor. The polymer product is recovered as a fine powder, which is then extruded and pelletised (Tellus Institute, 1992).

To produce 1 kg of PP, 1.05 kg of propylene is required. Energy requirements for the polymerisation process are 5.8 MJ steam and 6.3 MJ electricity per kg polypropylene produced. The total energy requirement, including material energy content, is 97.2 MJ/kg (Tellus Institute, 1992). By Boustead and Hancock's (1981) calculation, the energy required to produce polypropylene from crude oil is 114.12 MJ/kg. Franklin Associates (1992) give a value of 89.02 MJ/kg (38,353 Btu/lb).

As was the case for the polyethylenes, a major air emission during polymerisation is VOCs, in amounts of 36.7 g per kg PP produced in uncontrolled emissions. Other emissions include 36.7 g/kg NO_x and 1.50 g/kg total suspended particulates. Controlled air emissions were not available (Tellus Institute, 1992).

Water effluents are produced during diluent recovery, catalyst recovery, and filtering of wastewater. Water emission data are available only for solution polymerisation (which uses a solvent instead of a diluent), and are presented in Table 7–8. Gas phase reaction produces no water emissions. It is believed that emissions for suspension polymerisation are similar to those for solution polymerisation (Tellus Institute, 1992).

Table 7–8 Water Emissions from Solution Phase PP Polymerisation (g/kg PP)

Pollutant	*Untreated Water*	*Treated Water*
5-day biological oxygen demand	0.120	
Chemical oxygen demand	0.354	
Total organic carbon	0.154	
Benzene	4.02×10^{-4}	1.99×10^{-3}
Bis(2-ethylhexyl) phthalate	4.89×10^{-6}	5.84×10^{-5}
Chloroform	1.71×10^{-5}	2.05×10^{-4}
Ethylbenzene	2.34×10^{-3}	7.74×10^{-3}
1,1,1-Trichloroethane	1.10×10^{-5}	7.74×10^{-3}
Toluene	1.28×10^{-3}	6.19×10^{-3}
Chromium	1.95×10^{-3}	8.64×10^{-3}
Zinc	2.04×10^{-3}	8.99×10^{-3}

Note: Data from Tellus Institute, 1992; treated wastewater emissions not available for BOD5, COD, and total organic carbon, no explanation given for higher emissions for treated compared to untreated wastewater.

7.9.8 Polystyrene

PS is used in packaging in two distinct forms. A large amount of PS in expanded form (foam) is used as cushioning material, either in shapes or as loose fill. It is also used in egg cartons, produce trays, and so on. In non-expanded form (crystal), PS is used in lidding material, yoghurt cups, produce trays, and a variety of other applications. High-impact PS incorporates a comonomer (usually polybutadiene) to improve the impact strength of polystyrene, which, unmodified, is a relatively brittle material.

The basic chemical feedstocks used to produce PS are ethylene and benzene. First, ethylene and benzene are combined, in a process known as alkylation, to form ethylbenzene. Next, ethylbenzene is dehydrogenated to form styrene. This process, carried out at 650° C in the presence of a catalyst, also produces a small amount of toluene as a by-product. Manufacture of 1 kg of ethylbenzene requires 0.27 kg ethylene and 0.74 kg benzene. Manufacture of 1 kg styrene requires 1.077 kg ethylbenzene and produces 0.031 kg toluene by-product (Tellus Institute, 1992).

Polymerisation of styrene can be done with mass, solution, suspension, or emulsion techniques. Mass and suspension polymerisation are most common. In mass polymerisation, mineral oil is added as a lubricant and plasticiser. Production of 1 kg of PS requires 0.98 kg styrene and 0.03 kg mineral oil. General-purpose (crystal) PS is produced by both mass and suspension polymerisation. Expandable bead PS is generally produced by suspension polymerisation. The blowing agent can be added before polymerisation but is most commonly added after polymerisation (Tellus Institute, 1992).

Production of 1 kg PS from raw materials requires 88.6 MJ/kg, including the energy value of the materials, for production of general-purpose (crystal) PS by mass polymerisation. It is believed that suspension polymerisation has slightly higher energy requirements (Tellus Institute, 1992). The mass polymerisation process itself uses 2.28 MJ/kg. Dehydrogenation of ethylbenzene to styrene consumes 12.18 MJ/kg polystyrene produced. However, alkylation of ethylene and benzene to ethylbenzene produces a net energy gain of 0.73 MJ/kg polystyrene, in the form of steam (Tellus Institute, 1992).

Van Eijk et al. (1992) calculate energy requirements for the production of polystyrene granulate, consisting of 50% high-impact polystyrene and 50% general-purpose polystyrene, as 33.459 MJ/kg, in the Netherlands. Transportation of the oil required consumes an additional 1.045 MJ/kg PS, and transportation of the granulate to the forming facility uses 0.447 MJ/kg PS. Moulding of cups from the granulate uses 12.948 MJ/kg PS, and transportation from the moulder to the user requires 0.464 MJ/kg PS.

Franklin Associates (1992) calculate an energy requirement of 100.0 MJ/kg (43,101 Btu/lb) for manufacture of expandable polystyrene resin, using isopentane as the blowing agent. Crystal polystyrene is calculated to require 92.3 MJ/kg (39,796 Btu/lb), and high-impact PS requires 93.6 MJ/kg (40,333 Btu/lb). Fabrication of moulded foam polystyrene products is estimated to require 17.85 MJ/kg (7,691 Btu/lb), and manufacture of thermoformed foam PS products requires 17.96 MJ/kg (7,736 Btu/lb).

The air emissions of most concern in PS production are VOCs, especially styrene and benzene, which are toxic. Table 7–9 presents air emissions for production

Table 7–9 Air Emissions from Polystyrene Production (g/kg PS)

	Ethylbenzene/Styrene Production		*Polymerisation*	
Pollutant	*Uncontrolled*	*Controlled*	*Uncontrolled*	*Controlled*
NO_x	1.96×10^{-2}			
Volatile organic compounds	16.5		1.65	
Benzene	2.55	0.939		
Styrene	5.19×10^{-2}	5.19×10^{-2}		

Note: Data from Tellus Institute, 1992; controlled air emissions not available for all categories.

of ethylbenzene and styrene from ethylene and benzene, and for polymerisation of styrene to polystyrene. The compounds listed as VOCs are essentially all styrene. Batch-type polymerisation has styrene emissions of 1.55 g/kg, while for continuous polymerisation, the emissions are 1.75 g/kg. Suspension polymerisation of expandable bead PS produces air emissions of 9.29 g VOC/kg polystyrene (Tellus Institute, 1992).

Both mass and suspension polymerisation can produce a variety of water effluents. These are summarised in Table 7–10.

7.9.9 Polyvinyl Chloride

PVC is used widely in blister packaging and also has a number of other packaging applications, including bottles for potable water. Packaging consumes only a small portion of PVC production, since non-packaging applications such as pipe and siding are much larger markets.

Table 7–10 Water Emissions from Polystyrene Polymerisation (g/kg Polystyrene)

	Mass Polymerisation		*Suspension Polymerisation*	
Polymerisation	*Uncontrolled*	*Controlled*	*Uncontrolled*	*Controlled*
5-day biological oxygen demand	1.38		11.0	
Chemical oxygen demand	1.87		31.8	
Total organic carbon	0.794		10.5	
Total suspended solids	1.31×10^{-2}			
Aldrin	1.35×10^{-6}	1.39×10^{-6}	1.59×10^{-5}	7.94×10^{-6}
Benzene	1.04×10^{-2}	4.74×10^{-5}	1.30×10^{-3}	3.98×10^{-4}
Bis(2-ethylhexyl) phthalate	3.62×10^{-4}	1.39×10^{-4}	2.08×10^{-3}	3.98×10^{-4}
Butyl benzyl phthalate			8.74×10^{-5}	3.18×10^{-5}
1,2-Dichloropropane	9.19×10^{-5}	1.39×10^{-6}	3.98×10^{-4}	3.98×10^{-4}
Dimethyl phthalate	4.11×10^{-4}	1.39×10^{-4}	8.49×10^{-4}	3.98×10^{-4}
Ethylbenzene	2.38×10^{-2}	1.17×10^{-4}	9.10×10^{-4}	3.98×10^{-4}
Ethylene dichloride	2.70×10^{-5}	1.39×10^{-6}	1.59×10^{-4}	1.59×10^{-4}
Lindane (*R-bhc-γ*)	1.35×10^{-6}	1.39×10^{-6}	7.94×10^{-6}	7.94×10^{-6}
Phenol	6.04×10^{-4}	5.54×10^{-4}	7.14×10^{-4}	3.98×10^{-4}
Toluene	1.94×10^{-2}	8.09×10^{-5}	4.69×10^{-4}	3.98×10^{-4}
Chromium			4.61×10^{-4}	3.98×10^{-4}
Lead			2.39×10^{-5}	2.39×10^{-5}

Note: Data from Tellus Institute, 1992; controlled emission values not available for all categories.

Manufacture of PVC begins with chlorination of ethylene to yield ethylene dichloride. This can be carried out by direct chlorination, oxychlorination, or (the process used most frequently) a combination of these methods known as the balanced process. The chlorine used is produced from electrolysis of salt, which produces both chlorine and caustic soda, along with a small amount of hydrogen, from brine (Tellus Institute, 1992).

The next step in the process is dehydrochlorination of ethylene dichloride, yielding vinyl chloride and by-product hydrogen chloride (HCl). In the United States, over 80% of all ethylene dichloride produced is used for synthesis of vinyl chloride, which in turn is used primarily for production of PVC.

Polyvinyl chloride can be produced by mass, solution, suspension, or emulsion polymerisation, with suspension polymerisation being by far the most common. The reaction is carried out in a dilute water suspension under a nitrogen atmosphere, to prevent unwanted reactions with oxygen (Tellus Institute, 1992).

Production of 1 kg of PVC requires 1.02 kg of vinyl chloride and 84.0 MJ of energy, including the raw material energy. The polymerisation process energy is 4.32 MJ/kg PVC. Production of ethylene dichloride and vinyl chloride monomer consumes 0.46 kg ethylene and 0.58 kg chlorine, with a process energy requirement of 9.76 MJ/kg PVC. Production of the required amount of chlorine uses 12.52 MJ of process energy per kg PVC (Tellus Institute, 1992). Franklin Associates (1992) calculate total energy requirements for the manufacture of PVC resin as 70.88 MJ/kg (30,539 Btu/lb). Fabrication of PVC products by calendering is estimated to require 3.47 MJ/kg (1,495 Btu/lb).

Air emissions from chlorine production, production of ethylene dichloride and vinyl chloride, and from PVC polymerisation are listed in Table 7–11. Controlled and uncontrolled water emissions associated with PVC polymerisation are listed in Table 7–12.

7.9.10 Polyethylene Terephthalate

The major use of PET in packaging is bottles for carbonated soft drinks. However, PET use in other food and non-food markets, mostly in the form of bottles, has grown rapidly and continues to expand. PET films and coatings are also used in packaging.

PET is a condensation polymer formed from ethylene glycol and dimethyl terephthalate (DMT) or terephthalic acid (TPA). Manufacture from DMT is more common and produces methanol as a by-product. This polymerisation process requires 1 kg of DMT and 0.7 kg of ethylene glycol to produce 1 kg of PET and 0.33 kg of methanol.

DMT can be produced by oxidation of paraxylene or by esterification of terephthalic acid with methanol. Terephthalic acid is produced from acetic acid and

Table 7–11 Uncontrolled and Controlled Air Emissions from PVC Suspension Polymerisation Steps (g/kg PVC)

	EDC/VCM Production		*PVC Production*	
Pollutant	*Uncontrolled*	*Controlled*	*Uncontrolled*	*Controlled*
Total suspended particles			17.5	
SO_x			1.25×10^{-2}	
NO_x			100	
Volatile organic compounds	10.7		8.49	
1,3-Butadiene			2.10×10^{-4}	2.10×10^{-4}
Carbon tetrachloride	1.55	0.327		
Chloroform	1.82	3.65×10^{-2}		
Chloroprene			4.00×10^{-5}	4.00×10^{-5}
Ethyl chloride	4.89×10^{-3}	4.89×10^{-3}		
Ethylene dichloride	10.9	1.43	4.00×10^{-5}	4.00×10^{-5}
Hydrogen chloride			2.10×10^{-4}	2.10×10^{-4}
Propylene			6.99×10^{-5}	6.99×10^{-5}
Vinyl chloride	4.88	0.127	35.5	0.674
Vinylidene chloride			8.99×10^{-5}	3.24×10^{-4}

Note: Data from Tellus Institute, 1992; controlled emission values not available for all categories.

paraxylene by oxidation in air. Production of 1 kg of DMT requires 0.63 kg of paraxylene and 0.38 kg of methanol for the oxidation route (Tellus Institute, 1992).

Ethylene glycol is commonly manufactured by hydrolysis of ethylene oxide, although it can also be produced directly from ethylene. Ethylene oxide is produced by the direct oxidation of ethylene. Manufacture of 1 kg of ethylene oxide requires 0.9 kg of ethylene and 1.14 kg of oxygen. Manufacture of 1 kg of ethylene glycol requires 0.75 kg of ethylene oxide and 2 kg of water (Tellus Institute, 1992).

Production of PET from raw materials requires 122.7 MJ of energy per kg, including the energy content of the materials. Oxidation of paraxylene uses 13.14 MJ/kg of PET produced, while the condensation polymerisation reaction uses 6.16 MJ/kg (Tellus Institute, 1992).

Franklin Associates (1992) calculate an energy requirement of 88.74 MJ/kg (38,234 Btu/lb) for production of PET resin and 18.87 MJ/kg (8,131 Btu/lb) for fabrication of blow-moulded products.

PET can be recycled by either chemical or physical methods. Chemical recycling consumes only about one-third of the energy required to produce PET from virgin materials and can produce a product with properties identical to virgin, approved for food contact. Some physically reprocessed materials have also been approved for food contact. Mechanical recycling requires even less energy than chemical recycling (Gaines & Stodolsky, 1993).

Selected air and water emissions for production of xylene, production of ethylene oxide and ethylene glycol, production of DMT, and polymerisation of PET are shown in Tables 7–13 through 7–16.

Table 7–12 Uncontrolled and Controlled Water Emissions from PVC Suspension Polymerisation Steps (g/kg PVC)

	EDC/VCM Production		*PVC Production*	
Pollutant	*Uncontrolled*	*Controlled*	*Uncontrolled*	*Controlled*
5-day biological oxygen demand	1.78		0.989	
Chemical oxygen demand	4.73		0.719	
Total organic carbon	1.42		0.124	
Total suspended solids	5.89×10^{-2}		1.75	
Total dissolved solids			0.819	
Acrolein			1.61×10^{-5}	
Bis(2-ethylhexyl) phthalate	6.44×10^{-5}	6.34×10^{-5}	1.67×10^{-4}	
Carbon tetrachloride		2.90×10^{-3}		
Chloroethane	1.64×10^{-5}	1.64×10^{-5}		
Chloroform	1.74	0.434		
1,3-dichloropropene	8.44×10^{-5}	7.04×10^{-5}		
Di-n-butyl phthalate	4.04×10^{-4}	1.77×10^{-4}		
Ethylene dichloride	4.35	0.719		
Methylene chloride	2.82×10^{-2}	9.44×10^{-3}		1.32×10^{-4}
Pentachlorophenol	1.04×10^{-3}	2.68×10^{-4}		
Phenol			2.03×10^{-4}	
Tetrachloroethylene	3.66×10^{-3}	1.86×10^{-3}		
1,1,1-Trichloroethane				3.84×10^{-5}
Trichloroethylene	0.367	0.122		
Vinyl chloride	2.72×10^{-2}	6.84×10^{-3}	4.43×10^{-3}	5.49×10^{-4}
Chromium	1.22×10^{-3}	2.18×10^{-4}		
Copper	1.69×10^{-4}	3.62×10^{-5}		
Cyanide	2.52×10^{-5}	1.39×10^{-5}		
Lead				3.16×10^{-4}
Nickel	1.97×10^{-4}	6.99×10^{-5}		
Zinc				6.04×10^{-5}

Note: Data from Tellus Institute, 1992; controlled emission values not available for all categories.

Table 7–13 Uncontrolled Air Pollutant Emissions Associated with PET Production Steps (g/kg PET)

Pollutant	*Paraxylene Production*	*Ethylene Oxide/Glycol Production*	*Dimethyl Terephthalate Production*	*Polymerisation*
NO_x			3.00×10^{-2}	
Volatile organic compounds	0.346	5.24	19.8	9.69
CO			19.0	
Ethylene Oxide		1.83		
Triethanol			0.130	
Xylene	0.944		7.09	

Note: Data from Tellus Institute, 1992; controlled emission values not available for all categories.

Table 7–14 Controlled Air Pollutant Emissions Associated with PET Production Steps (g/kg PET)

Pollutant	*Paraxylene Production*	*Ethylene Oxide/Glycol Production*	*Dimethyl Terephthalate Production*	*Polymerisation*
NO_x				
Volatile organic compounds				0.534
CO				
Ethylene Oxide		4.84×10^{-2}		
Triethanol			0.013	
Xylene				

Note: Data from Tellus Institute, 1992; controlled emission values not available for all categories.

7.10 EFFECTS OF TRANSPORTATION

The cost and the environmental impact of transportation are affected by both the weight and volume of packaging materials. Which is more significant depends on whether the truck (the most common transportation vehicle) or other vehicle fills up first or reaches maximum weight limitations first. In general, vehicles will weight out first with heavy materials such as glass and fill up first with light materials such as plastics.

Table 7–15 Uncontrolled Water Pollutant Emissions Associated with PET Production Steps (g/kg PET)

Pollutant	*Paraxylene Production*	*Ethylene Oxide/Glycol Production*	*Dimethyl Terephthalate Production*	*Polymerisation*
5-day biological oxygen demand	4.99	20.3	0.339	4.36
Chemical oxygen demand	13.5	58.4	1.69	9.19
Total organic carbon	3.68	14.8	1.17	1.07
Oil and grease	3.11×10^{-5}	1.11		6.74×10^{-2}
Total suspended solids	0.939	15.9	3.39×10^{-3}	0.809
Total dissolved solids			2.35×10^{-2}	2.44
Acenaphthene		5.44×10^{-4}		
Acenaphthylene		1.12×10^{-4}		
Acrylonitrile				4.95×10^{-3}
Antimony				7.99×10^{-2}
Benzene	1.18×10^{-3}	0.123		
Bix(2-chloroisopropyl) ether		1.49×10^{-4}		
Chloroform	3.17×10^{-5}	2.13×10^{-5}		
2-Chlorophenol	4.48×10^{-5}			
2,4-Dimethylphenol	1.70×10^{-4}			
Dimethyl phthalate			5.09×10^{-3}	
Di-n-butyl phthalate			2.09×10^{-3}	
Ethylbenzene	5.34×10^{-4}	2.93×10^{-3}		6.74×10^{-3}
Ethylene dichloride		4.75×10^{-3}		
Fluorene		6.09×10^{-5}		
Methylene chloride	1.39×10^{-4}	1.15×10^{-4}		
Naphthalene	2.48×10^{-4}	1.32×10^{-3}		1.19×10^{-3}
Phenol	1.44×10^{-4}	1.07×10^{-3}	3.69×10^{-4}	0.704
Tetrachloroethylene		6.09×10^{-6}		
Toluene	2.42×10^{-3}	3.48×10^{-2}		
Vinyl chloride		4.13×10^{-4}		
Chromium	8.74×10^{-3}	1.21×10^{-5}		
Copper		1.18×10^{-4}		
Cyanide		6.99×10^{-5}		
Mercury			1.03×10^{-6}	
Sulfates		0.207		
Zinc	1.26×10^{-3}			

Note: Data from Tellus Institute, 1992; controlled emission values not available for all categories.

Table 7–16 Controlled Water Pollutant Emissions Associated with PET Production Steps (g/kg PET)

Pollutant	*Paraxylene Production*	*Ethylene Oxide/Glycol Production*	*Dimethyl Terephthalate Production*	*Polymerisation*
5-day biological oxygen demand				
Chemical oxygen demand				
Total organic carbon				
Oil and grease				
Total suspended solids				
Total dissolved solids				
Acenaphthene		1.53×10^{-6}		
Acenaphthylene		1.53×10^{-6}		
Acrylonitrile				1.55×10^{-3}
Antimony				3.43×10^{-2}
Benzene	3.29×10^{-4}	1.29×10^{-4}		
Bis(2-chloroisopropyl) ether		3.07×10^{-6}		
Chloroform	1.80×10^{-5}	1.53×10^{-6}		
2-Chlorophenol	3.66×10^{-5}			
2,4-Dimethylphenol	1.64×10^{-4}		1.29×10^{-5}	
Dimethyl phthalate			1.29×10^{-5}	
Di-n-butyl phthalate				
Ethylbenzene	1.65×10^{-4}	3.07×10^{-6}		2.10×10^{-3}
Ethylene dichloride		3.99×10^{-5}		
Fluorene		1.53×10^{-6}		
Methylene chloride	3.71×10^{-5}	1.53×10^{-6}		
Naphthalene	7.74×10^{-5}	3.07×10^{-6}		4.51×10^{-4}
Phenol	1.07×10^{-4}	7.94×10^{-4}	3.69×10^{-4}	0.270
Tetrachloroethylene		1.53×10^{-6}		
Toluene	1.98×10^{-4}	3.07×10^{-5}		
Vinyl chloride		1.53×10^{-6}		
Chromium	3.23×10^{-4}	1.23×10^{-5}		
Copper		1.16×10^{-4}		
Cyanide		3.68×10^{-5}		
Mercury			5.79×10^{-7}	
Sulfates				
Zinc	4.37×10^{-4}			

Note: Data from Tellus Institute, 1992; controlled emission values not available for all categories.

Boustead and Hancock (1981) present energy cost calculations for road vehicles of varying types in the United Kingdom for a full load, ranging from 9.61 MJ/vehicle-km (15.47 MJ per vehicle-mile) for vehicles of less than 1 tonne, to 27.5 MJ per vehicle-km (44.32 MJ per vehicle-mile) for articulated vehicles greater than 18 tonnes. When no load is carried, energy requirements can be reduced by a factor of 0.7. These energy requirements include energy consumed in supplying the fuel. Van Eijk et al. (1992) calculate an average energy cost for truck transport as 2.32 MJ/tonne-km. Cross, Welch, Hunt, and Park. (1974) estimated average truck transport energy consumption as 1.80 MJ/tonne-km (2.5×10^6 BTU/1,000 ton-miles). Franklin Associates (1992) calculate energy consumption for tractor trailer trucks as 1.40 MJ/tonne-km (1,945 Btu/ton-mile) for diesel trucks and 1.29 MJ/tonne-km (1,782 Btu/ton-mile) for gasoline trucks. For single-unit trucks, the values were 2.26 MJ/tonne-km (3,136 Btu/ton-mile) for diesel and 2.26 MJ/tonne-km (3,132 Btu/ton-mile) for gasoline.

For rail transport, Boustead and Hancock (1981) calculate a total energy requirement of 1.50 MJ per tonne-mile for general rail freight and 1.09 MJ per tonne-mile for circuit-working freight trains. Sea transport is calculated to require 0.17 MJ per tonne-mile. Cross et al. (1974) estimate average rail transport energy as 0.58 MJ/tonne-km (0.81×10^6 Btu/1,000 ton-miles). Franklin Associates (1992) give a value of 0.37 MJ/tonne-km (514 Btu/ton-mile).

7.11 PACKAGING TRENDS

It is generally agreed that a powerful way to reduce the environmental effects of packaging is to use less packaging. Source reduction has a long history in the packaging industry, although, until the 1980s, it was virtually always done for economic rather than environmental reasons. Nonetheless, many of the packaging reductions carried out in order to save money have also reduced environmental impact. This trend is continuing. Examples include removal of outer cartons, changes from rigid to flexible packages, no-frills refill packages, and product concentration permitting smaller containers.

Efforts to reduce the presence of toxic materials in packaging include removal of target heavy metals (lead, cadmium, mercury, and hexavalent chromium) from packaging materials, additives, pigments, and inks; these efforts have been spurred on in the United States by legislation in several states that forbids all nonessential uses of these materials in packaging. The paper industry is currently engaged in vastly reducing its dependence on chlorine as a bleaching agent.

Packaging is also undergoing changes designed to permit increased use of recycled materials in packages. Such developments include multilayer bottles, with a sandwiched inner layer of recycled HDPE or PET, for food contact, as well as specially processed PET for direct food contact. Other design changes are aimed at facilitating the recycling of packages, such as the removal of HDPE base cups from PET soft drink bottles.

It is not infrequent for packaging design changes that result in source reduction to also result in packages that are less easily recycled. Evaluation of the environmental trade-offs in such situations can be very complex. In general, it can usually be concluded that a significant degree of source reduction outweighs loss of recycling capability if recycling rates are low to moderate. If recycling rates are high, more specific information will be required.

It should be pointed out that the majority of the environmental impacts discussed in this chapter are listed on a weight basis. In comparing alternative packaging systems, the point of comparison should always be delivery of a standard amount of product. Thus, all per weight impacts need to be converted to a per package impact and then to a per-delivered-product impact. Only in this way can valid comparisons be made.

REFERENCES

Boustead, I., & Hancock, G.F. (1981). *Energy and packaging*. Chichester: Ellis Horwood.

Cross, J.A., Welch, R.O., Hunt, R.G., & Park, W.R. (1974). *Plastics: Resource and environmental profile analyses*. Washington, DC: Manufacturing Chemists Association.

Fenton, R. (1991). *Grocery Bag Comparison* (Report No. 1). The Winnipeg Packaging Product, University of Winnipeg, Winnipeg, Manitoba.

Franklin Associates. (1990). *Resource and environmental profile analysis of foam polystyrene and bleached paperboard containers*. Prairie Village, KS: Franklin Associates, Ltd.

Franklin Associates. (1992). *An energy study of plastics and their alternatives in packaging and disposable consumer goods*. Prairie Village, KS: Franklin Associates, Ltd.

Gaines, L.L. (1981). *Energy and materials use in the production and recycling of consumer-goods packaging*. U.S. Department of Commerce, National Technical Information Service, ANL/CNSV-TM-58.

Gaines, L.L., & Stodolsky, F. (1993). *Mandated recycling rates: Impacts on energy consumption and municipal solid waste volume*. Energy Systems Division, Argonne National Laboratory, U.S. Department of Energy, ANL/ESD-25.

Hunt, R.G., & Welch, R.O. (1974). *Resource and environmental profile analysis of plastics and non-plastics containers*. Midwest Research Institute, MRI Project No. 3714-D.

Little, Arthur D., Inc. (1984). *The life cycle energy content of containers: 1984 update*. Cambridge, MA: A.D. Little, Center for Environmental Assurance.

Little, Arthur D., Inc. (1991). *The life cycle energy content of containers: 1991 update*. Cambridge, MA: A.D. Little, Center for Environmental Assurance.

Tellus Institute. (1992). *CSG/Tellus packaging study*, vols. I and II. Boston.

van Eijk, J., Nieuwenhuis, J.W., Post, C.W., & DeZeeuw, J.H. (1992). *Reusable versus disposable: A comparison of the environmental impact of polystyrene, paper/cardboard and porcelain crockery*. Ministry of Housing, Physical Planning and Environment, The Netherlands.

CHAPTER 8

Disposal of Used Packaging

Peter Page and R. White

8.1 BACKGROUND

Modern packaging has given the public what it was demanding: quicker shopping, less frequent shopping, more choice, lower prices, and better hygiene. Modern supermarkets could not function without the sophistication of packaging today. This has led to a "throw-away" society, where much of the packaging has become one trip and then thrown away. Over two-thirds of packaging is used to protect food and drink. In these areas, packaging has had to be effective in complying with legal and health and safety requirements, while being kept to a minimum. A leading food packer describes the role of packaging as follows (Pré, 1992):

- protects the nutritional and sensory properties of the product
- identifies brand and provides required product information
- ensures optimum distribution costs
- provides convenience and "value for money" to the consumer
- provides sales appeal on the shelf
- minimises the possible negative impact on the environment

Over recent years, there have been major changes as governments everywhere endeavoured to reduce packaging consumption, control its disposal, and encourage its recovery, reuse, and recycling through legislation. It was in the United States in the 1980s, a growing concern about the amount of household waste being generated led to the rapid spread of mandatory recycling laws covering producers and users of all packaged goods. The waste disposal problem and the shortage of landfill sites for disposal of rubbish was highlighted when the mayor of New York arranged for a load of the city's garbage to be taken to the Caribbean and back. Although New York has a shortage of landfill sites, this is not generally the case elsewhere in the United States. In Europe, there are problems in the Netherlands and in countries where the geology is unsuitable; but, again, many countries have more than adequate suitable landfill sites and so could continue their present methods of disposal. Used packaging is perceived as one of the principal constituents of waste, mainly because of its high profile.

People are very aware of the high level of used packaging in their rubbish bins, but, in reality, used packaging represents only approximately 1.0% to 1.5% of the total tonnage of waste produced in Europe each year. About 150 million tonnes of waste go to landfill in the United Kingdom, of which 13 to 15 million tonnes are household waste. Of this household waste, about 3.5 to 4 million tonnes are packaging; the remainder comes from commercial, industrial, construction, and other waste sources.

A common perception is that packaging waste has increased over the years. While there are certainly more packaged goods available, the amount of packaging used for each pack is much less than it used to be. The total weight of material used for all packaging is probably no more than it was 20 years ago. Even if all packagings were removed, there would still be a waste problem.

However, the methods of waste disposal of the 1970s and 1980s were seen as a waste of valuable resources, and there was increasing pressure on governments to enforce reuse and recycling of used packaging. This started in Germany, where a minority, disaffected with what they saw as excessive materialism, was almost steered by the government towards packaging as an outlet for their frustrations. There were no easy answers to global warming, acid rain, the hole in the ozone layer, or use of nuclear energy; packaging offered a suitable outlet. The Greens identified packaging as a cause of consumerism, and boycotting plastics packaging was a way of demonstrating their feelings. Politicians who recognised the political threat from the Greens began to urge consumers to avoid packaging if they wanted to contribute to a solution to the "waste problem". Industry was unwilling to challenge the arguments. The European Court of Justice ruled in 1988 that limitations on the free movement of goods were permissible if there was an appropriate gain in environmental protection. This was seen as a way of protecting locally manufactured goods against imports and led to demands for legislation in the name of environmental protection.

In June 1998, the UK government issued a consultation paper on waste strategy called *Less Waste. More Value* (Department of the Environment, Transport and the Regions, 1998). In this paper, the government's key commitments were set out. After due consultation, the paper was expected to lead eventually to publication of a White Paper at the end of 1999.

The government's seven key commitments were:

1. substantial increases in recycling and energy recovery
2. engagement of the public in increased reuse and recycling of household waste
3. a long-term framework with challenging targets underpinned by realistic programmes
4. a strong emphasis on waste minimisation
5. using the waste hierarchy as a guide, not a prescriptive set of rules
6. creative use of economic incentives like the landfill tax
7. increased public involvement in decision making

8.2 LEGISLATION

Germany was the first country in Europe to introduce legislation requiring all used packaging to be collected and reused or recycled. Incineration is not permitted, although at the time of the introduction of the legislation, up to 30% of household waste in Germany was being treated in this way and much of the energy created recovered as hot water or electrical power. Subsequently, other countries have introduced or are introducing similar legislation, and the anticipated European Commission legislation was finally introduced in December 1994.

8.2.1 The European Commission Directive on Packaging and Packaging Waste

The objective of this legislation (European Commission, 1994) is to reduce the impact of packaging on the environment by reducing packaging at source; eliminating harmful materials in packaging waste; maximising the recovery of packaging waste for reuse, recycling, composting or energy generation; and minimising the amount going for ultimate disposal. The directive is implemented by individual countries, and it is the intention of the directive that national measures should be harmonised.

The directive specifies targets that national laws have to implement by 30 June 2001. These targets are:

- between 50% and 65% of packaging by weight to be recovered
- between 25% and 45% of packaging by weight to be recycled, with a 15% minimum of each material being recycled

These targets are likely to be "substantially" increased for the second 5-year period up to 30 June 2006 after a review of practical experience, which has to be carried out before 1 January 2001.

There has been considerable difference of opinion between member states on these targets. Some, like Germany, Denmark, and the Netherlands, considered the targets not tough enough and voted against the directive; however, they were outvoted.

The directive requires systems to be set up so that packaging can be recovered from the waste stream by the most appropriate method. In order to assist the process of collection, recovery, and reuse, packaging has to be marked to indicate what material has been used. Agreement on marking symbols had not been achieved at the time of writing (see Section 8.4.3).

Member states have to set up systems to gather information so that the Commission can monitor achievement of the targets and so that the public is aware of the targets and what is being done to achieve them. There have been various approaches to this problem. These range from the Netherlands' voluntary monitoring scheme under its Packaging Covenant (which collects sample data on what packaging each company has added through the supply chain) to the United Kingdom, where a comprehensive data collection system will be the responsibility of industry. There are many difficulties to be overcome in obtaining detailed, properly classified, and reliable data that do not have any double counting and that can account for all the problems of imports and exports from outside the EU and movements of packaging (filled and empty) within.

Individual countries in the European Union cannot create barriers to the use of certain forms of packaging, providing that the packaging meets the requirements of the directive. These "essential" requirements cover the minimisation of weight and volume, and suitability for reuse, recycling, energy recovery, or composting. Standards are being drawn up by the European standards organisation CEN (Comité Européen de Normalisation).

Member states are not allowed to adopt targets for the packaging of specific product categories and sectors; targets for recovery and recycling in the directive are not product-specific.

Implementation of the Packaging and Packaging Waste Directive has been slow, reflecting to some degree what are recognised, even by the European Commission itself, as fundamental flaws in the legislation. One of the main flaws is that packaging waste was singled out for specific legislation rather than being incorporated into legislation to set up integrated waste management systems.

There are significant differences among member states as to their ability to deal with packaging waste and achieve the required targets. There are also member states whose targets exceed those set by the Commission and who consider that their own environmental policies are being hampered by the EU legislation. Outstanding issues include questions as to whether packaging waste management hierarchies should be set or not, and whether in some cases precedence should be given to environmental considerations over those of the international market and the free circulation of goods, or vice versa.

8.2.2 Producer Responsibility

The Environment Act 1995 gave power for regulations to be made to implement the government's Producer Responsibility Initiative. This initiative, based on the polluter pays principle, aimed at encouraging business to recover value from products at the end of their lives.

The Producer Responsibility Obligations (Packaging Waste) Regulations 1997 (SI 1997:648) aims to:

- achieve a more sustainable approach to dealing with packaging waste
- reduce the amount of packaging waste going into landfill
- implement the recovery and recycling targets in the EC Directive on Packaging and Packaging Waste
- respond to the UK industry's wish to have
  - legislative support for recovery and recycling obligations so that "free riders" are deterred
  - business-led collection schemes to handle their obligations for them
  - an approach whereby the recovery and recycling obligations are shared across all parts of the packaging chain

These regulations require certain businesses to register with the Environment Agency (EA) or the Scottish Environment Protection Agency (SEPA), to recover and recycle specific tonnages of packaging waste and to certify that this recovery and recycling has been achieved.

8.2.2.1 Who Is Affected?

Any business that is involved in one or more of the following is affected:

- the manufacture of packaging raw materials
- converting materials into packaging
- using packaging to pack products or putting products into packaging
- selling packaging to the final consumer (this could include, for example, selling a jar of marmalade or selling meat on a plastic tray)

The regulations also affect any business that owns the packaging on which the above activities are carried out; any business that supplies to another stage in the packaging chain or to the final user; and any business that has a turnover of over £5 million (reducing to £1 million in 2000) or handles over 50 tonnes of packaging materials and/or packaging in a year.

8.2.2.2 How Is Packaging Defined?

The regulations define packaging as any product that is used to contain and protect goods or to aid in their handling, delivery, or presentation. Such products are regarded as packaging at all stages of the process of manufacture and distribution, starting as raw materials and ending when they are delivered to the final user or consumer.

8.2.2.3 What Are the Obligations?

1. To register either with EA or SEPA or to join one of the registered collective schemes.
2. To recover and recycle specified percentages of packaging waste for which the business has a responsibility.
3. To provide data on packaging handled, recovered, and recycled.

8.2.2.4 What Are "Recovery" and "Recycling"?

Recovery includes a number of processes that result in a net benefit being derived from used packaging. Recovery includes recycling, energy recovery from heat generated by incineration, and composting of packaging materials.

Recycling is the reprocessing of used packaging so that the material produced can be reused for its original or another purpose. Composting counts as recycling.

8.2.2.5 Further Information

Details on the levels of obligation, how to calculate obligations, and information on collective schemes are available from the EA or SEPA. Help is also available from relevant industry organisations, for example, Incpen (the Industry Council for Packaging and the Environment) or Pira International.

8.2.2.6 Review of the Producer Responsibility Obligations (Packaging Waste) Regulations

At the end of July 1998, the minister for the environment published a consultation paper for proposals for modification of the regulations. This consultation was expected to have been completed by the end of September 1998, with the government introducing changes later in 1998, but as of mid-1999, consultation is still going on and changes are still being drafted. The documents proposed a range of options, including, amongst others, increasing targets, some further data obligations, a household packaging waste recycling obligation, and a greater requirement for consumer information. Businesses obligated under the regulations will need to check any revision of the regulations made following this consultation.

8.2.3 Essential Requirements

As the United Kingdom had not fully implemented the EC Packaging Waste Directive, and to avoid action by the European Commission, the government issued regulations to implement the "essential requirements" provision of the directive. The Packaging (Essential Requirements) Regulations 1998 (SI 1998:1165) came into force on 31 May 1998 and covers heavy metal limits and essential requirements.

8.2.3.1 Heavy Metal Limits

The combined concentration of lead, cadmium, mercury, and hexavalent chromium in packaging or its components was limited to 600 ppm as of 30 June 1998 and to 250 ppm as of 30 June 1999; it will be limited to 100 ppm as of 30 June 2001.

Lead crystal glass is exempt from these limits, and there are derogations for recycled materials. Possible derogations for plastic pallets and crates and recycled glass are being negotiated.

8.2.3.2 Essential Requirements

All packaging placed on the market, other than reused packaging, has to comply with laid down essential requirements. The "responsible person" (as defined in the regulations) who places the packaging on the market has to ensure that it satisfies the relevant essential requirements. These requirements are detailed in the Schedule to the Regulations, and this Schedule is reproduced below:

> Essential requirements on the composition and the reusable and recoverable, including recyclable, nature of packaging
>
> 1. Requirements specific to the manufacturing and composition of packaging:
>   - Packaging shall be so manufactured that the packaging volume and weight be limited to the minimum adequate amount to maintain the necessary level of safety, hygiene and acceptance for the packed product and for the consumer.
>   - Packaging shall be designed, produced and commercialised in such a way as to permit its reuse or recovery, including recycling, and to minimise its impact on the environment when packaging waste or residues from packaging waste management operations are disposed of.
>   - Packaging shall be so manufactured that the presence of noxious and other hazardous substances and materials as constituents of the packaging material or of any of the packaging components is minimised with regard to their presence in emissions, ash or leachate when packaging or residues from management operations or packaging waste are incinerated or landfilled.
> 2. Requirements specific to the reusable nature of packaging—the following requirements must be simultaneously satisfied:

- Physical properties and characteristics of the packaging shall enable a number of trips or rotations in normally predictable conditions of use.
- Possibility of processing the used packaging in order to meet health and safety requirements for the workforce
- Fulfil the requirements specific to recoverable packaging when the packaging is no longer reused and thus becomes waste

3. Requirements specific to the recoverable nature of packaging:

(a) Packaging recoverable in the form of material recycling.

Packaging must be manufactured in such a way as to enable the recycling of a certain percentage of weight of the materials used into the manufacture of marketable products, in compliance with current standards in the Community. The establishment of this percentage may vary, depending on the type of material of which the packaging is composed.

(b) Packaging recoverable in the form of energy recovery.

Packaging waste processed for the purpose of energy recovery shall have a minimum inferior calorific value to allow optimisation of energy recovery.

(c) Packaging recoverable in the form of composting.

Packaging waste processed for the purpose of composting shall be of such a biodegradable nature that it should not hinder the separate collection and the composting process or activity into which it is introduced.

(d) Biodegradable packaging.

Biodegradable packaging waste shall be of such a nature that it is capable of undergoing physical, chemical, thermal or biological decomposition such that most of the finished compost ultimately decomposes into carbon dioxide, biomass and water.

Harmonised standards are being drafted by CEN. Enforcement of the regulations will be by Trading Standards authorities.

Questions still remain as to what the cost might be of implementing these regulations. It is assumed that changes in packaging design and manufacture attributable to these regulations will not incur significant additional costs for industry. According to an *ENDS Report* ("Packaging 'Essential Requirements'", 1998) article on the regulations, the Department of Trade and Industry's compliance cost assessment of the regulations assumes that each packer/filler will need to send a letter to a supplier to seek evidence of compliance and spend 4 hours updating packaging design and specification procedures and a further 3.5 hours designing new packaging—calculated at £54 for a typical firm's recurring compliance costs and £78 for non-recurring costs.

8.2.4 Other Controls

Largely as a result of pressure from consumer groups, some European countries have from time to time introduced restrictions on specific materials. Germany, for instance, banned polyethylene terephthalate (PET) for carbonated drinks, and Switzerland banned polyvinylchloride (PVC) bottles. Gentlemen's agreements between industry and government stopped the use of PVC in Sweden and led to restrictions on some forms of packaging in the Netherlands. Many of these restrictions were based on emotion rather than scientific fact. PVC, for instance, is seen as posing a carcinogenic hazard for the workers producing it. It contains heavy metal salts that can leak out and harm the environment if it is put into landfill, and it gives rise to dioxins (polychlorinated dibenzo-p-dioxins [PCDDs]) if it is burned.

The reality is that modern polymerisation production techniques do not put workers at risk, heavy metal salts are not used in the material used for packaging, and when the material is burnt there is no evidence of dioxins being produced (Freiesleben, 1992). However, the incinerated gases must be scrubbed to remove hydrogen chloride, and it is recommended that the material be burnt at very high temperatures—up to 1200° C.

PVC is one of the best packaging materials available, but there continues to be considerable debate about the environmental impact of PVC. Late in 1997, some retailers (Tesco, the Co-operative Wholesale Society, and The Body Shop) were warning PVC manufacturers about levels of emissions and other concerns and were saying that they must "clean up their act" or PVC would be phased out of their stores. The PVC industry maintains that this is completely unjustified.

At the same time, there has also been pressure on governments and the food industry for safer, cleaner food. Packaging has played an important role in helping to provide this but has also been subject to legislation that sets out what may and may not be used. In Europe, a directive on plastics in contact with foodstuffs (European Commission, 1990) sets out the monomers and additives that may be used and sets limits on the amounts that may migrate into the food under standard test conditions. These strict conditions have ruled out the use of post-consumer recycled plastics because their content cannot be guaranteed. The legislation, which states that the flavour or olfactory performance of the food may not be affected, also rules out the use of recycled paper products for similar reasons.

In the United States, the Food and Drug Administration does not allow the use of recycled materials in contact with foodstuffs, except in the case of PET, which has been broken down into its constituent monomers, purified and repolymerised. Thus, food safety laws are in direct conflict with recycling and reuse legislation, and it is against this background that the collection, reuse, recycling, incineration, and landfill of used packaging must be viewed.

8.3 COLLECTION OF USED PACKAGING

Used packaging is a lot of material, often in small deposits spread over a very large area, and is made up of a wide variety of materials, some of which are contaminated with food and products that can be hazardous to human health, thus making collection and sorting an expensive and difficult operation. The "packaging flow" throughout the life of a product, from manufacture to disposal, is important because waste is produced at various stages in this life cycle. Although waste arising at various stages is important in the recovery process, the main volumes arise at the household or industrial level, where the packed product is used. This environmental balance may be described diagrammatically (Figure 8–1). The materials flowchart for paper (Figure 8–2) illustrates how complicated this flow pattern is.

Most household waste is a mixture of used packaging (24%), other papers (29%), textiles, garden waste and waste food (25%), used nappies, and many other unsavoury products. In most countries, this is collected by local authorities and

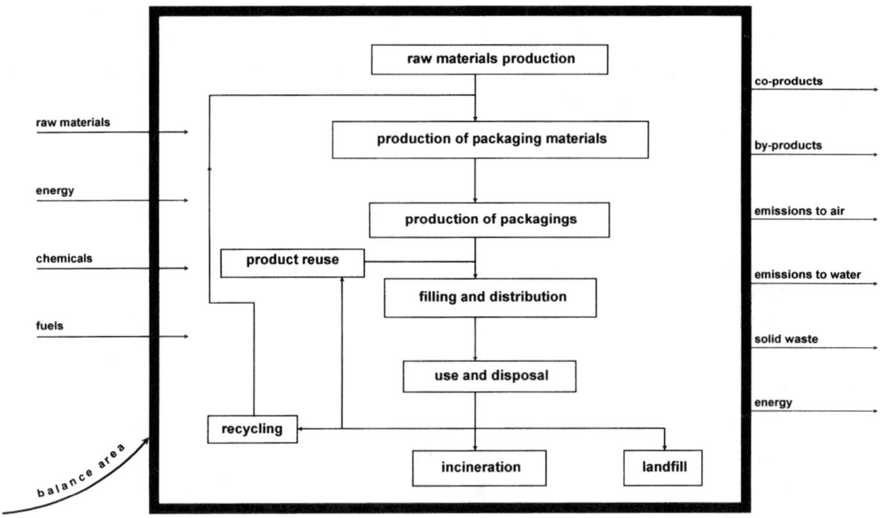

Figure 8–1 Packaging materials—environmental balance.

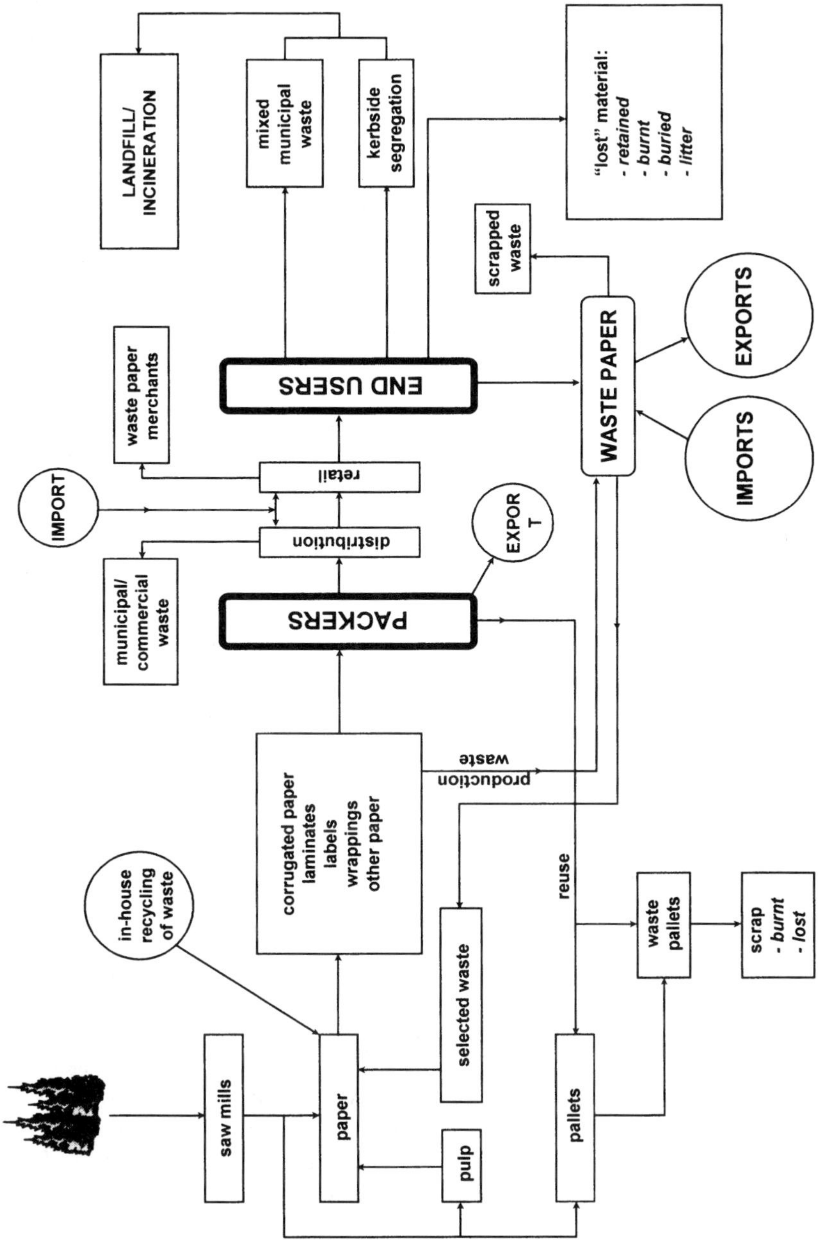

Figure 8–2 Materials flowchart for paper.

either incinerated or landfilled. Systems are in operation to sort material and to remove anything with value. In Dunkirk, France, waste is collected and dumped on a conveyer where people on each side remove bottles, plastics, paper, and cans for recycling. The stench is overpowering, and workers can only stay alongside the conveyer for an hour before having a break. It is significant that a qualification to work there is that people must have been unemployed for at least a year. This type of sorting poses health risks for the people, and the Danish government has banned the hand sorting of mixed household waste. The economics of these operations in Europe are such that they need a subsidy, because the value of the products recovered does not cover the operational costs (although this probably ignores the savings on landfill). In less developed countries, where labour is cheap, local industries arise where the unemployed scavenge the rubbish tips for anything that is recoverable; this has led to quite high recycling levels on plastic bottles, for instance, in Egypt.

There are many other options to use the mixed material arriving at a centralised waste treatment plant, from energy recovery alone to combined materials recovery, energy recovery, and compost systems. Many plants concentrate on producing waste derived fuel pellets and ferrous metals. The physical processing of the waste seeks to sort it into fractions by virtue of the different nature and characteristics of the various fractions. The material is first screened by passing it through a rotary screen and breaking it down into components. The second stage is magnetic extraction, which can remove 90% of the ferrous products present. Better extraction levels are obtained if the material passes through hammer mills before passing over a magnetic belt. The final stage is to use ballistic separators, density separators, and air classifiers to separate the waste into light and heavy fractions. Materials separated in this way tend to be heavily contaminated, so they only find low-grade applications.

Source separation schemes achieve higher levels of refinement, leading to higher quality output. To compete, recycled materials must be able to achieve the same performance specifications as virgin raw materials.

Legislation has therefore centred on separating waste before it is mixed and becomes rubbish. It has aimed at taking out the valuable and easily sorted components of the waste. Glass and plastic bottles and metal cans can be separated and reused or recycled. Deposit systems are employed whereby a small charge is made when the product is purchased and a refund made when the used empty package is returned. Practice has shown that a deduction from the deposit must be made to cover the cost of collection and recycling. When recovery rates are high (approximately 75%), the value of deposits retained is insufficient to cover the cost of the collection and sorting process. Also, the deposit has to be relatively high to overcome peoples' inertia and to persuade them to return the empty container. Where hazardous products are concerned, the container has a very high deposit to ensure that it is properly disposed of and does not get into the general waste stream of used packaging.

The other method of sorting and collection is known as kerbside collection, where households are asked to separate waste into separate bins—plastic, paper, glass, metal, and used batteries—and these are then collected and the products sent to recycling centres.

There are now many such schemes being developed throughout the United Kingdom. The cost of recovery may not be offset by the value of the materials recovered, especially at times of economic downturn, when prices of international products like plastics and pulps are low in order to keep plants operational. This makes it difficult to give any accurate costs for kerbside collection, but published data for the United Kingdom (Cooper, 1992) suggest they could be as high as £130 per tonne for kerbside collection, with the multi-materials collection schemes in the £40-£80 per tonne range. One scheme in Greater Manchester provided households with a second bin for mixed recyclables, which was then delivered to a centre for sorting. However, the costs of sorting were found to be higher than the revenue generated, and the scheme was abandoned. The actual cost of collection had only increased marginally, as the local authority changed collection schedules and the bins were supplied free to the council.

Although collection costs were minimised, the cost of sorting was well above the revenue from the segregated waste, even if the full cost of the landfill savings was credited to the scheme. The full cost of waste disposal is relatively high in Manchester; where disposal is significantly cheaper, this form of segregation would be even more uneconomic.

8.4 IDENTIFICATION OF MATERIALS FOR COLLECTION AND SORTING

8.4.1 Collection

Some European packaging collection schemes require that the used packaging materials be marked in some way to indicate that the manufacturer/user has prepaid a fee for the collection of the used packaging. Perhaps the best known is the green dot from Duales System Deutschland (DSD). DSD is a company that collects used packaging and sells it to recyclers. It initially charged on the number of packs per year and their capacity but, based on the initial experience, the system has been changed to charge on a weight basis rather than volume and to have a scale of charges for various materials, plastics and laminated materials costing the most and natural materials the least. Once packaging users have prepaid the fee, they are allowed to put a green dot on their packaging.

8.4.2 Sorting

A recent development is the use of tracers in plastics materials so that the different types of plastic can be sorted automatically (Simmons, Ahmad, & Overton, 1998). A very small amount (1–10 ppm) of a fluorescent tracer is put into the polymer at manufacture. At the separation plant, the waste is illuminated with ultraviolet (UV) light; sensors identify the type of plastic and automatically eject it into the appropriate bin. It is possible that the system could be adapted for other materials.

Many packs also carry a symbol indicating that the material can be recycled; rigid plastic containers carry the mark along with a number that identifies the polymer used in its manufacture and that aids in separation prior to recycling.

8.4.3 The European Union Marking Directive

The EU Packaging and Packaging Waste Directive has a framework for a materials identification system (Article 8). This system is designed to facilitate the sorting and recycling of packaging waste to help member states achieve the recovery and recycling targets of the directive.

Under Article 21 of the directive, a committee was set up (the Article 21 Committee) to undertake specific tasks in relation to the implementation of the directive; it also serves as a forum for member states to discuss issues concerning implementation of the directive. In 1996, the Article 21 Committee reached agreement on the issues of a Materials Identification System. The Commission Decision was adopted on 28 January 1997 with immediate effect. At present, the numbering and abbreviation system is voluntary, but the decision includes the means to make it mandatory at a later stage (Tables 8–1 and 8–2). Although composites are not covered by the original directive, they are defined in the Decision: "packaging made up of different materials, and which cannot be separated by hand, none exceeding a given percent by weight which shall be established in accordance with the procedure laid down in Article 21 of Directive 94/62/EC (Packaging and Packaging Waste)".

The draft Harmonised Marking and Identification for all European Packaging to Facilitate Collection, Reuse and Recovery Directive was published on 18 December 1996 to implement the obligations set out in Article 8 of the Packaging and Packaging Waste Directive. The directive introduces a voluntary system of marking based on two new symbols, one for reusable packaging (Figure 8–3) and the other for recyclable packaging (Figure 8–4).

By mid-1998, little progress had been made on the directive. The European Parliament's Environment Commission was querying with the European Commission why particular symbols had been chosen without appropriate information

Table 8–1 European Commission Numbering and Abbreviation Systems for Materials Identification

Material	*Abbreviation (Always Uppercase)*	*Numbering*
Plastics		
Polyethylene terephthalate	PET	1
High-density polyethylene	HDPE	2
Polyvinyl chloride	PVC	3
Low-density polyethylene	LDPE	4
Polypropylene	PP	5
Polystyrene	PS	6
Paper and fibreboard		
Corrugated fibreboard	PAP	20
Non-corrugated fibreboard	PAP	21
Paper	PAP	22
Metals		
Steel	FE	40
Aluminium	ALU	41
Wood materials		
Wood	FOR	50
Cork	FOR	51

Source: Adapted from Commission Decision of 28 January 1997 establishing the identification system for packaging materials pursuant to European Parliament and Council Directive 94/62/EC on packaging and packaging waste (97/129/EC), OJL 50, 20.2.97, p. 28.

campaigns. It also seems that European Parliament will be unlikely to move ahead on the directive until it has had a reply and until the ISO standards committee produces some conclusions from its own marking discussions.

8.5 LANDFILL

Landfill is the principal method of disposing of municipal (household) waste, with 90% of the UK waste being treated in this way. It is at the bottom of the EU hierarchy of waste management options. The word *landfill* broadly relates to any controlled operation in which waste is deposited onto land. Over the years it has gained a bad reputation because of dumping of rubbish on land that is unsuitable or not properly prepared, with the result that toxic materials have leached out, damaging the surrounding environment. There have also been cases where methane has built up in old landfill sites and been the cause of serious explosions. If properly exploited, this build-up of landfill gas can be tapped and used as a source

Table 8–2 European Commission Numbering and Abbreviation System for Composite Identification

Material	*Numbering*
Paper and fibreboard/miscellaneous metals	80
Paper and fibreboard/plastic	81
Paper and fibreboard/aluminium	82
Paper and fibreboard/tinplate	83
Paper and fibreboard/plastic/aluminium	84
Paper and fibreboard/plastic/tinplate	85
Paper and fibreboard/plastic/aluminium/tinplate	86
Plastic/aluminium	90
Plastic/tinplate	91
Plastic/miscellaneous metals	92
Glass/plastic	96
Glass/aluminium	97
Glass/tinplate	98
Glass/miscellaneous metals	99

Note: The abbreviation for a composite is C/ plus the abbreviation corresponding to the predominant material, e.g., C/HDPE

Source: Adapted from Commission Decision of 28 January 1997 establishing the identification system for packaging materials pursuant to European Parliament and Council Directive 94/62/EC on packaging and packaging waste (97/129/EC), OJL 50, 20.2.97, p. 28.

of energy. Modern landfill sites are now designed, engineered, and managed to minimise the risk of these problems arising. There are, however, many sites that became operational some time ago, and it is difficult to introduce controls retrospectively; to do so will incur costs that will be passed on to the waste producer.

Landfill sites are managed according to the permeability of the geology under the deposited rubbish. The preferred way is to contain the build-up of liquid in the site. The main sources of this build-up are rain, surface and ground water, and liquids in the waste or from its breakdown. Some water is lost through evaporation or taken up by surrounding plants, but it is necessary to prevent any water that could contain toxic materials from leaching away from the site. This is done by the geology of the site; for example, a clay layer may be covered with a plastic liner or, on sites where the underlying strata are permeable, clay is imported to form a base that is then covered with a plastic liner. It is good practice to regularly pump out the liquid that builds up in order to prevent problems, although on some sites it is not always necessary.

As part of good management policy, the materials placed in landfill sites should be controlled and toxic materials disposed of in other ways, such as controlled

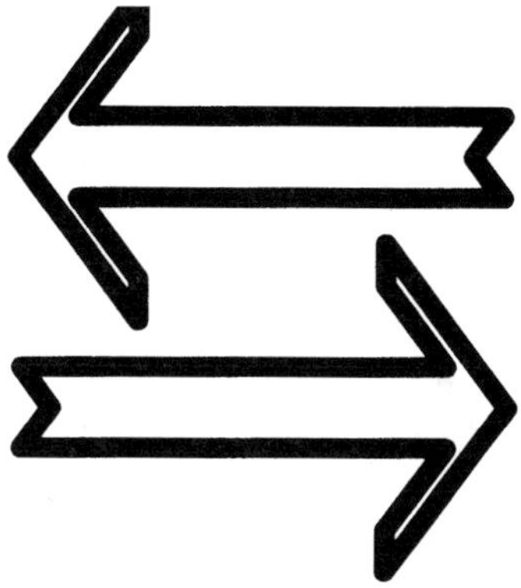

Figure 8–3 Reusable packaging symbol.

incineration. Used packaging does not present any hazard unless it has contained dangerous materials—medical waste, chemicals, or pesticides—and the safe disposal of these is now becoming a key issue, with landfill increasingly not being an option.

Household waste, which is a significant percentage of used packaging, can be compacted in the landfill site but will settle in excess of 10% of the compacted volume as a result of biodegradation. The waste provides a source of nutrients for micro-organisms that, in the initial stage, thrive in the presence of oxygen. As the oxygen is used up, the waste enters the methanorganic stage, where the principal

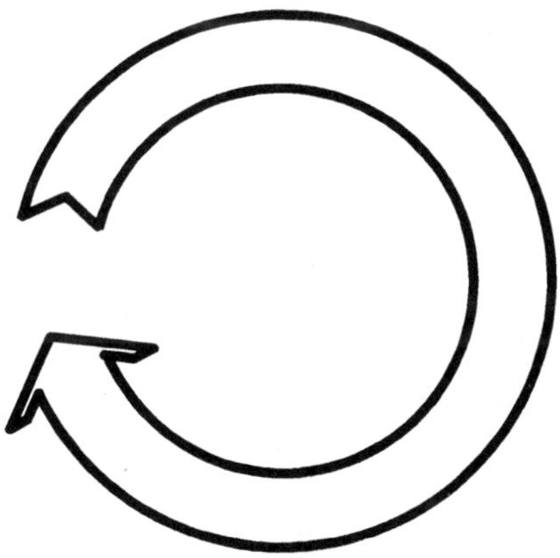

Figure 8–4 Recyclable packaging symbol.

by-products are methane and carbon dioxide. As the gas can cause serious nuisance and hazards, the site must be managed to allow it to vent to atmosphere, or a system of wells can be bored that draw the gas off for burning as an energy source.

Many governments are adopting the policy that the polluter must pay; as a consequence, landfill costs will increase because of the stricter controls and in an attempt to reduce the volume going to the sites. The average cost of landfill was about 32 Euros per tonne in urban sites and 20 Euros in rural sites (1993 prices).

The European Commission has adopted a new proposal on the Landfill of Waste. In fact this is a relaunch of a proposed directive dating from 1991 but rejected by the European Parliament in 1996 because of an insufficient level of environmental protection. The objectives of the new proposal are to prevent and reduce the adverse effects of existing and new landfills on the environment—in particular, the pollution of surface water, ground water, soil, and air—as well as the resulting risks to human health. The proposal also seeks to harmonise the environmental and technical standards for the landfill of waste.

Procedures are provided for controlling the opening, management, closure, and monitoring of a site and the acceptability of the waste in the landfill. To reduce the global greenhouse effect, there are provisions for reducing the landfilling of biodegradable municipal waste. The total amount (by weight) of biodegradable municipal waste must be reduced to 75% of 1993 levels by 2002, 50% by 2005, and 25% by 2010. Gases produced in new as well as existing landfills are to be collected, treated, and used.

All waste will have to be treated before landfilling, and landfills will become waste-specific, that is, non-hazardous waste will only be able to go into a non-hazardous waste landfill, hazardous waste only into a hazardous waste landfill, and inert waste only into an inert waste landfill. A ban on the landfilling of tyres is proposed. To ensure that all the costs of setting up, operating, or closing a landfill site are covered, member states will be required to increase the cost of landfilling.

8.6 COMPOSTING

Methods of composting household waste are being examined; this requires the putrescibles and other suitable materials to be separated and composted under controlled conditions. The problems associated with the process are to ensure that all unsuitable materials, such as glass, are removed and that no toxic contaminants, such as heavy metals, are included.

It is a process in which micro-organisms break down the materials in the presence of oxygen at elevated temperatures (<50° C) to give water, carbon dioxide, compost, and heat. The microbiological activity heats up the compostable material, and this composting stage is usually followed by a final stage at ambient temperature, in which the compost matures for several weeks. This is important to destroy any phototoxins formed in the initial stage that would affect any plants grown in the compost. Paper packaging products can be composted, but they are not ideal material for the process, and there are more appropriate methods for disposal of used packaging.

8.7 INCINERATION

In some countries, such as Denmark and Japan, incineration is the main method of disposing of household waste. Denmark incinerates about three-quarters of its waste and uses the heat generated to heat municipal buildings and homes. The amount of household waste incinerated varies throughout Europe: There are no incinerators in Ireland, and Germany burns 30% of its household waste, although the packaging legislation does not allow packaging to be burnt. The EU legislation favours incineration over landfill and allows for up to 30% of packaging to be incinerated, provided the energy is recovered and the calorific value of the materials is greater than 13 MJ/kg (e.g., paper and cardboard) and provided it has no greater impact on the environment than the fuel it replaces.

There are two main methods of combustion. The first is mass incineration, in which the collected waste is burned in specially designed and constructed incinerators. The combustion of waste leads to gaseous emissions that are closely controlled by law. There are also limits on the minimum residual lime, minimum oxygen levels, and a minimum furnace temperature of 850° C. The design of the furnace is important in controlling the emissions, which, following the combustion stage, are scrubbed to remove acid gases such as hydrogen chloride and sulphur oxides. Cyclones remove the coarse grit particles, and electrostatic precipitation and filter bags remove the fine dust. Rapid cooling of the gases to below 300° C is essential to prevent dioxin formulation in the post-combustion stage.

The estimated running costs of an advanced municipal waste incinerator with heat recovery are £45 to £50 per tonne.

The second method is the intermediate step in producing refuse-derived fuel (RDF) from household waste after it has been processed to remove noncombustibles such as metals and glass. RDF is fed directly to conventional power stations or made into pellets for use in modified boilers. This system has the advantage that materials are recovered from the waste for recycling; it is estimated that 7% of the reprocessed waste at an RDF plant can be recovered as ferrous scrap. Waste-derived fuel pellets have a typical calorific value of 18 MJ/kg, compared with 27 MJ/kg for coal. However, there are problems with their use:

- clinker formation and boiler fouling
- emissions of hydrogen chloride, which require scrubbing
- emission of particles high in metal chlorides, which need to be controlled

8.8 REUSE OF PACKAGING

The reuse of packaging rates second in the hierarchy of many governments when it comes to packaging legislation. Glass bottles have traditionally been collected and returned to the supplier for washing, inspection, and reuse. The closed-loop system used for milk and beer—in which used containers can be collected when fresh supplies are delivered—works well, and many return trips can be achieved. However, with modern supermarket retailing, the systems are not geared to taking returns, and the process has become less popular and the cost of recovery is often not justified. As a result, some governments have introduced legislation requiring retailers to take back empties and introduced deposit systems to encourage reasonable return rates.

Reuse systems focus on glass bottles, although plastic bottles are included in some countries, notably Germany and South Africa. Practice has shown that deposit schemes are successful, with return rates of 90% or more being common. Once returned, the packaging can be reused or recycled.

Bottles have to be washed and inspected to ensure that they are safe for reuse. The washing procedure includes caustic/detergent washes, and the effluent from such processes has to be disposed of safely. Energy requirements, including transport of returns, can be high.

Other packaging materials, such as pallets, are regularly reused, and international collection schemes exist. Bulk containers for liquids and solids are now widely used. They are generally designated for single product use to avoid contamination; the use of such containers is expected to grow.

There are now an increasing number of refillable packs. In the Netherlands, bulk packs of detergents (bag in box) are delivered to shops, and customers can either bring their own bottle for refill or obtain a new bottle from a dispenser alongside the bulk pack. When empty, the bulk pack is returned to the factory for refilling. This idea is not new—The Body Shop in the United Kingdom ran a similar project for shampoo. The main problem is product integrity, as one cannot ensure that the bottles used by the customer are free of contamination.

8.9 RECYCLING

Many processes for the manufacture of packaging recycle in-house waste as part of the production process. The plastics industry, for instance, recycles 99% of its production scrap. This is clean material, controlled in terms of purity, composition, and so on. Post-consumer packaging waste is a totally different problem because it is often soiled with decayed food or may contain toxic substances. It is because the history of such material is unknown that it creates problems in recycling. Metals and glass present the fewest problems, whereas plastics present the packaging industry with the greatest challenges.

The growing environmental awareness of society, particularly in the industrialised societies, has led to a demand for more recycling to husband the world's resources. The reasons for recycling need to be examined to be sure that there is a net environmental benefit. Recycling does conserve raw materials (mineral and energy) and reduces disposal costs and pollution effects such as emissions from incineration.

However, the collection of materials for recycling is a waste of effort (and may actually be harmful environmentally) if the economic balance does not favour recycling. This balance has to take into account the value of the reclaimed materials, whether there are markets for these materials, and other costs, such as storage or treatment.

A key factor is the amount of energy used throughout the life of the product. Consider the following sequence:

1. raw material extraction
2. purification
3. production of primary packaging materials
4. conversion
5. use for product protection in distribution, etc.
6. use by consumer
7. disposal
8. collection, cleaning, recycle, and reuse

If stage 8 uses fewer resources than stages 3, 4, 5, and 6 together, then recycling is resource-effective (but still not necessarily economically viable).

However, there is no agreed-upon method of evaluating these factors; until such time as a universally agreed-upon method of life cycle analysis or assessment is developed, there will always be arguments about the merits and demerits of recycling and other methods of packaging disposal.

8.9.1 Recycling Paper Packaging

Packaging paper and boards are used for a wide range of products: primary packaging such as bags, sacks, and cartons for frozen foods and dry goods, and secondary packaging such as corrugated cases and transit packaging. Paper is also used in combination with other materials, such as plastics or aluminium foil, when special properties or barrier resistance is required. Combining it with other materials makes it more difficult to recycle, as these other materials ("contraries") have to be separated from the paper fibres.

Paper packaging can be divided into three main categories:

1. corrugated cases
2. cartons
3. packaging papers

The material flows through the collection systems are complex, as Figures 8–5, 8–6, and 8–7 show. The collection of waste paper is dependent on efficient recovery and collection schemes in which the different grades of paper are kept separate. In all, there are 11 different grades of waste paper (Table 8–3).

Different grading schemes are used in the United States and Europe. Each grade in whatever system contains a heterogeneous collection of waste paper. Groups 7–11 cover the packaging grades; these tend to be the lower quality materials, and as there is a tendency to down-cycle paper as a raw material, waste packaging will tend to finish up in a lower quality paper or board.

After the waste paper is collected and sorted, it is transported to a paper mill, which will select the grade of waste required to make a particular type of paper or board. The material is pulped or slushed with large quantities of water to break the paper down into individual fibres; it is then passed through screens to remove contaminants and centrifugally cleaned to remove grit and small solid particles. It is often treated with steam or other methods to disperse plastic materials and adhesives, which cause problems in the paper-making process. After the material is cleaned, the fibres are refined and then flow onto the paper-making wire in the normal way.

The recycling process shortens the fibres each time they are processed; after about four cycles, the fibres are too short to stay on the paper-making wire and finish up as a sludge. Efforts are being made to find uses for this sludge, including feeding it into the middle ply in a multi-ply board where the first ply or layer of fibres on the paper-making wire helps retain the fine particles. Where the fibres are combined with materials such as polyethylene, a more complex pulping and cleaning process is required. For liquid packaging—board, for instance—the cartons are put into the pulper and soaked at 60° C and agitated for 30 minutes. The slurry passes through a multistage cleaning system to take out the polyethylene and aluminium. The polyethylene is screened off using a mechanical process, and the aluminium is removed using centrifugal force.

This stage takes around 20 minutes, depending how many times the slurry is screened, which can be up to three times to achieve the purest pulp possible. The pulp slurry is sent to the de-inking plant, where soaps are added to soften the remaining ink particles. Air is injected into the solution and rises to the surface, carrying with it the ink particles. The ink is removed with the soap and foam, and the clean fibre is concentrated ready for processing.

Although there is a very large use of waste paper worldwide, it is not recommended that packaging material made from it be used in direct contact with foodstuffs, as it cannot be guaranteed free from toxic contaminants. With stricter food laws being introduced worldwide, waste paper materials can only be used in non-food-contact outlets. As a result, there is an increasing lobby that recommends that waste paper packaging should be incinerated with energy recovery and fresh trees

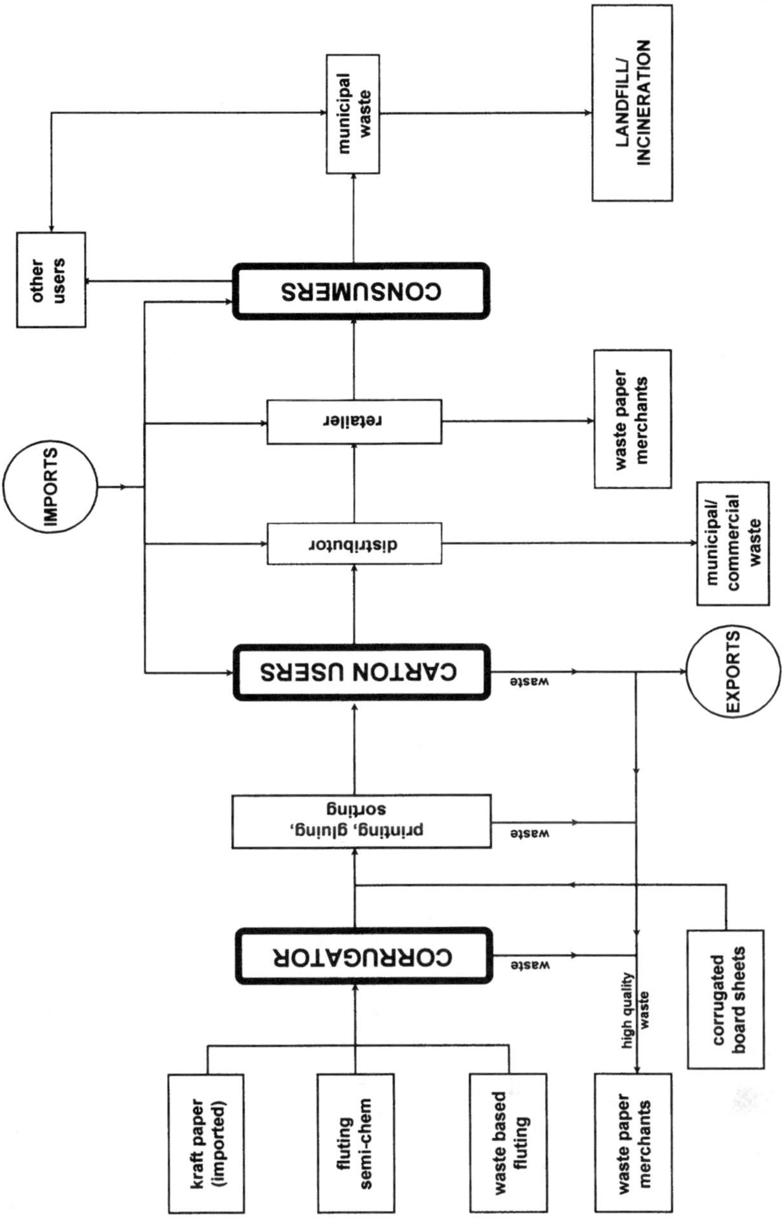

Figure 8–5 Materials flowchart for corrugated cases.

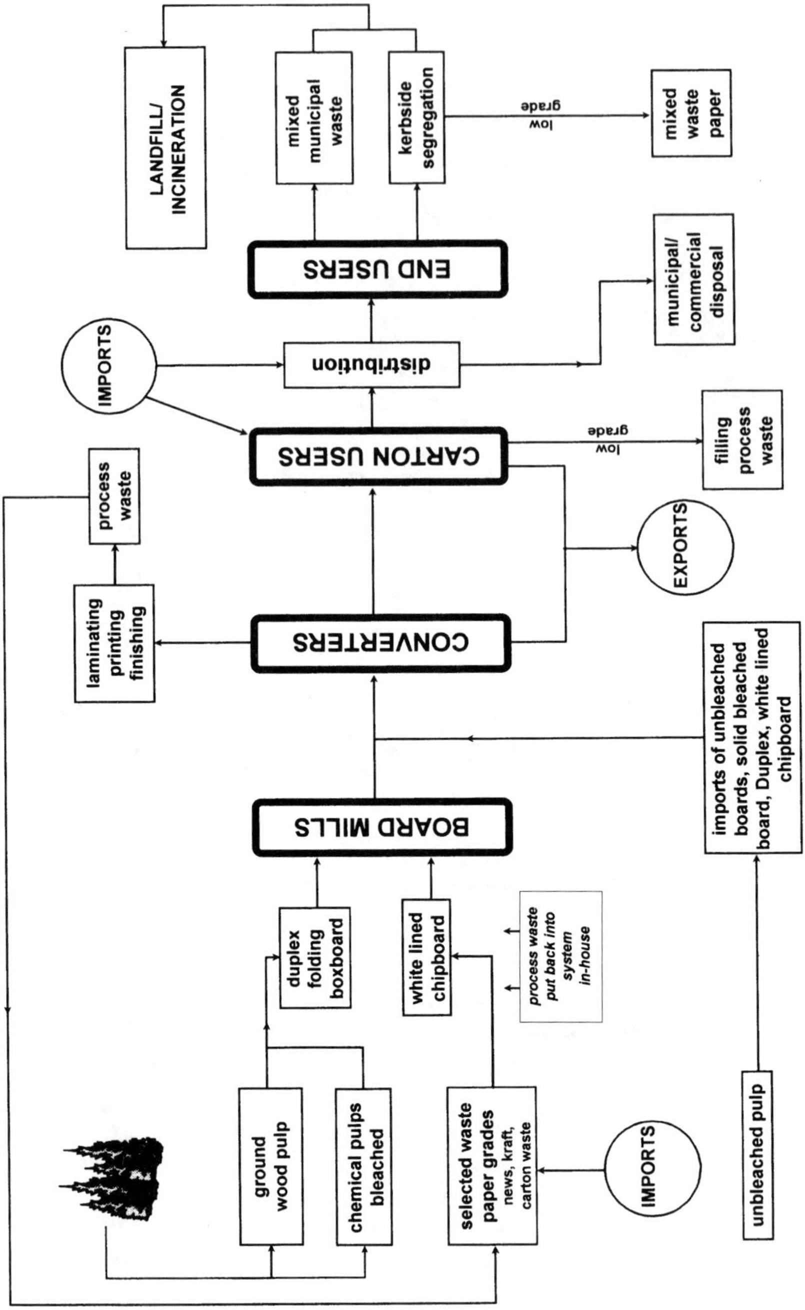

Figure 8–6 Materials flowchart for cartons.

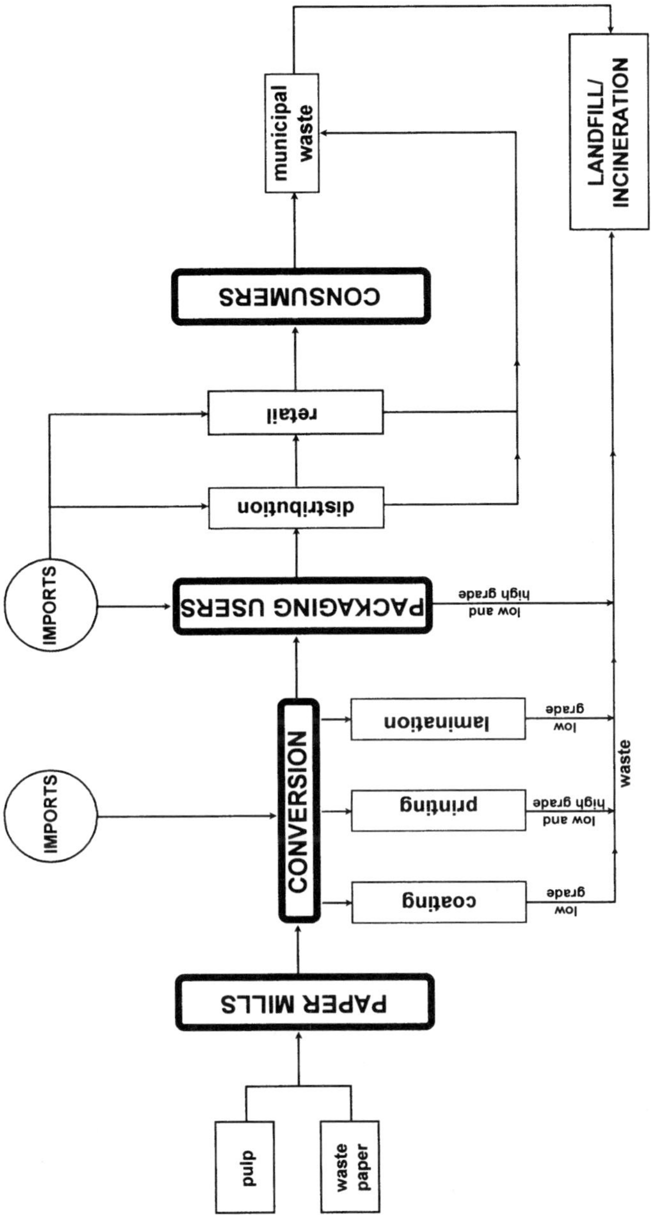

Figure 8–7 Materials flowchart for packaging papers.

Table 8–3 United Kingdom Waste Paper Grade Structure

Group	*Description*
1	Best white shavings, fine shavings, white and cream shavings, and white-coated shavings.
2	White unprinted, white duplex, and other mechanical wood pulp cuttings; slightly printed white card cuttings; best one-cuts; printed woody one-cuts; white and light tone shavings; white and coloured shavings.
3	White and lightly printed scanboard.
4	Ledgers, white and coloured heavy letter, white and coloured continuous stationery waste, white and coloured carbonless copy paper waste, quire, best white pams.
5	Over-issue news, once-read news, over-issue white and coloured woody pams, news and pams, telephone directories and soft covers, once-read woody pams.
6	Buff envelope cuttings, buff tabulating cards, coloured tabulating cards, light brown and buffs.
7	Double-lined kraft and new kraft liner cuttings.
8	Container waste (old kraft liners).
9	Mixed papers
10	Coloured card
11	Contaminated grades.

Source: BPBIF.

used to produce food packaging material, because trees are a renewable source of material and the forestry programme relies on harvesting the wood on a regular basis.

8.9.2 Recycling Aluminium

The aluminium used in packaging is principally either in the form of cans or foil, and the foil is usually combined with other materials to form flexible laminates used for either liquid or dry products. Figure 8–8 gives a brief outline of the flows of aluminium packaging.

The production of aluminium from bauxite, the naturally occurring aluminium mineral, requires large amounts of power—about 7.5 kWh for each pound of material produced (Aluminum Association, 1990). Most (95%) of the energy is saved when new metal is made from used cans instead of raw materials. It is important to the economics of the recycling process that there is a regular and continuous sup-

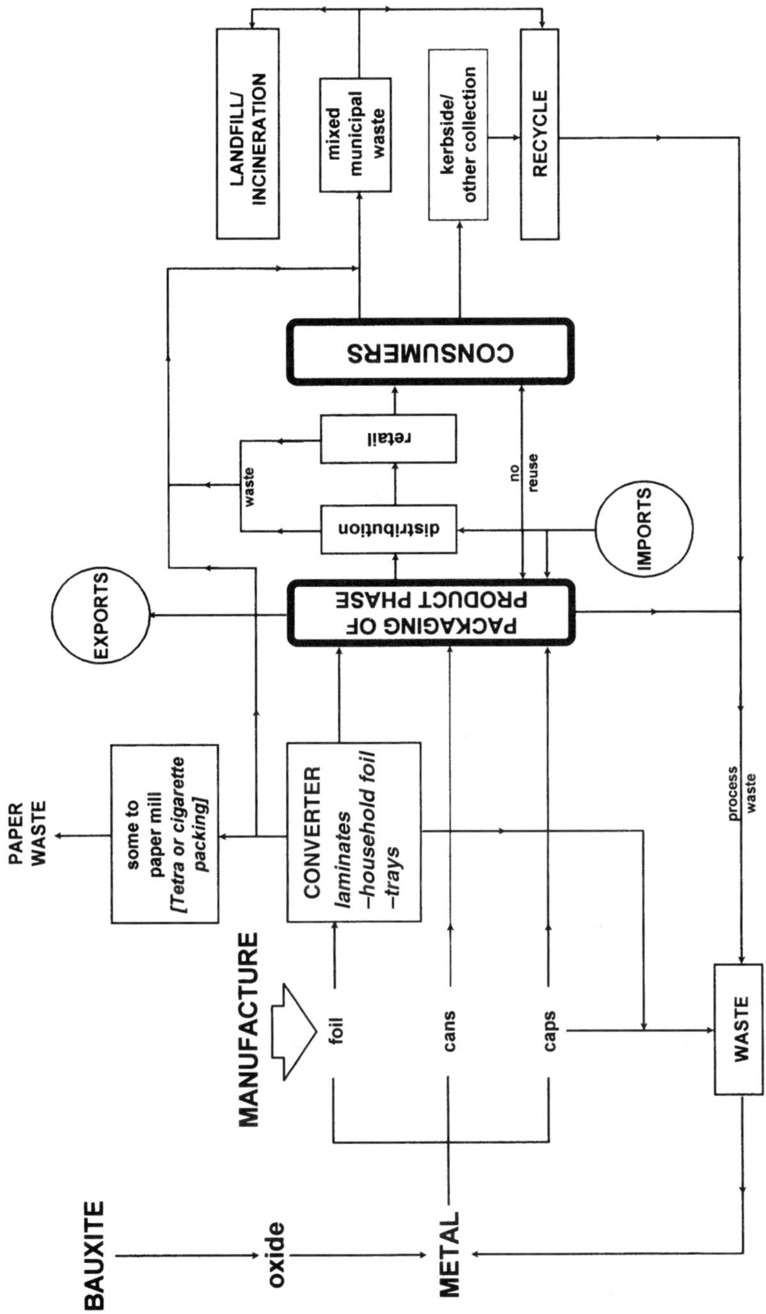

Figure 8–8 Materials flowchart for aluminium.

ply of used cans through effective and convenient collection systems that provide the closed loop for the recycling process.

Publicity and deposit schemes have been used to raise consumer awareness of the need to recycle aluminium cans, and this campaign has been a recycling success story. Recovery rates as high as 85% are recorded through these collection schemes. On collection, the cans are crushed and baled for shipment to the recycling plant *(The Can Makers*, 1992). The material is first shredded into small pieces and passed through a magnetic separator to remove any steel present. A decoating process burns off the paint and lacquer; the emissions from this process are recirculated and burned with more fuel to preheat further batches of shredded cans, improving the energy efficiency of the process and reducing emissions.

The hot shreds of aluminium are then fed into melting furnaces where the composition of the molten metal can be alloyed to obtain the correct composition before casting into ingots. The ingots then go forward for the manufacture of new cans.

Aluminium foil poses much greater problems for recycling because it is usually combined with other materials such as polyethylene, paper, and PET, to name but a few. Currently, there are no commercial methods available for separating the materials from these laminates, although research is being carried out. The only real alternative to landfill for these products is incineration. The one exception to this has been the separation of paper fibres from board foil polythene cartons for liquid packaging, which is discussed elsewhere in this chapter.

8.9.3 Recycling of Steel and Tinplate

Food cans are principally made from steel or tinplated steel. Metal packaging represents about 8% of household waste (*The Can Makers,* 1992). The recycling of both tinplate and steel cans from the domestic waste stream has steadily increased in recent years. Collection of the cans is arranged either through can banks or, more usually, by the use of electromagnetic separation at the waste treatment plants (*The Can Makers,* 1992). Up to 80% of the available steel can be recovered. Once collected, the tin cans are cleaned to remove dirt, paper, plastic, and aluminium, and the dense pellets are detinned using an electrolytic process (Figure 8–9). The steel produced needs no further processing before its use in high-specification steel making. The tin recovered can be reused to produce new tin plate for can making. The recycling of steel cans is also a recycling success story.

8.9.4 Recycling of Glass

The material flow in the glass packaging cycle is shown in Figure 8–10. The chemical composition of glass is largely unchanged by repeated reheating, so the

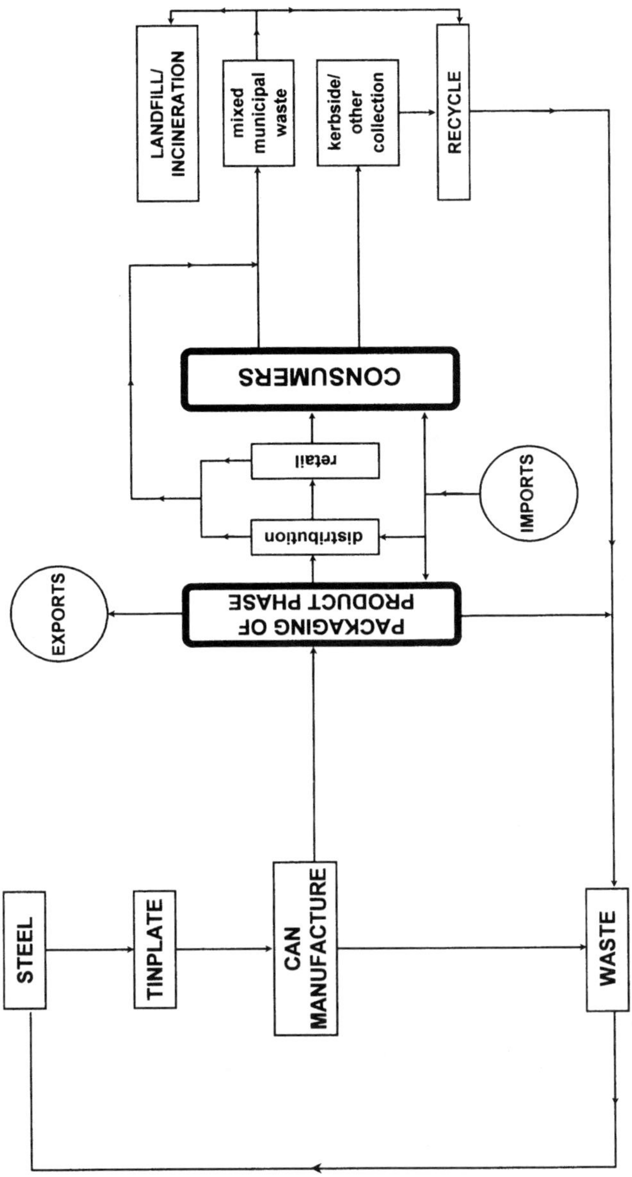

Figure 8–9 Materials flowchart for tinplate/steel.

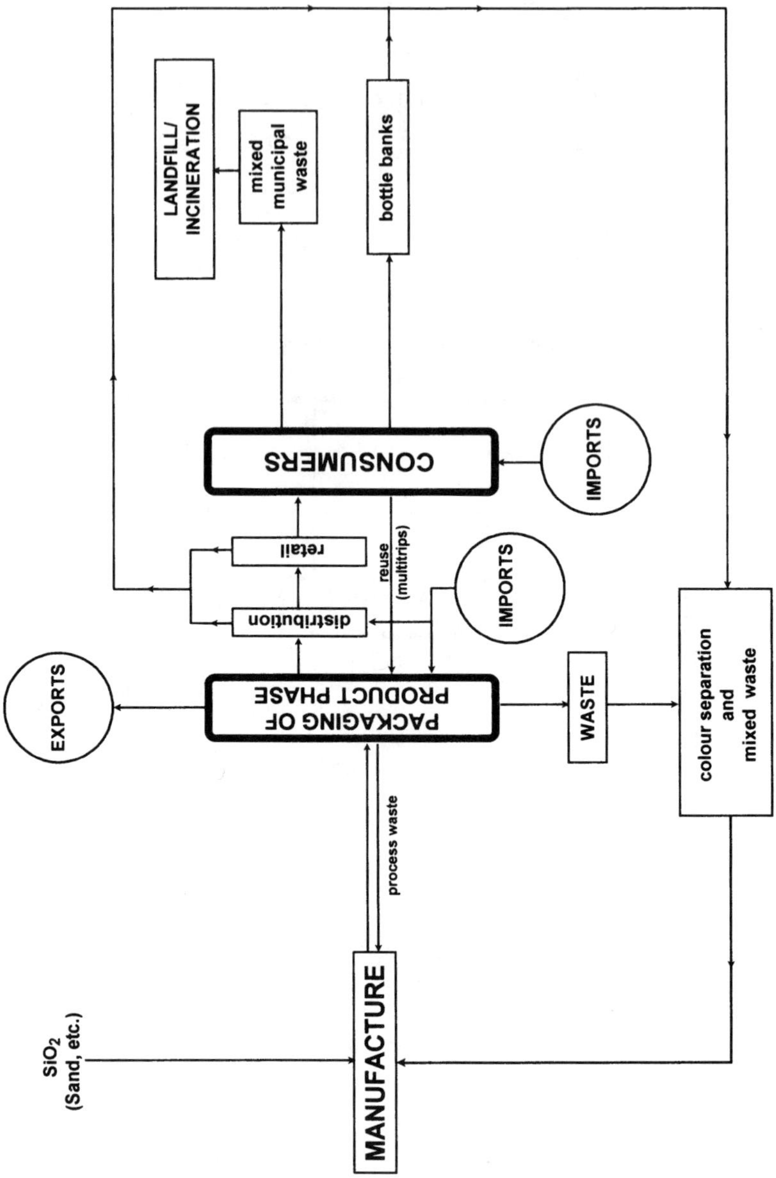

Figure 8–10 Materials flowchart for glass.

degradation processes that are associated with the reprocessing or re-melting of organically based packaging materials are absent.

Traditionally, glassworks have used 15% to 20% of broken glass, or "cullet", in their raw material mix, but this proportion has increased dramatically. This is a direct result of industry policy in relation to the recovery of broken glass from waste, a desire to save energy because broken glass melts more readily than the mixed raw materials, and the all-important attitude and response to customers.

The glass recycling industry in Europe has grown rapidly over the past 15 years. In the United Kingdom, the amount recycled has risen from 55,000 tonnes in 1980 to 372,000 tonnes in 1990 and 420,000 tonnes in 1996 (Table 8–4). However, the UK glass industry's target of a recycling rate of 50% by 2001 is not thought likely to be achieved ("Packaging Recovery Stagnates", 1997). Other recycling targets in the EU are shown in Table 8–5.

From much of the world come reports of steady increases in recycled glass. Switzerland already uses 75% recycled glass in new bottles and jars (Cornaz, 1989), and some 1.5 million tonnes have been collected by its Bottle Bank scheme ("Farewell to Swiss Glass Recycling", 1989), which saves the Swiss government at least SFr20 million per year.

All containers produced by the German glass industry contain 50% cullet ("Increase in Recycled Glass", 1989). The Netherlands reports the same figure and claims to be Europe's leading glass collector ("Dutch Bottle Bank 10 Years Old", 1988). Japan claims that 9.2 billion containers incorporate 55% cullet (Ichiko, 1988). Average figures such as these indicate that many containers have in excess of 65% cullet, and Swiss green container production is manufactured entirely from cullet ("Farewell to Swiss Glass Recycling", 1989).

The glass collected is generally contaminated with a variety of materials: metals, stone, cork, plastics, and other materials not suitable for use by the glassworks. So it is necessary to clean the material, which has traditionally included hand sorting, although with increased volume, automatic methods are being installed.

In the older method, the glass is put onto a conveyor and hand sorted to remove obvious contraries and oversized material; ferrous materials are then taken out by magnetic separator. Then the material is crushed in a hammermill followed by magnetic separator. The crushing operation gives cullet of a specific size. The material is then screened with an air classifier, which removes light material such as paper, aluminium foil, cork, and plastic. Some glassworks employ a final hand sorting operation. As the volume of cullet used and the specifications required by the glassworks have both increased, hand sorting has been replaced by electronic methods of detection and selection.

Separation of the coloured glass automatically presents the major problem to the industry. To do it efficiently requires the glass to be in large pieces—that is, whole bottles—which can then be separated into brown, green, clear, and opaque

Table 8–4 European Glass Recycling Levels, 1996

Country	*Tonnes Collected ('000)*	*National Recycling Rate (%)*	*Collection Method*
Austria	206	—	100% bottle banks
Belgium	224	66	85% bottle banks; 15% other systems
Denmark	122	66	50% bottle banks; 50% other systems
Finland	33	63	17% bottle banks; 83% bottlers
France	1,400	50	90% bottle banks; 10% multimaterial collection
Germany	2,839	79	95% bottle banks; 5% multimaterial collection
Greece	39	29	22% bottle banks; 78% multimaterial collection
Ireland	43	46	33% bottle banks; 5% multimaterial collection; 62% cullet dealers
Italy	894	53	99% bottle banks; 1% multimaterial collection
Netherlands	380	81	95% bottle banks; 5% multimaterial collection
Norway	40	75	
Portugal	120	42	95% bottle banks; 5% multimaterial collection
Spain	456	35	100% bottle banks
Sweden	120	72	90% bottle banks; 10% multimaterial collection
Switzerland	259	89	
Turkey	44	13	
United Kingdom	420	22	100% bottle banks

Courtesy of FEVE/Glass Gazette, September 1997.

by measuring the effect of a beam of visible wavelength light on the bottle, measuring either the transparency or the reflectance. Once cleaned and sorted, the cullet is fed into the furnace for reuse.

8.9.5 Recycling Plastics Packaging

Plastics packaging is lightweight and functional. It has replaced many traditional materials in recent years, and the use of plastics has grown rapidly. The term

Table 8–5 Recycling Targets for Glass in the European Union for 2001

Country	*Target (%)*
Austria	93
Belgium	†
Denmark	80
Finland	48
France	75
Germany	75
Greece	25*
Ireland	55
Italy	†
Netherlands	90
Portugal	25
Spain	50
Sweden	70
United Kingdom	15

* = recovery
† = Packaging Directive targets
Courtesy of FEVE/Glass Gazette, September 1997.

plastic covers a wide range of polymeric materials made from different base chemicals. They are divided into two types: thermosetting resins and thermoplastics. Packaging is mainly composed of the latter, which can be recycled. Some thermoset resins are used for bottle caps; these cannot be recycled, and their use will decline as a result of legislation. At least 13 different thermoplastic polymers are used on their own or in combinations in packaging. The flow patterns are complicated, as Figure 8–11 shows.

For the inexperienced person, it is difficult to identify the materials, and this is a major problem. A conventional numbering system (Figure 8–12) has been set up for plastic containers so that the material can be identified. This covers the main materials used but does not take into account laminated materials. Also, the system can only be applied to rigid containers or printed film.

For a material to be successfully recycled, it should not be mixed with other plastics. If polypropylene (PP) is mixed with more than 10 percent of high-density polyethylene (HDPE), the impact strength and other properties are substantially reduced. Other materials are incompatible if processed together, for example, low-density polyethylene (LDPE) and polystyrene (PS). Others, like polyethylene terephthalate, have to be dried before processing. Thus, it is essential for any recycling scheme to keep each plastic material separate to be successful. The most

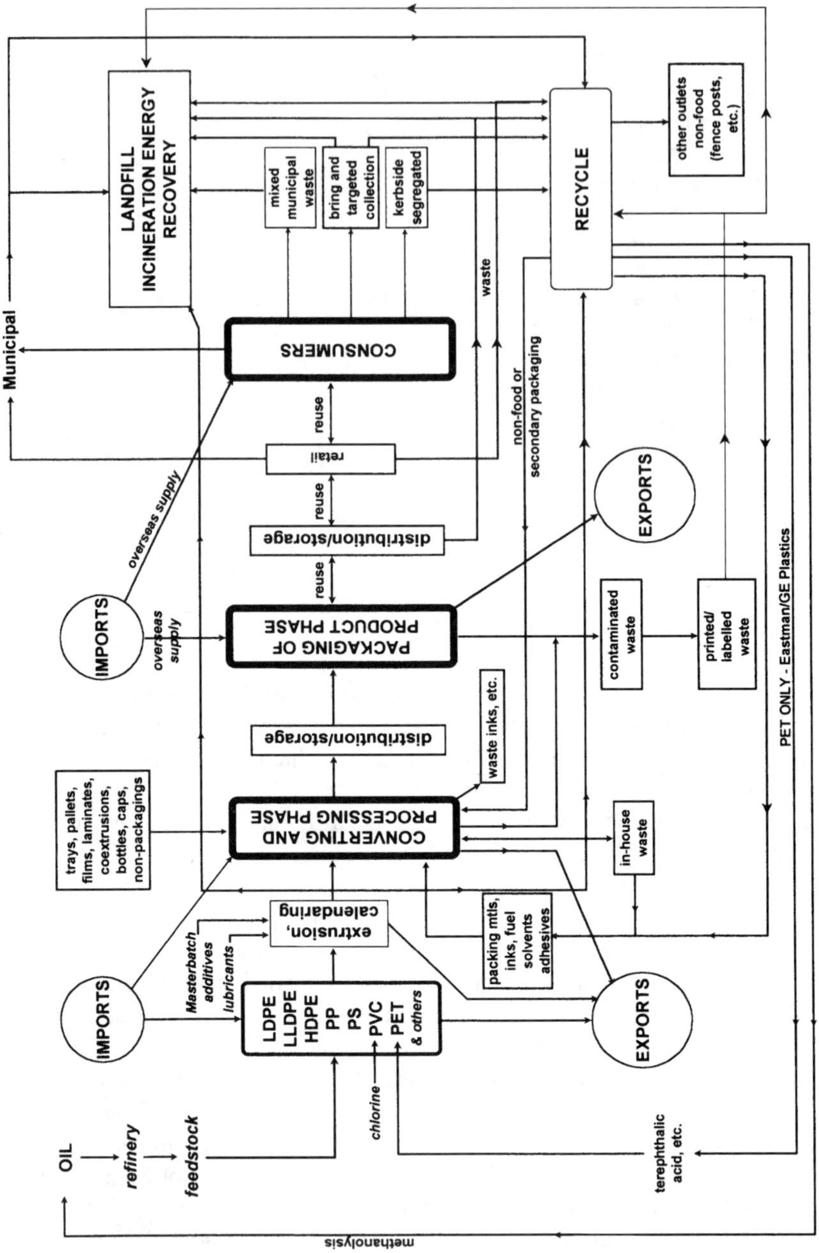

Figure 8–11 Materials flowchart for plastics.

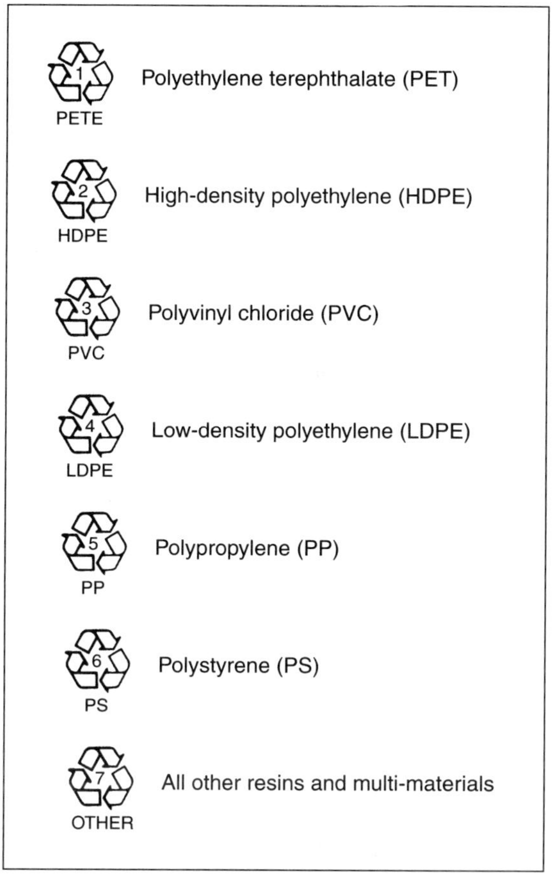

Figure 8–12 Society of the Plastics Industry plastics container material code.

successful schemes have been the ones in which the material can be easily recognised by the pack design, for example, plastic Coke bottles and high-density milk bottles in the United States. The emphasis during the collection process is to keep each material separate.

Mixed plastics can be recycled, but it is necessary to add a compatibiliser, a rubber-based compound that acts as a binder for the plastic materials. This can add up to £200 per tonne to the cost of the material. The product can be used for fence posts and as a general wood substitute. One problem with the material is that it can have variable properties because of variation of the constituents; recyclers of such material try to make it in large batches in order to establish a minimum performance level.

In principle, the same recycling procedures are used for all the materials. They are sorted and, in general, plastic bottles are kept separate from films, as the grades of polymers used for each outlet differ.

The bottles are granulated and the resultant material washed to remove labels, adhesives, and other contaminants. The material is dried and passed through a cyclone to remove any metallic particles and grit. It is then extruded and chopped into small pellets and can then be mixed with virgin polymer or used on its own to make new bottles or other plastic components. A recent development has been to co-extrude the recycled material between two layers of virgin material (Figure 8–13), which gives a bottle indistinguishable in appearance from one made of virgin material. The inner layer of virgin material also protects the bottle contents from possible contamination from the recycled material.

A similar process can be used for film, in which it is chopped into small pieces, washed, cleaned, and then re-extruded. As films are often heavily printed, the colour of the resultant product will be variable and can only be used for low-grade outlets such as rubbish bags and carrier bags.

8.9.5.1 Low-Density Polyethylene

The main uses of LDPE are plastic squeeze bottles, bags, and carrier bags. If uncontaminated, the material can be recycled many times (tests up to 20 times have shown no deterioration).

8.9.5.2 High-Density Polyethylene

The uses for HDPE include rigid bottles and bags. The materials can be recycled many times, provided they are not contaminated. Laundry detergent bottles and milk bottles are a major source of materials for recycling, although it is diffi-

Figure 8–13 Section through wall of co-extruded bottle.

cult to remove the perfume of the detergent or the smell of stale milk from the polymer.

8.9.5.3 Polypropylene

PP is used for closures and dairy pot bottles, but probably the main outlet is biaxially orientated film for wrappers and bags. The material can be recycled without problem provided it is not contaminated. Because it is mainly used in film and bags, it is costly to collect—little material spread over a large area.

8.9.5.4 Polyethylene Terephthalate

PET is widely used for carbonated beverage bottles, ovenable trays, and film—one of the recycling success stories. The material is clearly identified from the bottle labels and distinctive design, so it can be separated from the waste stream, granulated, washed and cleaned, dried, and then reprocessed into base cups for new carbonated drinks bottles but more often into fibre fill for padded clothing and duvets.

8.9.5.5 Polyvinylchloride

PVC has been the focus of the criticism of plastics by many of the militant 'green' organisations. Twenty years ago the debate concentrated on the health hazards to which workers in PVC plants were exposed. Later, the attention switched to the disposal of PVC and the dioxins that may be formed when it is burnt. Current scientific evidence shows that incineration of all chlorine-containing materials can give rise to dioxins, but that if PVC is incinerated at high temperature and the flue gases washed and quickly cooled as previously described in this chapter, it is a safe material to use. There will always be people who see PVC as an environmentally harmful material, however carefully it is produced, however inert the chlorine content, and however beneficial the application to which it is put.

PVC is one of the best packaging materials available—it is widely used for bottles, storage trays, and thermoformed trays for modified-atmosphere packed foods. It uses much less of our non-renewable petroleum resources than other thermoplastics; the chlorine used is derived from salt, of which there are high natural reserves. PVC formulations can be tailored to different applications, and it is more tolerant of contamination by other materials in the recycling process than is the case with some commodity polymers such as polyethylene and PP.

The main source of post-consumer used PVC packaging is plastic bottles, although there are some closed-loop systems in which trays used to collate and display goods on supermarket shelves are collected and returned for recycling. Recycled bottles are collected and separated from other plastic bottles manually using the polymer identification circles (see Figure 8–12). The chlorine atoms in

PVC also give it a unique characteristic that enables it to be identified by automatic methods of sorting. The PVC is cleaned and granulated and then moulded into long-term applications such as pipes, where it can replace new PVC.

8.9.5.6 Polystyrene

PS is widely used in packaging: thermoformed for dairy pots and tubs and injection-blown for talc bottles/dispensers. It is also used for caps and closures, and recent developments include biaxially orientated film. These outlets have attracted little attention from the "green" lobby. However, another form of the material that has been the focus of green protest is foam PS, which is used as a protective packaging for electronic and other products to prevent impact damage. It is also used for take-away disposable food containers, and the litter these produce has been the subject of legislation. Originally, foam PS was made using chlorofluorocarbons (CFCs) as the foaming agent. These are now banned under the Montreal Protocol to protect the ozone layer, and other foaming agents are used. As a result of the litter problem from fast food chains and the high profile of the material, recycling methods have been developed for foam PS. The material is collected, cleaned, crushed, and re-extruded for non-food outlets.

8.10 MULTILAYER MATERIALS

Co-extrusion technology has developed over the past 10 years to produce packaging materials with specific barrier properties. It has enabled producers to tailor-make materials using thin layers of expensive materials with good barrier properties, and to produce a laminate that has good barrier properties and is economically viable. Some of the commercially available materials have up to seven layers of material and can contain up to four or five different polymers.

Opinion is divided on the feasibility of recycling such materials. Oxygen barrier is one property that these materials are required to exhibit, and this is often achieved by using ethylene vinyl alcohol (EVOH) or polyvinylidene chloride (PVdC). The tendency is to use EVOH; PVdC cannot be recycled and as it contains chlorine it is not considered acceptable by some organisations.

Because of the relatively small quantities of these materials, they could be recycled with films made of the commodity polymers; however, as the market widens and because of the diverse range of materials, it may be necessary to find alternative ways of recycling these materials. The cost of recycling gives materials that are more expensive than virgin materials, and whilst no overall accurate figures can be given, recycled materials will be more expensive unless subsidies or tax incentives to use them are introduced.

All the preceding recycling methods depend on mechanical methods of reprocessing the polymers. Because of the strict food laws in Europe and the United States, post-consumer plastics waste cannot be used in direct contact with foodstuffs because it cannot be guaranteed to be free of contaminants. The difficulties of sorting plastics and mechanically recycling mixed polymers has encouraged the industry to look at chemical methods of recycling used packaging. Although mechanical recycling should be seen as a relatively low-cost, low-energy solution (as long as markets for the recycled materials can be found) and should be seen as the first priority, the alternatives being developed would substantially contribute to increasing the amount of plastics recovered from plastics packaging.

Mechanical recycling produces similar material with generally a lower quality, whilst chemical recycling leads to new materials with no quality compromise. The first commercially approved process is for PET: The material is broken down into its components, which are then purified and repolymerised. At present, this process is only being operated on a relatively small scale, but it is being claimed that on scaling up the process it becomes economically viable. Recently, DuPont indicated that it may be possible to extend the process to polyethylene and polyamide (nylon) film structures ("Getting the Chemistry Right", 1993).

Techniques have also been developed on a pilot scale to convert mixed plastics into an acceptable hydrocarbon feedstock suitable for mixing with the feedstock currently used in polymer production. Contamination remains one of the problems of the process, including contamination with heavy metals and chlorine from PVC. Work is also going on to convert the mixed plastics from post-consumer waste back into oil, which can then be used to replace virgin mineral oil.

8.11 DEGRADABLE PLASTICS

The industry has been pressured to find ways of making plastics biodegradable because plastics do not break down into basic components or rot when put in landfill. None of the plastics in general use for packaging are biodegradable. As yet, there are no universally accepted definitions of terms such as "degradable" and "biodegradable".

Biodegradable is generally defined as being capable of being broken down by micro-organisms, within a reasonable period of time, to carbon dioxide, methane, water, and humus-type materials. Biodegradation results in the total loss of the chemical identity of the original material. Photodegradable, on the other hand, refers to a significant loss in mechanical properties within a reasonable time after exposure to sunlight. Although there is a significant reduction in the molecular weight of the polymers, photodegradation does not itself destroy the chemical identity of the material.

Table 8–6 (Selke, 1996) shows the degradability of plastics and some other packaging materials.

The polyethylene used by Hi-Cone to make can carriers (the strips of plastic that are used to hold four or six cans together to form a multipack) has an additive incorporated in it so that after about 6 weeks' exposure to ultraviolet light, the plastic becomes brittle and breaks into small pieces. This is an embrittlement rather than biodegradation and is the result of various green pressure groups accusing Hi-Cone of killing wildlife. Supposedly, there were instances of seabirds who were killed because they were strangled in the plastic rings. The claim was disputed by the company, but it incorporated the additive in the material so that even if the can carriers were discarded as litter there would be no risk to wildlife.

The plastics industry has been careful to stress that if biodegradable plastics were available they could become a problem if they became part of the recycling systems, because there would then be no guarantee on the life of the products made from recycled plastics.

Much research effort has gone into producing thermoplastics from naturally occurring materials. The first one to be marketed was ICI's Biopol, which is completely biodegradable and is made from naturally occurring materials. It can be made into film and bottles. It was not a commercial success, and ICI has sold the technology to Monsanto.

Several companies have developed biodegradable polymers from starch. These materials should not be confused with starch-filled polyethylene, which is used for carrier bags. These starch-filled materials are not biodegradable, but when they are exposed to the elements, the starch is leached out, leaving the basic polymer.

Table 8–6 Degradability of Packaging Materials

Material	*Biodegradable*	*Photodegradable*
Polyester (PET)	No	Very slow
Polyalkenes	No	Very slow
Polystyrene (PS)	No	Very slow
Polyvinyl chloride (PVC)	No	Very slow
Cellophane	Yes	Very slow
Paper and paperboard	Yes	Very slow
Aluminium	No	No
Steel	No	No
Glass	No	No
Wood	Yes	Very slow

Courtesy of Pira International, Leatherhead, United Kingdom.

Biodegradable materials were principally seen as a solution to the litter problem, but the answer to litter must lie in the education of people to take rubbish home rather than dumping it in the towns and countryside.

Two important issues remain for biodegradable materials. First, there has to be an assurance that the products of degradation are not more hazardous than the original materials. Second, and significantly, is the issue of cost. On the whole and with the exception of paper, biodegradable packaging materials are more costly than the nondegradable alternatives. Thus, they are unlikely to be used unless there is some stimulus, such as perceived consumer preference or legislation, providing the necessary driver.

8.12 THE WAY AHEAD

The field of recovery and disposal of used packaging is very dynamic because of the effects of legislation and restriction being imposed by governments and other organisations. The pressures are to reduce the volume of waste packaging and to recycle as much as possible. As already stated, this assumes that suitable outlets can be found for recycled materials. The economics of collection recycling and reuse are confused. There is no doubt that glass, aluminium, and tinplate are viable materials to recycle, and that greater quantities will be collected and used.

Paper recycling is an established industry, but it is easily overwhelmed by the amount of material available. Capital investment is required to use more material and the economics—compared with other methods of disposal, such as incineration—may in the long run make this a more attractive proposition.

Plastics present the largest problem. Recycling is uneconomic compared with the use of virgin material, so incineration may well be the long-term solution. The debate on the need to recycle, incinerate, and landfill will continue until such time as an acceptable and viable method of total life cycle assessment is available for packaging that will give numerised data that will enable the correct choice to be made to produce a sensible balance of economics, environmental protection, and packaging function.

8.12.1 Life Cycle Assessment

Life cycle assessment (LCA) is defined as "a process to evaluate the environmental burdens associated with a product, process or activity by identifying and quantifying energy and materials used and wastes released to the environment, throughout the life cycle and to assess the impact of those energy and material uses and releases on the environment" (Dobson, unpublished).

LCA is one of a number of environmental tools. Others include:

- *EIA—environmental impact assessment.* A site-specific tool used to assess the environmental impact of a specific planned investment or facility.
- *EA—environmental audit.* A site-specific tool used to assess the environmental impact of an existing facility. Usually includes consideration of the communication and management of environmentally related information.
- *RA—risk assessment.* Examines the risk presented by a material or facility, including consideration of the potential hazard and its likelihood of recurring.
- *EVA—environmental valuation analysis.* The cost to society of an emission or emissions is calculated by assessing the environmental consequences of the emissions and placing economic values on these consequences.

The difference between these methods and LCA is that they do not attempt to cover the whole product, process, or activity life cycle.

The EU Directive on Packaging and Packaging Waste states that LCAs should be used to justify a clear hierarchy among reusable, recyclable, and recoverable packaging, and that the European Commission should promote the creation of European standards for the criteria and methodologies for life cycle analysis of packaging.

It is generally thought that LCA is best used on a case-by-case basis and that it cannot be used to identify a "best" waste management system. Comparisons made using LCA tend to show the environmental implications of different choices rather than producing clear winners and losers. An LCA study covers one specific system at a defined point in time—it will not indicate whether the option will still be valid at some point in the future; this would require a new study with new data.

LCA methodology consists of four generally accepted parts (Jönson, 1996):

1. *goal and scope definition:* the purpose of the study, its scope, and the procedure for quality assurance of the results
2. *inventory analysis* (sometimes referred to as LCI): the quantitative listing of all material and energy inputs and outputs
3. *impact assessment:* aggregation of the inventory data to make it easier to draw conclusions and make decisions
4. *interpretation*

Methodologies for impact assessment and interpretation of results are the subject of discussion within the International Standards Organisation (ISO) and elsewhere.

For packaging waste, LCA studies are unlikely to be able to create a universal hierarchy between reusable and one-way packaging, since the preferred option for any given product will vary by product category and will be dependent on several variables (trippage, kilometrage, etc.). LCA can be used to optimise the life cycles of individual packages and could be used to make a judgement on the best choice environmentally between two new packs for the same product. It can also help to identify where improvements can be made to reduce the environmental impact of a product in terms of energy and raw materials used and the wastes produced.

8.12.2 Being Environmentally Responsible

Incpen provides advice on taking into account environmental issues when choosing a pack. These are in the form of guidelines as follows (Incpen, 1998):

1. In choosing a pack, take into account all environmental factors, especially:
   - energy and raw material requirements throughout the distribution chain
   - waste arisings at all stages throughout the distribution chain
   - expected impact on post-consumer waste
2. Avoid the use of excessive packaging. Use the optimum resources to provide adequate protection and performance.
3. Take positive action to reduce litter. Put the Tidyman logo and/or anti-littering message on packaging likely to be consumed outdoors; sponsor litter bins.
4. Examine opportunities for using secondary materials. Allow specifications where possible to encourage the use of secondary materials; for example, a higher level of recycled fibre for corrugated cases, low-quality recycled paper for internal memos, secondary plastics for non-food-contact containers.
5. Support resource-efficient use where appropriate. Consider whether material would help.
6. Encourage repeated use where appropriate; for example, pallets and carrier bags (which are thought of as packaging and seen as a wasteful use of resource).
7. Assess potential for supplying your own waste to a recycler.

DO NOT CLAIM ENVIRONMENTAL SUPERIORITY FOR YOUR PACK. No pack has a monopoly of environmental virtues. In solving one environmental problem, do not cause another.

REFERENCES

Aluminum Association. (1990). *Aluminum recycling* (p. 6). Washington, D.C.

The Can Makers. (1992). Update No. 8. London: Can Makers.

Cooper, J. (1992). *Packaging and the environment* (p. 54). Leatherhead, UK: Pira International.

Cornaz, J.D. (1989). The practical repercussions of the beverage regulation for glass. *Verpacking (Zurich), 44*(5), 20–22.

Department of the Environment, Transport and the Regions. (1998). *Less waste. More value. Consultation paper on the waste strategy for England and Wales.* London.

Dobson, P.J. *Life cycle assessment (LCA): An introduction to the methodology.* Leatherhead, UK: Pira International, unpublished.

Dutch bottle bank 10 years old. (1988). *Voedingsmiddelentechnologie, 21*(13), 11.

European Commission. (1990). Directive 90/128/EEC, Plastics in contact with foodstuffs.

European Commission. (1994). Directive 94/62/EC, Packaging and packaging waste.

Freiesleben, W. (1992). *Plastics in Municipal Incineration.* Brussels: European Centre for Plastics in the Environment.

Farewell to Swiss glass recycling. (1989). *Verpacking (Zurich), 44*(5), 22–25.

Getting the chemistry right. (1993, February 18). *Packaging Week*, 15.

Ichiko, T. (1988). Recycling of glass bottles. *JPI Journal, 26*(12), 12–22.

Increase in recycled glass. (1989). *Brauwelt, 129*(16), 631.

Industry Council for Packaging and the Environment (Incpen). (1998). "Environmental Guidelines". http://www.incpen.org/html/enviroguides.htm; accessed June 1998.

Jönson, G. (1996). *LCA–a tool for measuring environmental performance.* Leatherhead, UK: Pira International, 25.

Packaging "essential requirements." (1998, June). *ENDS Report,* (281), 42–43.

Packaging recovery stagnates as EC deadlines loom. (1997, September). *ENDS Report,* (272), 19.

Pré, G. (1992). *Packaging and the environment* (p. 151). Leatherhead, UK: Pira International.

Selke, S. (1996). *Biodegradation and packaging* (p. 5). Leatherhead, UK: Pira International.

Simmons, B., Ahmad, R., & Overton, B. (1998, June). Fluorescent tracers enable automatic identification and sorting of waste plastic. *British Plastics and Rubber*, 4–6, 8–10.

Glossary

Alumina: Purified aluminium oxide

Bauxite: Aluminium ore consisting of 40% to 60% aluminium oxide

BOD: Biological oxygen demand, biochemical oxygen demand, the oxygen consumed by aerobic microorganisms to degrade organic matter in effluent

BOD5: Five-day biological oxygen demand

COD: Chemical oxygen demand, a measure of the amount of potassium dichromate needed to oxidise reducing material in a water sample. It is expressed as the concentration of oxygen gas in parts per million chemically equivalent to the amount of dichromate consumed.

Cullet: Recycled glass used as a feedstock for glass production

Degree day: A measure of the cumulative difference between the outside temperature and a base temperature in a period of time for all occasions when the outside temperature is below (heating degree day) or above (cooling degree day) the base temperature. There are two definitions of degree day. One is a summary indicator of the weather and uses a formula based on the daily maximum and minimum outside temperatures. The formula used in the United Kingdom for this purpose is different from that used in other countries. The other definition, recognised only since 1984, is a coefficient derived from the physical law of cooling for a building. These are often used interchangeably, but care needs to be exercised in assuming that they are identical.

Dioxins: A group of chemicals with the chemical name polychlorinated dibenzo-p-dioxins that are ubiquitous environmental contaminants

DMT: Dimethyl terephthalate

Eutrophication: The process of nutrient enrichment of an aquatic system

EVOH: Ethylene vinyl alcohol

Gene: The fundamental physical and functional unit of heredity. The portion of a DNA molecule that is made up of an ordered sequence of nucleotides that produce a specific product or have an assigned function.

GWP: Global warming potential

HDPE: High density polyethylene

IPC: Integrated pollution control

Kraft pulping: The most common pulping process for packaging papers—a chemical process using sodium sulphate

LDPE: Low-density polyethylene

LLDPE: Linear low-density polyethylene

LNG: Liquefied natural gas

LPG: Liquefied petroleum gas

MAC: Maximum admissible concentration

NO_x: Oxides of nitrogen

NSSC: Neutral sulphite semi-chemical pulping

ODP: Ozone-depleting potential

PE: Polyethylene

PET: Polyethylene terephthalate

POM: Particulate organic matter

PP: Polypropylene

PS: Polystyrene

PVC: Polyvinyl chloride

PVDC: Polyvinylidene chloride

SEC: Specific energy consumption

SO_x: Oxides of sulphur

TOC: Total organic carbon

TPA: Terephthalic acid

TSP: Total suspended particles

TSS: Total suspended solids

VOC: Volatile organic compound

Index

F

G

P

Y

Z